Efficiency in Irrigation

A World Bank Symposium

Efficiency in Irrigation

The Conjunctive Use of Surface and Groundwater Resources

edited by
Gerald T. O'Mara

The World Bank
Washington, D.C.

Manufactured in the United States of America
First printing March 1988

The findings, interpretations, and conclusions expressed in this study are entirely those of the authors and should not be attributed in any manner to the World Bank, to its affiliated organizations, or to members of its Board of Executive Directors or the countries they represent. The maps that accompany the text have been prepared solely for the convenience of readers; the designations and presentation of material in them do not imply the expression of any opinion whatsoever on the part of the World Bank, its affiliates, or its Board or member countries concerning the legal status of any country, territory, city, or area, or of the authorities thereof, or concerning the delimitation of its boundaries or its national affiliation.

Library of Congress Cataloging-in-Publication Data

Efficiency in irrigation.

(A World Bank symposium)
Includes bibliographies.
1. Irrigation—Economic aspects. 2. Irrigation efficiency.
I. O'Mara, Gerald T. II. Series.
HD1714.E34 1988 333.91′3 88-195
ISBN 0-8213-1030-5

Contents

Preface

The papers in this volume are from a conference on the effects of externalities on the efficiency of irrigated agriculture in developing countries. The conference, which was held at the World Bank in Washington, D.C., May 11–13, 1983, examined the economic effects of the physical externality that is created when agricultural producers rely on a common aquifer or stream-aquifer system for their water supply. Although this seems at first glance to be a special case, it applies to more than half the irrigated acreage in the developing countries.

Such a physical externality is known to lead to inefficiency. The additional costs producers impose on one another in pumping water from the aquifer can destroy the social efficiency of individuals' decisions on resource allocation. But all of the suggested remedies present difficulties in implementation that restrict their application. For example, corrective taxes or subsidies are an attempt to adjust private costs to reflect the full social costs. But in many circumstances it is difficult to monitor the activities of economic agents, who thus have an incentive to cheat in reporting their activities. The remedy of cooperative management of the resource through a centralized operation suffers in many circumstances from a similar monitoring problem that again creates incentives to cheat. The assignment of legal property rights, whereby the span of physical effects is made to coincide with legally recognized responsibility, also suffers from severe implementation problems in most contexts. There is therefore a need to consider other approaches to the problem of efficiency in irrigated agriculture when such externalities are present. Recognition of this need led to the conference on which this book is based.

The consensus of the conference participants was that the available remedies for coping with the gap between private and social costs induced by such an externality may perform tolerably well in most situations and that several innovative approaches may improve the allocation of resources in more difficult cases. Thus O'Mara (chapter 1) argues that farmer participation in the planning and management of water resources—through farmer organizations empowered to monitor water use and redress abuses—will eliminate, or at least reduce the scale of, inefficient irrigation practices. In a similar vein Randall (chapter 2) contends that prospects for stable cooperative approaches have been underrated and that better designed incentives may induce irrigation bureaucracies to improve the services they provide their farmer clients. Radosevich (chapter 3) takes the position that the legal rights remedy could be made more effective through reform of the legal institutional structure surrounding water use.

These upbeat perspectives, based largely on conceptual grounds discussed in the initial chapters, receive empirical support from the case histories described in chapters 4–6. Coe (chapter 4), for example, reports that in several groundwater basins in California the adverse effects of an externality among the users of a common aquifer led farmers to work out a solution among themselves—in this case, a tax on water withdrawals. Johnson's description (chapter 5) of the experience with public tubewells in Pakistan is less encouraging, but farmer participation was absent in this case and thus the incentives for bureaucratic efficiency were attenuated. Although Huang, Cai, and Nickum (chapter 6) indicate that significant inefficiency still exists in the North China Plain, they also describe a search for greater efficiency that is remarkably flexible considering the difficult institutional arrangements for water use in China. As O'Mara notes (chapter 1), California has already chosen an institutional solution to the inefficiency induced by the externality among groundwater users, but Pakistan and China have yet to find a stable solution to the problem.

Participants in the conference reviewed various analytical methods for evaluating policy alternatives and providing guidelines for eliminating inefficiency in water allocation. Gorelick (chapter 7), for example, reviews a set of models which simulate the behavior of economic agents under conditions including complex groundwater–surface water interactions and specific institutional arrangements.

Although these models do not explicitly select a best policy, they can be used to evaluate policy alternatives. In addition, Gorelick examines more complex multilevel optimization models linked to agent simulation submodels, which are designed to generate a specific optimal allocation policy.

Rogers, Harrington, and Fiering (chapter 8) examine the dialogue between policymaker and modeler. The true objective of this process is to find an *acceptable* policy, which is not always the same thing as the optimal solution to a model. Since a model abstracts from many factors that are not considered relevant to economic or technical analysis, it is likely to abstract from political tradeoffs that are relevant to policymakers. When this happens and the policymaker rejects the model's optimal solution as "unacceptable," a review of alternative solutions in the neighborhood of the optimal solution may yield an "acceptable" one that trades off some economic inefficiency in favor of a significant political gain. Rogers, Harrington, and Fiering ascribe this circumstance to the common occurrence of a relatively flat solution surface in the neighborhood of the optimum and present several illustrative examples.

O'Mara and Duloy (chapter 9) discuss an application of multilevel policy evaluation modeling to the Indus Basin of Pakistan. Their model is designed both for investment appraisal and to help address the difficult problem of coordinating the tubewell pumping of hundreds of thousands of well-owning farmers with canal diversions by a large irrigation bureaucracy. Their simulation experiments suggest a potential gain on the order of 20 percent from more efficient conjunctive use. Zapata (chapter 10) uses econometric methods to estimate the difference between private and social costs of pumping in Mendoza Province, Argentina. He finds the difference to be between 20 and 30 percent.

Finally, the conference examined several special cases that illustrate significant aspects of conjunctive use of aquifer and stream-aquifer resources. One is dynamic conjunctive use. In dynamic programs, fluctuations in production and agricultural income are minimized through a policy of (a) increasing groundwater withdrawals to offset diminished surface supplies in times of drought and (b) using excess surface water to replenish the aquifer when surface supplies are greater than normal. If tubewell pumping and the surface water system are controlled by different agents, the coordination of both aquifer management and investment planning will pose a problem.

Bredehoeft and Young (chapter 11) examine conjunctive use in the South Platte River Valley in Colorado. They find that actual well capacity was approximately sufficient to irrigate the entire area. Using a simulation model of optimal conjunctive use under conditions of uncertain water supply, they find that what seemed to be an overinvestment in wells was in fact optimal because it maximized the net benefit and reduced income variance to essentially zero. In this case the farmers used tubewell pumping to stabilize the water supply completely, no matter what the surface water supply.

This policy utilizes the first part of the dynamic program described above, but says nothing about the second. Perhaps the recharge from streambanks is sufficient, or perhaps the farmers do manage to apply excess water in periods of greater than normal surface supplies.

In some dry environments water is so scarce that the resource will be destroyed unless aquifer withdrawals are constrained to a low safe level that severely limits agricultural production. Thomas (chapter 12) looks at two such situations in which rents from petroleum production significantly extend the range of feasible policy options. He finds that restricting withdrawals to the safe level and relying on food imports for the remainder of the nation's food supply is the least-cost solution. However, importing water from abroad and desalinizing seawater in amounts sufficient to support national self-sufficiency in food supply are also feasible alternatives.

Often, conjunctive use develops as an unintended consequence of the introduction of inefficient surface water irrigation. Basu and Ljung (chapter 13) examine such a case. They study a canal system which after fifteen years had not achieved its original objectives because it was only 20 percent efficient. Large seepage losses had induced a rapid secular rise in the water table, and farmers had installed shallow tubewells to exploit the readily available groundwater. Concerned over the efficiency and equity of the situation, Basu and Ljung estimated the additional quantity of groundwater that could be extracted safely and then projected a program of exploitation. Deep public tubewells were to discharge into surface water canals and farmers were to be given incentives to utilize the shallow aquifer directly.

In summary, the conference examined the effects of physical interdependence among agents on their economic behavior in several institutional contexts. Although the situation can be difficult when different sets of agents control surface and groundwater allocation, there are conceptual and practical grounds for expecting that the problem can be resolved even in difficult cases. The top-down philosophy of investment and water resource allocation that has dominated the decisions by governments in the developing countries is seriously flawed, however, and this mind-set needs to be changed before the substantive issues can be resolved. In particular, the exclusion of farmers from active participation in the process has doomed past efforts to solve the problems. Farmers must be brought into the picture if the requisite cooperation and coordination are to be achieved. This entails encouraging farmer organizations—water user associations, for example—by giving them the right to monitor performance and to seek redress when this is indicated by clear and reasonable standards. In effect, this means sharing political power with farmers for the good of all.

Since neither farmers nor bureaucrats will seek change unless confronted by unacceptable consequences or certain rewards, irrigation professionals have an important role to play. Both farmers and irrigation bureaucrats need to learn that present practices can be significantly improved. Sensible people do not regard an unpleasant situation as a problem unless and until they learn that a better outcome is feasible.

Irrigation professionals have tools at their disposal today that previous generations did not even dream of, but these tools will be of little use until the glacial resistance to institutional change in irrigating societies thaws perceptibly. This resistance stems from the fear on the part of groups capable of blocking change that change may be detrimental to their interests. The irrigation professional is well situated to demonstrate that this need not be the case.

G. Edward Schuh
Director
Agriculture and Rural Development Department
The World Bank

Participants

D. N. Basu
Operations Research Group, Baroda, India

John D. Bredehoeft
U.S. Geological Survey, Menlo Park, California

Nathan Buras
University of Arizona, Tucson

Jack J. Coe
Department of Water Resources, State of California, Los Angeles

John H. Duloy
Economics and Research Staff, The World Bank

Fei Jin
Regional Hydrogeology Research, Ministry of Geology and Minerals, Beijing, China

Myron B. Fiering
Harvard University, Cambridge, Massachusetts

Kenneth D. Frederick
Resources for the Future, Washington, D.C.

Steven M. Gorelick
U.S. Geological Survey, Menlo Park, California

F. L. Hotes
Agriculture and Rural Development Department, The World Bank

Huang Ronghan
Ministry of Water Resources and Electric Power, Beijing, China

Jiang Ping
Chinese Academy of Agricultural Sciences, Beijing, China

Sam H. Johnson III
University of Illinois, Champaign-Urbana

Max K. Lowdermilk
Colorado State University, Fort Collins, and U.S. Agency for International Development, New Delhi, India

David M. Newbery
Cambridge University, Cambridge, England, and Development Research Department, The World Bank

James E. Nickum
Cornell University, Ithaca, New York

Gerald T. O'Mara
Agriculture and Rural Development Department, The World Bank

Robert Picciotto
Europe, Middle East, and North Africa Projects, The World Bank

George Radosevich
Colorado State University, Fort Collins

Alan Randall
University of Kentucky, Lexington

Peter P. Rogers
Harvard University, Cambridge, Massachusetts

Mark W. Rosegrant
International Food Policy Research Institute, Washington, D.C.

Bruce Stone
International Food Policy Research Institute, Washington, D.C.

Robert G. Thomas
United Nations Food and Agriculture Organization, Rome, Italy

Juan Antonio Zapata
Economics Institute, Catholic University, Santiago, Chile

Contributors

Joseph J. Harrington
Howard University, Washington, D.C.

Cai Lingen
Ministry of Water Resources and Electric Power, Beijing, China

Per Ljung
South Asia Projects, The World Bank

Robert A. Young
Colorado State University, Fort Collins

Definitions

Technical Terms Relating to the Management of Stream-Aquifer Systems

Energy cost of pumping Cost of the energy required to efficiently pump a unit of water from the aquifer.

Evapotranspirative withdrawal Effect that occurs when the water table is near the surface and whereby withdrawals of water are induced through capillary action and plant suction.

Groundwater quality Defined by the level of dissolved salts or other contaminants. Quality also depends on the intended use.

Groundwater table Level of water storage (above mean sea level) in an aquifer; hence, the point at which the soil is fully saturated with water.

Hydraulic head Force vector caused by the effect of gravity on a liquid body.

Long-term storage capacity Total capacity of the aquifer to store water.

Mean long-term recharge An average annual recharge, exclusive of net pumping, in which the average is taken over enough years to obtain a tight confidence interval on the estimate of the population mean.

Mining of an aquifer Persistent withdrawals in excess of mean recharge.

Recharge Change (per unit of time) in water stored in an aquifer, which may be negative. It is defined by the identity: recharge = additions − withdrawals + net underflows.

Rim station A flow measurement point—where a river carrying runoff from mountains enters the plains of an alluvial basin.

Secular reserves Volume of groundwater capable of being mined.

Stochastic steady state, stochastic equilibrium An equilibrium that exists only in terms of the long-run average; periodic observations will differ from the long-run average because of random (stochastic) variation.

Underflow Underground flow or movement of water in an aquifer.

Underground transmission Water distribution by means of underflows within an aquifer system.

Waterlogging and salinization Unproductive soil condition that occurs when the water table is very near the surface.

Acronyms and Abbreviations

ACZ agroclimatic zone
ET evapotranspiration
SCARP Salinity Control and Reclamation Project
FAO Food and Agriculture Organization
FGW fresh groundwater
MAF million acre-feet
MBO management by objectives
MBR management by results
MCM million cubic meters
SGW saline groundwater
UNDP United Nations Development Programme
USAID United States Agency for International Development
WAPDA Water and Power Development Authority
WTP willingness to pay
WUA water user association

Editor's Notes

Dollars are current U.S. dollars unless otherwise noted. *Billion* is 1,000 million. *Tons* are metric tons, equal to 1,000 kilograms, or 2,204.6 pounds. The pronouns "he," "his," and so on are used generically in referring to indefinite persons and are not meant to imply sexual bias. Numbers in tables may not add up to given totals because of rounding.

1

The Efficient Use of Surface Water and Groundwater in Irrigation: An Overview of the Issues

Gerald T. O'Mara

The use of irrigation for crop production in arid and semiarid regions dates back to the dawn of the Neolithic agricultural revolution (about 6000 B.C.). But the intensive development of water resources to permit greater agricultural development in virtually all regions of the world is of relatively recent origin, dating back only as far as the past century. For not until progress in engineering had reached the point that structures could be placed astride sizable rivers was it possible to extend perennial canal irrigation to large alluvial basins. And similarly, not until efficient and compact engines and pumps were developed could groundwater resources be exploited on a significant scale. These developments have made large-scale agricultural development possible in certain water-short basins, such as the Gangetic Plain, the Indus Basin, and the North China Plain. Now that the limits of water resource endowments are being reached in many regions, the emphasis has shifted away from development toward more efficient utilization of these scarce resources.

Conjunctive Use

In particular, research in irrigation has focused attention on the potentially large benefits to be gained from efficient joint use of surface water and groundwater in those large alluvial basins where the physical interdependence of the two water resources complicates allocation. For example, Duloy and I estimate that a 20 percent increase in output is feasible in Pakistan through more efficient conjunctive use of surface and groundwaters (see chapter 9). A related problem is the environmental deterioration and loss of resource productivity that come with sustained irrigation development when salt accumulates in confined or slowly draining alluvial basins. This chapter will review several approaches that are being used to resolve these problems and will consider some schemes that have been proposed but have not yet been implemented.

Some Important Technical Concepts

To convey to the reader the technical complexity of the problem of efficient management of conjunctive use in a surface and groundwater system, some preliminary discussion of essential concepts will be useful. Readers may also want to refer to the list of definitions at the front of this book.

A groundwater aquifer may be visualized as a giant sponge that is buried underground.[1] Water that is absorbed by the soil at the land surface above the giant sponge will (after some losses to evaporation) seep downward by the force of gravity to be stored in the sponge after reaching either a volume of water already stored—in other words, by reaching the groundwater table—or an impermeable boundary. Water may also move horizontally in the sponge, in a process known as underground transmission, if it is subjected to a gravitational or hydraulic force vector in that direction. In addition, water may enter and leave the sponge horizontally, by means of underflows deriving from subsurface sources or sinks that connect with the aquifer boundary. If the groundwater level in the sponge rises close to the surface, withdrawals due to evapotranspiration (ET) will become significant. Undisturbed by human intervention, a groundwater aquifer will tend to reach a steady state in which the change in storage over a sufficiently long period is zero. The main adjustment mechanism is through changes in the groundwater level which vary ET losses so as to achieve a kind of stochastic steady state.[2] If the inflow from seepage and underflows is negligible, then the outflow from ET and underflows must also be negligible and the aquifer is essentially an exhaustible resource. The deep aquifers of desert regions are examples. In the more typical case, inflows are significant, and the water table is likely to be fairly close to the surface if outflows from ET and underflows are to match inflows in the statistical sense. Since alluvial aquifers in large river basins are usually quite large and typically contain many times the annual inflows, a relatively small adjustment in the water table level may be sufficient for them to accommodate the large positive or negative recharge of exceptional years. Alluvial aquifers are clearly renewable resources for which the prudent yield is the mean long-term recharge.[3]

The introduction of surface water irrigation in a river basin with an alluvial aquifer disturbs the stochastic equilibrium. Since canals will leak and not all the irrigation water applied will go to crop consumption, an increase in inflows via canal and field seepage will occur. Thus, mean recharge becomes positive, and the water table commences a secular rise that ends when increased withdrawals from ET match the increased flows. By the time this happens, the water table is probably so close to the surface that significant waterlogging and salinization of the soil are noted. The process may take decades or just a few years, depending on environmental factors and the pace of development. If the groundwater quality is adequate for cropping, drainage of the excessive accumulation of secular reserves may be achieved by using tubewells as a supplementary irrigation source. If the groundwater quality does not meet cropping standards, pure drainage is probably required. In any event, at least some of the excess water stored in the aquifer can probably be mixed with incoming canal water and used for irrigation.

The interdependence of surface and groundwater supplies means that controls on groundwater withdrawals depend on the allocation of surface water. While this allocation is typically controlled by the government, the quantity to be allocated in any one season is stochastic. The dynamic, stochastic nature of the problem affects the behavior of both the government officials who allocate the supply and the farmers who use it. The ability to efficiently manage joint utilization of surface and groundwater supplies makes it possible to pursue policies that both decrease exposure to risk and significantly increase mean agricultural production. The stochastic nature of naturally occurring flows makes it desirable to define the safe yield in a long-term statistical sense; this permits greater withdrawals in times of drought to minimize the impact on agricultural production, and the mined quantities can be restored during periods of greater than normal flows. In drought years a greater proportion of the diminished naturally occurring flows can be diverted to those parts of an alluvial basin that do not have access to good groundwater—the saline groundwater areas of the Indus Basin, for example. In such years areas that do have access to good groundwater will depend on tubewell pumping for a greater proportion of their water supply. Unfortunately, the efficient use of secular reserves in an aquifer to smooth out the effect of fluctuations in naturally occurring flows on agricultural production is very rare.[4] Although hydrologists and systems engineers have known for some time that more efficient use of groundwater is possible, a comprehensive specification of the problem that encompasses the institutional, legal, and political constraints has yet to find general acceptance.

"Rational treatment demands that systems engineering be based not so much on the methodologies sometimes mistaken for systems engineering but on a broad, scientifically firm foundation encompassing the very nature of the system concerned, historically, politically, legally, physically, productively, technically and economically" (Hall and Dracup 1970, pp. 3–4). This quotation from a water systems engineer accurately states the nature of the problem. In practice, however, as a review of the literature on conjunctive use demonstrates, there is a strong tendency to treat legal and institutional factors as outside the system. A political scientist notes that "water planners and engineers, in search of constraints to simplify their task of system design, have found it convenient to read inflexibility into governmental institutions and to treat them as immutable, if irrational, restrictions. On the contrary, there is and should be considerable flexibility in legal and administrative forms, which are quite adaptable in the face of demonstrated economic, technological, social or political needs" (Maass and others 1962, p. 565).

The Diffusion of Tubewell Technology

Historically, the development of the technology of surface water irrigation preceded that of tubewells based on compact diesel or electric power. In fact, the introduction of tubewells in the Indus Basin, and perhaps in the North China Plain as well, was motivated by concern over waterlogging and salinization that occurred when canal irrigation caused the water table to rise. The demonstration effect of public drainage wells set off a boom in tubewell investment by individual farmers (in South Asia) and individual communes (in North China). In Pakistan, the number of tubewells increased from less than 5,000 in 1960 to more than 200,000 by 1980. In the Indian states of Punjab and Haryana, there were more than 400,000 wells (of somewhat smaller capacity than those in Pakistan) by 1980. It is estimated that more than 2 million shallow, small-capacity private tubewells have been installed in the Gangetic Plain of India. In the 1960s and 1970s more than 2 million small-capacity tubewells were installed in North China. These massive investments in tubewells have completely transformed the use of water resources in these regions and raise problems of resource management that are beyond the grasp of existing irrigation bureaucracies. For example, in the Indus Basin of Pakistan tubewells now supply more than half of the water actually available for crop consumption in the fresh groundwater areas but of course supply none of the water for crop consumption in the saline groundwater areas, which account for about one-third of the total irrigated land of the basin. However, the bureaucracies concerned with the distribution of water do not attempt to achieve an efficient joint allocation of surface and groundwater supplies. In fact, the legal basis for doing so is unclear and open to challenge.

Although the massive investments in tubewells pose a challenge to resource management, such investments are a precondition of efficient conjunctive use of surface and groundwaters. Once the governments of these regions meet the challenge, a significant improvement in water productivity and agricultural output can be achieved.

The Need for Sustainable Policies

The evolution of irrigated agriculture in large alluvial basins—from inundation to modern perennial canals to the tubewell explosion of recent decades—traces the interaction of human society with the environment in arid and semiarid regions. As the scale of development increases, the pressure on the natural resource base increases, and the environment is changed in significant ways by human intervention. The introduction of modern perennial canals induces positive net recharge to alluvial aquifers, forcing a rise in the water table until increased ET losses establish a new equilibrium level. The required level of ET losses is likely to push the water table so close to the surface that significant waterlogging and salinization occur. If certain facilitating investments are undertaken, such as construction of drainage works, the productivity of the salinized and waterlogged lands can be restored. The point is, as initial investments change the state of the aquifer system, further investments are needed to make the system more conducive to human welfare. Thus, a proper appraisal of initial investments requires the recognition that intervention will change the state of the system and that subsequent investments are really a necessary complement. Unfortunately, such foresighted appraisals are quite rare in the history of irrigation. In the past, when the level of development was much lower and many countries had unexploited resources, such myopic investment planning—although inefficient—did not present intolerable consequences for future generations. At present, fewer countries still have unexploited resources, and most developing countries can no longer tolerate shortsighted planning which fails to adopt sustainable strategies and policies that can bequeath to future generations a resource base whose productivity is unimpaired.

Institutional Constraints

Physical Externalities

In the microeconomic theory of industry supply, an important determinant of supply sometimes consists of effects external to the providing firms but internal to the industry. These external economies and diseconomies—or externalities—affect the cost functions of individual firms as the output of the industry changes. Externalities are classified as "pecuniary" or "physical" (or "technological"). For a pecuniary externality, the interaction between industry output and firm costs comes about solely through changes in input prices. For a physical externality, the interaction is through industry output effects that directly operate on the physical production possibilities of the firm.

Interdependence through water supply is such a classic example of physical externality that it is used illustratively in many economics textbooks. For example, farmers who depend on a common aquifer have related costs. If some farmers pump their tubewells sufficiently to lower the water table in the vicinity of their neighbors' wells, the increased lift with its increased energy cost will raise the average and marginal costs of production of the neighboring farmers. If the drawdown is sufficient to put wells out of service, the required investment in deeper wells will further increase costs. Thus, overpumping by some farmers imposes a physical external diseconomy on their neighbors. Conversely, if a situation of positive recharge leads to waterlogging and salinization of soils, overpumping by some farmers may lower their neighbors' costs through the increase in yields (for given variable inputs) that comes with the improvement in soil productivity that a lowered water table will bring. In this instance, overpumping of the aquifer leads to physical external economies for neighboring farmers.

Physical externalities lead the invisible hand of the market mechanism astray by driving a wedge between private and social costs: the private agent does not recognize as a cost the effect of his actions in raising or lowering the costs of others. This is certainly true of the pumping decisions of individual well owners. Another example much used by textbook authors concerns firms that depend (in serial fashion) on the same river for water supply. Viewed as a productive input, water as a commodity has dimensions that include place, time, quality, and degree of certainty of supply. If an upstream user is free to degrade the river by discharging an effluent into it, downstream users who need water of a certain quality will suffer an increase in costs as they either expend resources to purify the polluted river water or seek more expensive alternative sources. Once again, the problem is that the cost to the downstream user is not recognized by the upstream user, who supplies his output at a price that does not reflect its true social cost. Thus, a physical externality has long been recognized by economists as a potential source of the failure of markets to achieve an efficient resource allocation.

It has been suggested that the term "externality" contributes nothing to analytic insight and that the terms "nonexclusive" or "nonrival" describe the essence of the failure of the market mechanism to deal with the phenomena subsumed under the rubric of externality. Clearly, as Randall points out in chapter 2, it is the nonexclusivity of groundwater resources and the technical difficulties in assigning meaningful property rights (exclusivity) to groundwater that constitute the essence of the externality among groundwater users. Nevertheless, this chapter will follow the accepted practice in economics and use the term "externality" to assign a label to the phenomenon of nonexclusivity that lies at the heart of the matter.

Since physical externalities represent sources of social gain or loss that are not reflected in market signals to private agents, economic theorists have considered various solutions to the problem. Mainstream economic analysis has focused on the following remedies:

- Corrective taxes or subsidies so that private agents are penalized or rewarded to the extent necessary to adjust private costs to reflect social costs
- Centralized control over the resource embodying the physical externality so that a single management will fully internalize the external effects in its calculations of costs—for example, the unitized operation of an oil field, in which individual well (or land) owners receive pro rata shares of overall profits
- Assignment of legal property rights so that the span of actual physical effects always coincides with legally recognized responsibility.

Theorists tend to prefer the last solution since it minimizes the need for state intervention. It has been generalized into a proposition known in the literature as Coase's theorem: given well-defined and nonattenuated initial legal property rights and zero transactions costs, the market allocation will be efficient.[5] According to this argument private agents offer payments to owners of legal rights to exercise or to refrain from exercising their rights, depending on whether the cost of the externality is negative (and thus a benefit) or positive. In the resulting market equilibrium, with the price of, say, nonexercised rights driven to equality with the marginal cost of the externality, private costs are equal to social costs.

The problem with the actual application of Coase's theorem lies in the limiting assumptions required to establish this result. Legal rights must be well defined and nonattenuated, and transactions costs must be at least small relative to benefits. Perhaps the most serious drawback is the elusive nature of *well-defined* legal rights when physical interdependence is in any way complex. Less serious but also troubling is the fact that the cost of the required legal services is usually small in relation to benefits only when the benefits are absolutely large.[6] Thus, the comparative simplicity of the physical interdependence inherent in an irrigation system based purely on surface water has permitted precise definition of legal rights in the form of the doctrine of prior appropriation, or rights based on historic use. That is, earlier users have preference over late users. This means that in times of low natural flow the rights of senior users take precedence over those of junior users, who may in fact receive no allocation at all. Note that often such rights are usufructs and may be lost if not exercised. Since water rights are legally well defined in this simple situation, they are marketable (if the sale of rights does not undermine the legal rights of the seller), and junior users are free to negotiate for the temporary purchase of rights from senior users. If the value of a unit of water is greater to the junior user, he will be able to offer a price that exceeds its value in use by the senior user, and a mutually beneficial transaction will occur. In this fashion, the marginal value of water is equalized among users (after allowing for transmission losses and transactions costs) and an efficient allocation can be achieved. It should be observed that this scenario is rarely followed in actual practice; water rights are usually not legally defined in such a manner as to permit transactions in rights that are encompassed by legally enforceable contracts. Moreover, the "use it or lose it" nature of many legal rights provides a perverse incentive to use water inefficiently.

Once a more complex physical interdependence is introduced in the form of an alluvial aquifer, the precise definition of legal rights becomes problematical. For instance, tubewell pumping that lowers the water table adjacent to a streamflow will induce increased seepage to groundwater or interrupt return flows to the stream, with the result that downstream flows are reduced to the detriment of the holders of appropriated surface flow rights downstream. Thus, where groundwater is an unrestricted common property resource (as it is in Pakistan and to a large extent in the western United States), a very junior upstream user who invests in a tubewell can obtain an implicit senior right over downstream users. In response to litigation stemming from such a situation, the Supreme Court of the State of Colorado ruled that groundwater (with certain exceptions) would eventually reach the surface stream and thus was subject to the doctrine of prior appropriation. In consequence, well owners with junior rights cannot pump unless all senior rights holders downstream have satisfied their rights.[7] In contrast to the previous situation, which placed a zero value on these costs, in this case the court has placed an infinite value on these costs. In a simulation study of the Colorado case, Daubert and Young (1982) found that either extreme situation was grossly inefficient and that efficient resource management required both joint use and an appropriate price for the externality.

Clearly, a working market in transferable water rights would allow upstream well owners to purchase rights from senior holders downstream and, by setting a market price for the externality, produce an improvement in the efficiency of resource use. Note that the gain comes from treating surface and groundwaters symmetrically. This implies establishing pumping rights on the same basis as rights to surface water. It implies further that water rights are fungible irrespective of source. It should be noted, however, that the hydrologically sensible principle of symmetry of water sources specified implicitly by the Colorado court is not accepted in almost all other jurisdictions.

The principle of symmetry of water sources—that is, that water is water—would seem to be fundamental to either efficient or equitable conjunctive management of surface and groundwaters. For example, consider the dynamic use of the secular reserves in an aquifer to smooth out the impact on agricultural production and rural welfare of abnormally low natural flows in an alluvial basin in which only some farmers have access to good groundwater. How can this be accomplished unless well owners pump more water and accept relatively smaller allocations of surface water? Clearly, it is not possible to utilize total water resources efficiently if surface and groundwaters are segregated into separate compartments.

Furthermore, a workable market in transferable water rights requires that sellers be able to transfer title to a given quantity of water with certainty, and this is often possible to only a limited degree with many irrigation systems. For example, in the Warabandi rotation system of South Asia, watercourse deliveries are passive consequences of diversions to distributory and minor canals, where such diversions are rotated among a number of such canals from flows along major or branch canals.[8] For this system, rights transfers can be effected with certainty only along a watercourse, and even here the amount of water transferred remains uncertain.

More generally, the introduction of supplementary groundwater supply via private tubewells destroys the simplicity which facilitates the establishment of the well-defined legal rights specified by Coase's theorem. Some form of conscious management would seem to be necessary.[9] In addition, in the context of surface and groundwater irrigation, private tubewells have an agricultural significance that goes beyond this complication. A tubewell can supply water on demand, and this capacity permits use of tubewell-supplied water to eliminate the risks of failure or inadequacy of canal supply; augment water supplies for greater cropping intensity (for example, a second crop cycle per year); and optimize water inputs over the cropping cycle to secure larger yields. The risk-reducing effect of tubewell supply also induces greater use of other yield-enhancing inputs, such as fertilizer. But tubewell water is available on demand only when the farmer controls the well. Thus, when tubewells are introduced into a region with usable groundwater, significant investments in private tubewells are probable. The farmer's incentives for making such investments are the prospects of increased profit and decreased risk. He neglects the drainage effects of his tubewell pumping since these are negligible for any one farmer. Moreover, even if he controls his pumping for environmental reasons, his neighbors have an incentive to pump more as long as the marginal revenue exceeds the marginal cost of pumping. When tubewells are highly profitable, pumping in excess of net inflows (that is, a negative recharge) is highly probable. Sustained mining of the aquifer will eventually lower the water table to the point at which many wells will go out of service. Then, either artificial recharge (via collective action) or investment in deeper and more expensive wells with greater energy costs will be required to sustain the supply of tubewell water. Eventually, even if the deeper wells are built, sustained mining will totally exhaust an aquifer.

The preferred solution (well-defined, legally enforceable property rights) fails in the absence of public interest restraint on well pumping to achieve efficiency or even equity when farmers rely on a common aquifer. This failure leaves the centralized operation and tax-subsidy alternatives as prospective remedies.[10] But public or private schemes to manage aquifer operation on a cooperative basis suffer from the disadvantage that individual agents have an incentive to cheat, and detection of cheating is difficult since the well owners are free to choose when to pump—and may do so in the dead of night, for example. Taxation of pumping suffers from the same agent incentive to cheat. Public ownership solves the problem of cheating, but introduces an even greater problem. Control of the wells is now out of the farmers' hands and entrusted to a large number of low-level (and often poorly paid) public employees, and this circumstance provides a multitude of opportunities for rent-seeking behavior on the part of the tubewell operators. As a result, publicly owned tubewells have acquired an evil reputation and are widely viewed as an inferior solution to the problem (in the sense that large inefficiencies in groundwater utilization are seen to be inevitable).[11]

Equity and Legality in Water Allocation

In an agrarian, preindustrial society, control over the essential input to agricultural production determines the social class structure and income distribution. This input is usually good agricultural land. In an arid or semiarid environment, however, water is the factor input that determines the scale and intensity of agricultural production. Thus, control over water is equivalent to control over income and wealth, to say nothing of survival. For this reason, the mechanism for allocating water in agrarian societies situated in arid regions is intensely political and is typically sanctioned by institutions dominated by the ruling elite. In fact, this is usually the case even in industrial societies located in arid areas. It follows that any significant change in the mechanism for water allocation must meet with the approval of the ruling elite (at least if not accompanied by bloody revolution). Therefore, the basic principle of all institutional mechanisms for implementing improvements in water utilization and water resource conservation is that existing allocations must not be reduced. To put this in the language of welfare economics, the improvement must be Pareto-efficient—that is, the improvement while increasing total welfare must not adversely affect any agent's economic welfare. In practice, this proposition usually assumes a somewhat different form, which might be called the first principle of equitable water allocation:

> PRINCIPLE OF PRIOR APPROPRIATION. *Sustained historic access to irrigation water confers property rights enforceable by either law or custom and tradition. These may be lost with sustained disuse.*

This quasi-legal principle says nothing about source, transferability, or association with other attributes (as when water and land are linked as a productive unit). It can be, and often is, extended to order of priority so as to cope with the stochastic character of naturally occurring flows. Other distributive arrangements may fix proportionate shares of available supply, turns in a circle, reserved time in a rotation, seasonal allocation, crop priorities, or salable water rights. The common feature is that all such arrangements are invariably based upon precedent—historic water use.[12]

The first principle is backward-looking and Pareto-safe.[13] It says nothing about future increments to irrigated water supply resulting from development of water resources through public investment or better public management. This prospect is of great interest in many developing countries in arid regions and is covered by the following principle:

> PRINCIPLE OF DISTRIBUTIVE JUSTICE. *Allocation of uncommitted water resources should not violate the accepted canons of equity for the irrigation society.*

Readers quite naturally will tend to interpret this principle in accordance with the rules of equity that characterize their values—for example, that the allocation of uncommitted water should not worsen the distribution of income or blatantly discriminate against any minority. But the principle specifies only that the canons of equity for the irrigating society be observed, and this might entail distributing the water entirely to some preferred group such as the ruling class. Since water can be made fungible, such an action is not necessarily inconsistent with efficient utilization of the resource.

Since water allocation norms must cope with the stochastic character of natural water flows and the variation in these can be so great as to threaten the stability of the society, a rule for this situation is needed:

> PRINCIPLE OF SYMMETRIC LOSSES. *The burden of extreme stochastic variation in total water supply should not be placed asymmetrically on a subset of users.*

While it might be argued that the power structure of a society would operate so as to shift disproportionate burdens to those with the least power, the degree of variation in supply is specified as great enough "to threaten the stability of the society." Under such a condition, only a society in the process of sustained dissolution would permit gross asymmetry of burden.

Last, but by no means least, when water is scarce and valuable, there is a strong need to reconcile the norms of custom and legality with economic efficiency in such a way that efficiency is not badly impaired. This need is particularly strong in developing countries and is met by the following rule:

> PRINCIPLE OF SYMMETRIC TREATMENT OF SOURCES. *Rationalization of rights to water resources with efficient resource utilization requires that rights be specified in terms of water of common characteristics. That is, rights should cover all sources, with appropriate allowances for differences in cost and characteristics. Rights need not be identified with a source.*

This principle has a somewhat different thrust from the preceding three, which are concerned with the stability and continuity of the irrigating society. The principle of symmetric treatment of sources is concerned with reconciling economic efficiency with the prevailing value system of the irrigating society. Hence, it presupposes that economic welfare is a significant value to the society. Where this condition is not met, the principle will be rejected.

The four quasi-legal principles of equitable water allocation enunciated above are sufficient to cope pragmatically with equity problems in a variety of cultural contexts as long as water is scarce and an abiding concern for the stability of the society motivates the leadership.

These principles are formally expressed in the laws of the irrigating society that apply either explicitly or implicitly to the use of water and the development of water resources. This body of law, called water law, is the basis for administrative rules and regulations concerned with water resource allocation and development. The origins of water law, the varieties observed internationally, the failures of water law, and possible remedies are discussed by Radosevich in chapter 3.

Guidelines for Efficient Water Utilization

In the previous discussion of externalities, the physical interdependence of farmers utilizing tubewell water from a common aquifer was discussed at length, as was the physical interdependence of users along a stream with respect to water quality. There is another form of interdependence among users along a stream, and in this instance the externality is pecuniary. Suppose the users along a stream are partitioned into upstream and downstream users, and some water is taken from downstream users and given to upstream users. The increase in water available to upstream users lowers the opportunity cost of water as perceived by these users, inducing more water-intensive cultivation and increasing gross agriculture supply for a given level of output prices.[14] That is, when more water is allocated to a region, the value of its marginal product (the appropriate input price for the farmer) drops for all of the farmers, lowering costs across farms and inducing an increase in gross output.[15] The pecuniary external economy is induced by the action of the public agency that allocates water supply along the stream. If the additional water is taken from another region, as was hypothesized, a reverse effect, or a pecuniary external diseconomy, is induced in the region losing water. These opposed effects need not cancel each other out, given a fully utilized water endowment. If the value of the marginal product of water in the gaining region is greater than the value of the marginal product of water in the losing region, the effect on agricultural output will be positive. Since water is seldom allocated by a market, there is no presumption that the value of the marginal product of water is equalized across regions, and gains from the reallocation of water are typically possible among users along a stream. Moreover, if the increased allocation of water comes from an increase in total diversion for irrigation, or from diversions that achieve greater efficiency in delivering water at a specified time and place, it is possible for all users along a stream to benefit. Opportunities for such Pareto-efficient gains are not as rare as they should be.

To summarize this discussion, several basic propositions on efficiency in water utilization will be presented. Before that is done, it will be useful to define a standard commodity unit of water. Our definition of a standard commodity requires that the water commodity be distinguished by location, time, quality, and probability of supply. If two quantities of water differ in any of these characteristics, they are different commodities. Different water commodities often can be transformed into the same commodity, although perhaps with some loss in the transformation. For example, transporting otherwise equivalent quantities of water to the same location will transform them into the same commodity, but there will be losses to seepage and evaporation during transport. Similarly, different water commodities can often be substituted for one another within limits. For example, water that is too saline for crop use by itself can be mixed with water of better quality and in that fashion be used for cropping. Thus, rates of exchange between different water commodities can be established. When discussing water utilization, it is convenient to use a standard commodity of water, but it should be understood that the argument generalizes across water commodities by means of rates of exchange between them.[16]

The basic propositions on efficiency in water utilization assume no risk aversion by policymakers and are to be interpreted in terms of the expected values for stochastic quantities. They are:

Proposition 1. *The standard water commodity is the same, whatever the source.*

Proposition 2. *Efficient spatial operation of an irrigation system using only surface water requires that the social opportunity cost of water—the value of its marginal product at efficiency prices, at a common source (rim station)—be equalized across farms and regions at each point in time.*[17]

Proposition 3. *Efficient intertemporal operation of an irrigation system using only an aquifer requires that the discounted social opportunity cost of water at the wellhead be equalized across time periods.*

Proposition 4. *Given the optimality of preserving the productivity of the resource, renewable aquifer resources have an efficient mean annual yield equal to mean annual recharge.*

Proposition 5. *Efficient operation of an irrigation system encompassing both surface and groundwaters requires that the discounted social opportunity cost of surface water equal the discounted social opportunity cost of groundwater across farms and regions at each point in time. Both social opportunity costs must be measured from a common reference point, for example, field level.*

These propositions can be derived heuristically, in the manner of the discussion above on pecuniary external economies, or more formally by using a model. Proposition 1 is trivial, but in water resource utilization, for historical reasons, it is necessary to explicitly state the obvious. Proposition 2 converts all water units to a standard commodity and thus is really a standard proposition on efficient static allocation of one input in any one period. Proposition 3 is simply the well-known Hotelling (1931) result on efficient allocation of an exhaustible resource. Proposition 4 is the necessary condition for a sustainable policy with respect to a renewable aquifer resource. Proposition 5 generalizes proposition 2 to the conjunctive use case. Since both its proof and its implementation in practice require dynamic utilization of secular reserves, neither proof nor practice is simple. The proposition can be proved using nonlinear programming or optimal control methods.

Each of the efficiency propositions is commonly violated in actual irrigation practice. For example, many water laws treat surface and groundwaters as separate, distinct categories, thus violating proposition 1. Where the principle of prior appropriation is applied to water allocation, typically the rule of proposition 2 does not hold. Because of physical externalities, it is often the case that the exploitation of aquifer systems deviates from what proposition 3 specifies. Similarly, earlier discussion has already noted that propositions 4 and 5 are often not descriptive of actual practice, although in the long run proposition 4 has to hold if the productivity of the resource is to be sustained. The obvious moral of all this is that there exist significant opportunities to reap output gains from improved utilization of irrigation waters. The skeptic may respond to this assertion by asking, "If there are so many opportunities for unrealized gains, why have neither private agents nor governments acted to realize them?" The answer to that question involves several points:

- Both governments and private agents naturally prefer to exploit easily realized opportunities first, and historically this has been possible in many regions.
- Many of the possible gains in efficiency involve an institutional change that is not, or is not perceived to be, Pareto-safe, and thus conflict among interested groups prevents adoption of the output-improving institutional changes.
- Many of the potential gains require significant investments as a necessary condition for realization (large-scale drainage, for example), and other investments have seemed to be more attractive to governments in the past. Now that some of the more obvious and easily realized gains have been exploited in many countries, the less obvious, or apparently more difficult, systemwide gains constitute the frontier for improvements in water resource productivity.

Farmer Preference for Local Control

Since access to water is crucial to income and subsistence in water-scarce regions, farmers in such areas naturally prefer to have control over the disposition of the available water. As previously noted, the apparent certainty of control over groundwater supply that a private tubewell brings is a powerful inducement to a farmer to invest in a well. The control over water supply that a tubewell brings can often both increase farm income and decrease exposure to risk. This combination of advantages usually more than offsets the higher cost of water and the perceived risk of collective overpumping by well owners. Hence, the introduction of tubewell technology tends to set off a tubewell investment boom.

Similarly, farmers dependent on surface water or public tubewell supply prefer to have control over these supplies. Although it is not possible to achieve individual control as in the case of private tubewells, some control can often be achieved via collective action. There is much to be said for local control by farmers, since they have the most to gain from efficient water allocation and are the best informed about water requirements, local hydrology, and so on. Moreover, the information requirements for efficient centralized control of an irrigation system of significant size are formidable, and actual practice tends to fall palpably short of maximum efficiency. In fact, large centralized systems, such as those of South Asia, often must delegate significant autonomy to district managers and engineers. As Wade (1982) has documented, when relatively low-level functionaries have significant discretion in decisionmaking, the desire of farmers for local control tends to result in significant rent collecting by the functionaries. That is, farmers are willing to pay for some degree of control over the disposition of available water, and the functionaries are willing to accept payment in return for allowing farmer participation in decisions about water allocation. Of course, the farmers would prefer a less costly route to some control over their own destiny, but they usually must deal with the political and administrative environment as they find it. This characteristic pragmatism of farmers in their drive for control over water was considered by Maass and Anderson (1978, p. 366) to be an important finding of their study of irrigation systems:

> The most powerful conclusion that emerges from the case studies is the extent to which water users have controlled their own destinies as farmers, the extent to which the farmers of each community, acting collectively, have determined both the procedures for distributing a limited water supply and the resolution of conflicts with other groups over the development of additional supplies. With important variations to be sure, local control has been the dominant characteristic of irrigation in these regions, regardless of the nationality or religion of the farmers, the epoch, whether formal control is vested in an irrigation community or in higher levels of government, the forms of government at the higher levels, and perhaps even the legal nature of water rights. In this realm of public activity . . . formal centralization of authority, where it has occurred, has not meant substantial loss of local control *de facto.* General administrative, legislative, and judicial norms laid down by higher authorities have not negated customary procedures. The norms have been either too general to accomplish this or they have been ignored by local organizations.

This quotation somewhat overstates the extent to which local control is achieved in large irrigation systems—Maass and Anderson studied only relatively small systems—but accurately reflects the farmers' preference for control over the water supply so crucial to their welfare and their determination to achieve some control by whatever means they can. Moreover, it is true that farmers are better informed about crop water requirements than irrigation bureaucrats or engineers. It follows that a system responsive to farmer demand is far more likely to achieve an efficient allocation than a system which presupposes superior information and decisionmaking capacity on the part of the irrigation bureaucracy and neglects feedback from farmers.

Some Institutional Solutions

California

Although California is one of the most developed regions of the world, the early development of groundwater utilization there and the variety of the resulting institutional responses provide instructive lessons for the developing countries. Because large-scale groundwater exploitation in developing countries did not begin until relatively recently (within the past three decades), it is difficult to find experience similar to California's among the developing countries. But throughout recent decades farmers and other local interests have been forced to consider solutions to deal with the adverse effects of overexploitation of groundwater.

In California, permits and licenses are not required for use of groundwater as they are for use of surface water. As Coe points out in chapter 4, "Landowners are free to drill wells on their property and pump water unless the water rights have been adjudicated by the courts. Groundwater is considered to be appurtenant to the land, and the right to its use is analogous to a riparian surface water right."

The combination of rapidly growing demand for water and an early lack of public concern over the external diseconomies from uncontrolled pumping produced the expected adverse effects: higher energy costs because of increased groundwater pumping lifts, higher investment costs because of well deepening and pump lowering, land subsidence, degradation of water quality (including intrusion of seawater) and exclusion of some pumpers from the aquifer supply. As the effects of overdraft became apparent, a groundswell of response by users typically emerged. Coe gives the following description:

> When overdevelopment gave rise to detrimental effects, the typical response was for those experiencing problems to create a water association to provide a forum for discussion. Usually consultants were retained to provide advice. Once a plan had been agreed upon and it was necessary to levy taxes, condemn property, or contract to import water, a public agency was created. In some cases the groundwater rights were adjudicated to ensure equitable allocation of the scarce resource. But adjudication of water rights does not provide additional water, and the importation of supplemental supplies has been the solution to overdraft in California whenever feasible.

California's experience is examined in greater detail in chapter 4, where Coe discusses the case histories of four areas of the state.

Pakistan

Although surface waters have been used for irrigation in the Indus plains of Pakistan for millennia, large-scale irrigation dates from the development of perennial canals supplied from low weirs or barrages across the Indus and its tributaries, which commenced in the 1860s. The distribution across the plains of large quantities of waters that formerly had flushed down the rivers to the sea disturbed the dynamic equilibrium of the natural drainage system, as seepage from tens of thousands of miles of leaky canals caused an enormous increase in recharge to the groundwater aquifer underlying most of the basin. This change induced a secular rise in the groundwater table, which generated widespread salinization and waterlogging of the soils—a problem that attracted significant attention in the late 1950s and early 1960s. With funds from the U.S. Agency for International Development, a successful demonstration project for vertical drainage using public tubewells was initiated in 1960 and completed in 1963. Reports on the problem of salinization and waterlogging by several expert groups differed on a number of points, such as the level at which the groundwater table should be stabilized, but they all agreed on the need for horizontal drainage (to remove salt accumulation) in the long run and on the efficiency of vertical drainage by means of public tubewells in the intermediate run. An exception to the general drift of expert opinion was expressed by Ghulam Mohammad of the Pakistan Institute of Development Economics, who argued that public tubewells should be installed only where the groundwater is too saline for direct use by farmers. The demonstration project, Salinity Control and Reclamation Project (SCARP) I, did show that the water table could be successfully lowered by distributing tubewells uniformly over a large area. Apparently the project also demonstrated to thousands of farmers that there was usable water a few feet below the surface that could be made accessible on demand for a relatively modest investment in a tubewell. The 1960s and 1970s saw a sustained and still continuing investment boom in private tubewells: the number increased from less than 5,000 in 1960 to about 200,000 by 1980. As yet, there is no clear evidence that the tubewell boom has produced costly and unacceptable decreases in the depth to water table in the fresh groundwater areas of the basin. The tubewell boom has, however, rendered moot the strategy of using public tubewells to control the aquifer in fresh groundwater areas. Control over the aquifer now demands control over the pumping by hundreds of thousands of farmers.

The tubewell investment boom has been confined to the fresh groundwater areas of the basin, which account for about two-thirds of the command area. In the other third of the command area, public control over the aquifer is intact in principle, but a past reluctance to undertake the necessary enabling investments in drainage facilities has nullified the potential for public control thus far.

The failure of the projected public control of the aquifer in Pakistan by means of publicly owned and controlled tubewells is not to be taken lightly. Until the reasons for that failure are thoroughly understood, the public tubewell approach to conjunctive use must be considered unreliable. In this respect the review of the SCARP program by Johnson in chapter 5 is instructive and important. Johnson identifies a number of sources for the disappointing performance of the SCARPs: engineering and design problems, operations and maintenance problems, lack of integrated management at the local level, lack of effective supervision of individual well operators, and lack of planning to encourage farmer organizations along the watercourse.

The North China Plain

The North China Plain is the alluvial basin of three major rivers—the Huang (Yellow), Huai, and Hai—and covers 30 million hectares, some 24 million of which are cultivated. The Yellow is the longest (5,460 kilometers) of the three rivers and has the greatest mean flow (56 billion cubic meters), but in some important respects it is the least significant. The Huai and Hai have lengths of 1,080 and 1,090 kilometers respectively and mean annual runoffs of 50 billion and 29 billion cubic meters respectively. The diminished significance of the Yellow is due to the extraordinarily heavy load of silt it carries, some 1.6 billion tons in an average year, most of which originates in the loess plateau of the middle reaches. As a result, large storage reservoirs are impractical in the river's lower reaches, and diversions for canal irrigation onto the North China Plain are limited to relatively short canals. Approximately 400 million tons of sediment are deposited annually on the riverbed of the relatively flat reach across the plain. Lifted over time, the riverbed is now four to eight meters above the level of the surrounding plain.[18]

The North China Plain, or the Huang-Huai-Hai, as the plain is called in Chinese, has a temperate, semihumid monsoonal climate and a highly variable rainfall, which declines from the south (mean of 700–900 millimeters) to the north (mean of 500–600 millimeters), with a coefficient of variation ranging from 0.5 to 0.8. More than 70 percent of the annual rainfall is concentrated in the monsoonal months of June to September. During the twenty-four years from 1949 to 1972, flooding was relatively serious in eight years and drought was relatively serious in seven. In an average drought, agricultural output is reduced by one-third, and losses under serious conditions range from 50 to 100 percent. The lack of storage capacity on the plain means that much of the precipitation that falls in storms is not usable for agriculture, although it can and often does create drainage problems. Much of the plain is a water-deficit area in the sense that mean precipitation does not exceed crop requirements. Thus, on average, sustaining high agricultural yields has meant constructing access to irrigation supplies, and the history of irrigation in the region goes back millennia.

The need for supplemental supplies of water and the difficulty of exploiting the water of the Yellow River led to an early interest in developing groundwater supplies. The flat (slope of 1 in 10,000), slow-draining basin is underlain in most areas with a shallow aquifer 10 to 30 meters in depth. In the western part of the basin, in a region of piedmont and alluvial fans, there are multiple aquifers of greater thickness and less salinity. Moving east, the thickness and number of aquifers decrease and salinity increases. The capacity of the shallow aquifer in fresh groundwater areas has been estimated at 48 billion cubic meters. Modern tubewell technology reached the North China Plain about the same time as it did the Indus and Gangetic Plains in South Asia. In the 1960s and 1970s more than 2 million shallow, small-capacity wells were constructed. Total pumpage from the North China Plain was estimated at 27 billion cubic meters in 1978, or 13,500 cubic meters per shallow well per year, and there are reports of many wells running dry in years of severe drought. The problems of the North China Plain are discussed in illuminating detail by Huang, Cai, Nickum, Fei, and Jiang in chapter 6 in Part II of this book.

Review of Policy Responses to Externalities

The case histories that have been reviewed show a wide diversity of policy responses to the problems of externalities. It is useful to compare these responses in the context of the quasi-legal allocation principles, efficiency guidelines, and institutional options discussed above and by Randall (chapter 2) and Radosevich (chapter 3). Such a comparison is provided in Table 1-1. A question mark following an answer in the table indicates that there is insufficient evidence for a definitive answer. A "yes and no" answer indicates that each of the two answers is applicable to some subregion.

With respect to quasi-legal allocation principles, an affirmative answer (with some uncertainty) can be given for all cases on all principles except source symmetry. This was to be expected since the principles for which there is unanimity are essentially rules for maintaining the stability of the irrigation society. Source symmetry, which is essentially an allocation rule concerned with economic efficiency, shows a mixed picture—with China a definite yes, Pakistan a definite no, and California yes and no.

Answers with respect to the efficiency guidelines exhibit more ambiguity (yes and no answers) and more uncer-

Table 1-1. *Matrix of Policy Responses*

Item	California	Pakistan	North China Plain
Quasi-legal allocation			
Historical water rights	Yes	Yes	Yes?
Distributive justice	Yes	Yes	Yes?
Loss symmetry	Yes	Yes	Yes?
Source symmetry	Yes and No	No	Yes
Efficiency guidelines			
Spatial efficiency	Yes and No	No	Yes
Temporal efficiency	No	No?	Yes and No?
Resource sustainability	Yes and No	No	Yes and No?
Spatial and temporal efficiency	No	No	Yes and No?
Choice of institutional solution			
Fungible legal rights	No	No	No?
Efficiency tax or subsidy	Yes	No	No?
Centralized control	Yes and No	No	Yes and No
Legal local control	Yes	No	Yes?

tainty. As with the responses to source symmetry, however, China shows mostly affirmative answers, Pakistan negative answers, and California intermediate answers. A tentative (owing to uncertainty) conclusion is that China has been more alert to the efficiency gains possible from conjunctive use than Pakistan and possibly California.

With respect to the institutional options, neither China nor Pakistan has yet selected a solution that is both efficient and workable. California's choices favor local control and pumping taxes. Given China's preference for local administration and the ease of administering taxes or quotas through the commune system, pumping taxes (or quotas) would seem to be a likely ultimate choice. Pakistan has effectively rejected the centralized control remedy and is confronted with a dilemma regarding the other alternatives. Neither a tax-subsidy program nor a legal rights scheme is likely to be feasible until some form of water users' association is in place along each of the country's 90,000 watercourses.

Some Analytical Methods for Managing Efficient Conjunctive Use

By now it should be clear that management of efficient conjunctive use is both complex and skill-intensive. As a result, the potential for efficient management is constrained by the availability of skilled managers and technicians and the institutional restrictions which define the range of choices available to water management. In many developing countries the scope for immediately implementing a Pareto-efficient move toward more effective management is quite limited. Over the long run, however, skills can be acquired and institutional restrictions can be removed. Often all that is needed is a demonstration of the potential gains to be had. Once a strong, development-oriented leadership is convinced of the desirability of removing institutional and skill constraints, effective action usually follows. Without such leadership, effective and purposive action is unlikely.

In the positive spirit that affirms the possibility of change once the feasibility of obtaining desired results from well-defined actions has been established, this section previews the chapters in Part III of this book, which is devoted to an examination of recently developed analytical methods that facilitate the efficient management of conjunctive use. These methods all depend on developments arising from the parallel revolutions in mathematical modeling and digital computation of the past several decades. The ongoing development of ever less costly and more powerful computer hardware makes it certain that this technology will ultimately be extended to nations at all levels of development. The only real constraint to this diffusion is the shortage of human skills, and this is a removable constraint.

Groundwater Modeling Methods

Following Gorelick's lead (see chapter 7), studies which link aquifer simulation with management decision models may be divided into two classes: hydraulic management models and policy evaluation and allocation models. Models aimed primarily at managing groundwater stresses such as pumping and recharge are included in the first class. These models treat the stresses and hydraulic heads directly as variables in management decisionmaking. Models which simulate the behavior of economic agents, where the environment includes complex groundwater–surface water interactions and specific institutional content, are included in the second class. Although these models are not explicitly designed for policy selection, they can be used to evaluate policy alternatives. More complex, multilevel optimization models linked to agent simulation submodels do generate specific optimal allocation policies. Both classes of model utilize linear or quadratic programming methods. In both classes of model, the component simulating aquifer response is based on the equation of groundwater flow in saturated media (Pinder and Bredehoeft 1968; Remson, Hornberger, and Molz 1971).

Groundwater Hydraulic Management Models. Groundwater hydraulic management models incorporate a simulation of a particular groundwater system as a constraint in a management decision model. To quote Gorelick (chapter 7):

> The embedding method treats finite difference or finite element approximations of the governing groundwater flow equations as part of the constraint set of a linear programming model. Decision variables are hydraulic heads at each node as well as such local stresses as pumping rates and boundary conditions. The response matrix approach uses an external groundwater simulation model to develop unit responses. Each unit response describes the influence of a pulse stimulus (such as pumping for a brief period) upon hydraulic heads at points of interest throughout a system. An assemblage of the unit responses, a response matrix, is included in the management model. The decision variables in a linear, mixed integer, or quadratic program include the local stresses such as pumping or injection rates and may include hydraulic heads at the discretion of the modeler.

Groundwater Policy Evaluation and Allocation Models. Groundwater policy evaluation and allocation models analyze water allocation and investment problems from the viewpoint of economic efficiency. These models are of three types:

- *Hydraulic-economic response models* extend the response matrix approach to include agricultural supply response and surface water allocation. They are formulated as single optimization problems in which both hydraulic and economic target and instrument variables are included.
- *Linked simulation-optimization models* use the output from an external aquifer simulation model as an input to an economic optimization model. The linked model allows more economic and institutional content in the decision model, while the hydraulic nonlinearities are treated separately to avoid the need for either linearization or nonlinear programming. These models are concerned with economic objectives and have been used to evaluate alternative policies (for example, quotas and taxes on pumping).
- *Hierarchical, or multilevel, models* consider multiple optimizing agents in a hierarchical decision structure. In the simplest case, this would be the government (considered as monolithic) and farmers (considered collectively, using a representative farmer). Typically, such models involve some type of area decomposition, although this is not always necessary. The aquifer model tends to be simplified—for example, a two-dimensional asymmetric, polygonal finite difference system—since only broad, areawide hydraulic variables are considered. Large and complex systems can be decomposed into subsystems and decomposition methods used in solution. But if objectives differ between groups of agents, the resulting problem will be nonconvex, and only local optima may be obtained. Thus, some structural simplification is necessary if such problems are to yield reliable, policy-relevant results. (See Bisschop and others 1982 for a discussion of this problem.)

Finding an Acceptable Policy

Although it is neglected in the literature, all analysts who use modeling methods are familiar with the extended dialogue between modeler and policymaker that is intrinsic to any direct application of results from simulation experiments. Often in such dialogues critical political tradeoffs will emerge when the policymakers are confronted with recommendations (derived from model experimentation) that they find unacceptable. At such junctures it is useful to explore alternative policies that involve less-agonizing political tradeoffs. Rogers, Harrington, and Fiering point out in chapter 8 that programming models often generate flat response surfaces for which alternative feasible solutions differ little from the optimal solution with respect to the value of the criterion function (for example, agricultural income) but represent quite different values for key decision variables. When the simulation model generates such a flat response surface, a solution which differs from the optimal solution by, say, only 1 percent of the value of the criterion function may specify decisions which are much more acceptable politically. For example, changing the sequence of construction for several reservoirs may enable a policymaker to initiate construction in a politically sensitive region at quite low real cost.

An Application of Multilevel Policy Evaluation and Allocation Modeling

The modeling of the Indus Basin Irrigation System of Pakistan (chapter 9) is perhaps the earliest application of modeling methods to an analysis of efficient conjunctive use for a large alluvial basin that preserves enough of the structural characteristics of the system modeled to provide direct answers to practical policy questions. Background information on the development of conjunctive use in Pakistan is discussed in considerable detail by Johnson in chapter 5. As that discussion shows, at present the Pakistani government controls, through provincial irrigation departments, the annual allocation of some 100 million acre-feet (MAF) of surface water diverted through the largest integrated irrigation system in the world. But 75 percent of the some 30 MAF of annual tubewell withdrawals is controlled by hundreds of thousands of individual farmers with private tubewells. Given an estimated overall canal system efficiency of 40 percent, it is clear that canal seepage losses are a principal source of recharge to what is essentially a single unconfined aquifer, albeit one with substantial local variations in water quality. Coordination of the tubewell pumping of large numbers of farmers with the canal water diversions of the government so as to achieve efficient overall resource use has thus far eluded policymakers and administrators in Pakistan. The Indus Basin model was designed not only as an analytical instrument for broad policy evaluation but, most particularly, as a tool for addressing the difficult problems of coordination and control of efficient conjunctive use in the Indus Basin.

To evaluate existing and potential water allocation policies for the basin, a sequence of simulation experiments was performed. Each experiment tested the steady-state response of agricultural production and employment to a different policy rule for water allocation. In all but one case, the groundwater balance constraint was included for each region, and the value of the dual variable (accounting or shadow price) in the solution indicated the tax or subsidy on tubewell pumping required to induce farmers to pump at the desired levels. In addition, endogenous drainage investment variables were included for the saline groundwater areas to test the efficiency of increased water allocations (and associated drainage) for these areas.

Results from these experiments showed overall gains ranging from 17 to 20 percent above base level for agricultural production (value added) and from 14 to 16 percent for agricultural employment. When decomposed between fresh groundwater (FGW) and saline groundwater (SGW) areas, however, the results show FGW increasing only 2 to 4 percent for both measures, while SGW posted gains of 55 to 65 percent in production and 45 to 54 percent in employment. That is, the several alternative allocation rules showed little difference among themselves, but indicated significant gains were possible in comparison with existing policy. The reasons for the striking difference in the output and employment responses of FGW and SGW areas become clearer when the results with respect to land and labor intensity are analyzed. The base level data show divergent levels of input intensity for both inputs between FGW and SGW areas, with higher intensities prevailing in the FGW areas. For the entire basin, the effect of alternative system management policies is to bring the levels of input intensity much closer to equality between the two regions of groundwater quality.

Measurement of the Costs Caused by an Externality

The externality between farmers that is induced by dependence upon a common aquifer or stream-aquifer system for water supply is susceptible to measurement, and all of the policy remedies proposed for removing the inefficiency or inequity caused by the externality depend upon the feasibility of measurement. In practice, the complexity of the physical systems usually precludes the derivation of an analytic solution that can serve as a guide to policy in a variety of cases. Thus the measurement of the externality in any given aquifer or stream-aquifer system is usually accomplished by constructing a numerical simulation model that characterizes the particular system under study. This is the approach that Duloy and I take in chapter 9 in modeling the Indus Basin of Pakistan. Although this approach is both straightforward and flexible, in that it permits simulation of the effects of a number of policy remedies, it remains skill-, data- and time-intensive. For this reason its application tends to be limited to situations in which potential benefits justify the costs.

Zapata has developed another approach to the measurement of the cost of an externality in an aquifer or stream-aquifer system (chapter 10). This approach uses econometric methods and has the virtue of being less skill-, data-, and time-intensive than a numerical simulation model if the time series data it requires are available. It has the disadvantage that simulation of the effects of alternative policy remedies is not feasible. However, where an estimate of the difference between social and private costs can be used to determine key policy parameters (for example, the optimal tax on pumping), Zapata's method can be used to advantage.

In an empirical application of his method to two regions of Mendoza Province in Argentina, Zapata found the excess in social costs over private costs to be 20 percent in one region and 30 percent in the other.

Some Analytical Results

Dynamic Conjunctive Use to Minimize Income Fluctuations

One of the most attractive benefits from efficient conjunctive use of surface and groundwaters is the possibility of using the groundwater aquifer as a massive underground reservoir. It can be managed so as to minimize fluctuations in total water supply caused by random variation in ambient rainfall and rim station inflows (for well-defined alluvial basins). Efficient dynamic programs for conjunctive use increase groundwater withdrawals in times of drought and permit temporary mining of the aquifer to offset diminished surface supplies; in times of heavy surface supplies, such programs call for the greater than normal application of surface water (and diminished groundwater pumping) to replenish aquifer supplies. With a stabilized annual water supply, agricultural production and incomes are also stabilized. Of course, sufficient capacity in both surface delivery and tubewell pumping is needed to meet peak requirements in an efficient, dynamic conjunctive use program. If the surface water delivery system and tubewell pumping are controlled by different agencies, problems of coordination are likely in both aquifer management and investment planning for surface water delivery and tubewell pumping capacity. In practice, as already noted, these two sources of water supply are usually managed by separate agencies. The outcome is usually inefficient conjunctive use and, frequently, overinvestment by one agency owing to the absence of coordination.

In a study of efficient intertemporal use of surface and groundwaters in the South Platte Valley of Colorado, Bredehoeft and Young (chapter 11) found that actual installed well capacity was approximately sufficient to irrigate the entire area, which would appear to be an overinvestment in well capacity. Using a simulation model which coupled the hydrology of a conjunctive stream-aquifer system to an economic model simulating farmer production and investment decisions, they investigated the pattern of conjunctive use in an environment where surface flows are allocated by a system of historic water rights ("first in time, first in right") and tubewell investments and pumping are controlled by individual farmers. Their findings suggest that, given the existing institutional arrangements with respect to allocation of surface and groundwaters and prevailing prices and technology, the optimal groundwater pumping capacity for an individual farmer is that which permits irrigation of his entire acreage. Installing this level of capacity not only maximizes the expected net benefit, but also reduces the variance in annual income to essentially zero. Of course, heavy pumping by upstream users can reduce the return flow available to downstream users from surface supplies; this problem has given rise to litigation in Colorado. Thus the optimal investment response by individual farmers upstream, given prevailing institutional arrangements, imposes an externality on users downstream. In short, in an institutional context of flawed overall coordination, what is optimal for one group of agents may impose losses on others. In the Colorado case adjudication forced the upstream tubewell pumpers to tax themselves to secure additional supplies to make good on the losses to downstream users.

Externalities That Forced Reexamination of Policy

In some arid regions the only natural source of water is groundwater, and if external developments induce rapid population and economic growth, the pressure on the resource may threaten its continued productivity. In chapter 12, Thomas reports on two case studies of countries in arid regions, Qatar and Libya, where the discovery and subsequent development of petroleum resources resulted in rapid population and income growth. In both cases the increase in demand for water led to exploitation of aquifer systems beyond the safe yield and the predictable consequence of seawater intrusion. Advisers from the Food and Agriculture Organization (FAO) have participated in plans to explore the policy options for dealing with the situation, and Thomas briefly describes the results of simulation experiments that model the policy options. Since both governments place a priority on an extreme form of food security, full self-sufficiency, the policy options range from total reliance on food imports to total reliance on domestic production. In both cases Thomas finds that keeping to the safe yield of the local aquifer and relying on imports for the remainder of the food supply is the policy of least cost. At the other extreme, nearly total self-sufficiency is approximately ten times as costly in both cases. The large increase in costs arises in both cases from the need to desalinize seawater or import freshwater from a great distance to achieve the requisite levels of agricultural production. Since both countries have sufficient income from petroleum production to consider the full range of options, the choice of the policymakers (which Thomas does not report) clearly depends on the value weights attached to such policy objectives as full self-sufficiency in food. As in the case of California, however, policymakers apparently want to defer taking positive action on a water resource problem caused by an externality until the consequences of further delay become critical.

Case Study of Planning for Conjunctive Use

In chapter 13 Basu and Ljung analyze a traditional surface irrigation system in India for which actual performance of fifteen years had deviated significantly from the original objectives of the project. Total canal system efficiency was only 20 percent, and this low efficiency was largely responsible for the project's poor performance. However, the area was underlain by both shallow and deep aquifers, and the increase in recharge through leaks in the canal system had induced a secular rise in the water table (for the shallow aquifer) of 0.14 meters a year. With a depth to water table of 6 to 16 meters, a number of farmers had installed shallow tubewells; an estimated 44 million cubic meters (MCM) were pumped annually, about 30 percent of estimated gross additions to aquifer storage. With median canal head releases of 230 MCM yielding only an estimated 46 MCM at the root zone, it was evident that water utilization was significantly conjunctive. The issues were the efficiency and equity of the existing arrangement. Basu and Ljung estimated on the basis of preliminary hydrological balances that an additional 50 MCM of groundwater could be extracted within the safe yield of the shallow and deep aquifers. They projected the exploitation of the deep aquifer from a battery of deep public tubewells, whose discharge would be to the existing canals, and the institution of a program of incentives to encourage farmers to exploit the shallow aquifer.

Summary

Quite clearly the difficulties in achieving efficient conjunctive use of surface and groundwaters discussed in this chapter are problems of middle-aged irrigation systems in alluvial basins. Eventually, however, such difficulties will arise in more than half of the irrigated areas of the world. Moreover, they are already painfully evident in three great alluvial basins—the Indus and Gangetic Plains of South Asia and the North China Plain—that dominate the irrigated agriculture of three nations with close to half of the world's population. Thus the issue of efficient conjunctive use is important to the welfare of a majority of the population of the developing world.

The difficulty is due on the one hand to the physical interdependence of surface water distribution and the groundwater aquifer and on the other hand to the need to coordinate the activities of two sets of agents—individual farmers (or production teams in China) who control tubewell pumping and irrigation system managers who control canal diversions. In the 1960s it was thought that the government could easily coordinate pumping and canal diversions by keeping control of both in state hands, but this concept has been superseded by the reality of farmer control of pumping in the alluvial basins reviewed. In fact, the evidence from the Pakistani experience suggests that even when the government starts with control of both sources in state hands, the management problem becomes too difficult for existing irrigation bureaucracies to resolve satisfactorily. In short, the focus has shifted from blind faith in a benevolent, omnipotent state to skeptical scrutiny of all too fallible and shortsighted irrigation system managers.

In point of fact, the top-down philosophy of the planners has always been conceptually as well as practically flawed because neither the objectives and preferences of the farmers nor the particular moral hazards to which irrigation managers are exposed were taken into account. It is now apparent that a feasible plan for conjunctive use must recognize not only farmers' insistence on making only Pareto-safe policy changes but also the farmers' objective of having some control over the water allocation process. An allocation policy which provides for at least limited farmer participation in the process is also likely to diminish the exposure of irrigation managers to moral hazard. Since farmer participation will inevitably require cooperation among farmers, development of water users' associations becomes a high priority. A program of research on cooperative use of water could lead to improved understanding and more effective irrigation projects.

Along with better understanding of cooperation among water users, there is a need for better understanding of irrigation management. Improvement in irrigation operations has long been neglected in the search for technical (design and investment) solutions. Since an irrigation system may have a service life of a century, with only relatively minor changes in structure, it is perhaps not surprising to find areas where irrigation management has been technically static for decades. As the experience of Taiwan shows, however, significant improvement in performance is possible by designing operations around the constraints of older structures. Given that efficient conjunctive use is information- and skill-intensive, large improvements will be necessary in both the skills of irrigation managers and the quality of the management information and control systems at their disposal.

The review of ongoing experience in conjunctive use for several prominent irrigated regions has yielded several propositions that may serve as working hypotheses until they are confirmed or rejected by further experience. For example, California's experience suggests that farmers will act to cope with external diseconomies only when they are actually confronting serious adverse consequences. Pakistani experience confirms this thesis, since farmers in FGW areas both accept the status quo and have yet to suffer significant losses from externalities, whereas farmers in many SGW areas have suffered significant losses from waterlogging and salinity and have spearheaded a demand for drainage investments.

Another obvious proposition is that farmers (and bureaucrats) will seek a remedy only when they believe a remedy exists. An important implication is that neither farmers nor bureaucrats will accept proposed solutions that are beyond their understanding or experience. For this reason, irrigation professionals have a responsibility to communicate in simple language to all concerned parties the techniques that can improve irrigation efficiency, particularly if they require action on the part of those concerned. A somewhat negative confirmation of this proposition comes from the experience in the North China Plain, which is dominated by the problems presented by the heavy load of silt that the Yellow River picks up in the loess plateau region of China. Both the design of canals and the location of tubewell fields near the Yellow are heavily influenced by the need to cope with the enormous mass of silt that the river carries. But there are no large-scale efforts to control erosion in the loess plateau region. Apparently this action is not believed to be a practical solution to the problem.

Finally, some perspective on the recent advances in analysis and simulation of complex, conjunctive use regimes is in order. The mathematical modeling and computer methods required for the more sophisticated conjunctive use schemes are now being rapidly diffused; they are part of the engineering curricula of universities around the world. But although these methods are being taught at regional universities in India, at present it is difficult to find any of them being applied in India. This will change as a new generation of graduates replaces a generation accustomed to other methods. The real problem is neither analytical methods nor the shortage of people with relevant skills. The real problem is a skepticism about prospective gains combined with a glacial resistance to institutional change—because at least some participants believe that changing the status quo is not Pareto-safe.

Notes

1. To focus attention on some important physical linkages, this discussion of aquifers is limited to a special case. It is possible for an aquifer to be confined in several dimensions, so that natural recharge is from distant sources or else is nonexistent. For such aquifers, evapotranspirative withdrawals are either negligible or nonexistent. Multiple aquifers in several strata separated by impermeable barriers also occur.

2. A stochastic steady state is an equilibrium that exists only in terms of the long-run average; periodic observations will differ from the long-run average because of random (stochastic) variation. Also called stochastic equilibrium.

3. Mean long-term recharge is average annual recharge, exclusive of net pumping, where the average is taken over enough years to obtain a tight confidence interval on the estimate of population mean.

4. Secular reserves are the volume of water that can be mined, or withdrawn from the aquifer, without causing severe damage to the environment.

5. The concept of nonattenuated rights originated with Cheung (1970) and specifies property rights that are exclusive, transferable, and enforced. Compare the discussion by Randall in chapter 2.

6. Knapp and Vaux (1982) present evidence from California which indicates that "fifteen years is not an inordinate period for the settlement of groundwater litigation and the costs of adjudication can run as high as $76 per acre-foot of annual groundwater use."

7. Note that this requirement forces the well owner to get a waiver of assignment of rights from all downstream senior rights holders before a transaction transferring rights is valid. This is an impossible requirement and defeats the mechanism which validates Coase's theorem.

8. For a good description of the Warabandi system, see Reidinger 1974 or Van der Velde 1980.

9. Gisser (1983) argues that the market can perform efficiently in a stream-aquifer system, at least in the case of the Pecos Basin of New Mexico. His argument, however, presumes assignment of exclusive water rights by a state engineer; this is conscious, public interest resource management and not the operation of the invisible hand.

10. Not all resource economists agree that well-defined nonattenuated property rights are impractical for water resources, nor do all resource economists rule out remedies other than those mentioned here. Compare the positions taken by Randall in chapter 2 and Radosevich in chapter 3.

11. See chapter 5 for an illuminating discussion by Johnson on the administrative problems attendant upon the large-scale introduction of public tubewells in Pakistan.

12. For a discussion of distributive arrangements for water, see Maass and Anderson (1978), especially chap. 9.

13. A principle can be called Pareto-safe if it admits only Pareto-efficient improvements.

14. Strictly speaking, this is true only if the shadow price to the farmer is also the social opportunity cost. This may not be true if the farmer is induced to expand output of a heavily subsidized crop. In such cases, the change in gross output in social welfare terms may be negative.

15. Since water charges are subsidized almost universally and water supply is allocated by nonmarket means, water users typically are not in an equilibrium such that the value of the marginal product of water is equal to the water charge. Rather, farmers allocate their given water allocation so that the value of the marginal product of water is equalized across crops at a given point in time. Since there is no assurance that water can be stored at no cost, the value of the marginal product of water is not necessarily equalized over time even on the same farm.

16. The rate of exchange may be zero if substitution is not possible; where substitution is possible, the rate of exchange may change discontinuously when a bound is reached.

17. A rim station is a flow measurement point at which a river carrying runoff from the mountains enters the plains of an alluvial basin.

18. Historical records show 1,600 shifts in river courses for the plain. The Yellow River has had six major changes in course since 602 B.C., affecting an area from Tianjin in the north to Huaiyin in the south. In a very real sense, most of the plain is largely the product of alluvium from the Yellow.

References

Bisschop, Johannes, Wilfred V. Candler, J. H. Duloy, and G. T. O'Mara. 1982. "The Indus Basin Model: A Special Application of Two-Level Programming." *Mathematical Programming* 20:36–38.

Cheung, S. N. 1970. "The Structure of a Contract and the Theory of a Non-Exclusive Resource." *Journal of Law and Economics* 13:49–70.

Daubert, J. T., and R. A. Young. 1982. "Ground-Water Development in Western River-Basins: Large Economic Gains with Unseen Costs." *Ground-Water* 20 (1): 80–85.

Gisser, Mischa. 1983. "Ground-Water: Focussing on the Real Issue." *Journal of Political Economy* 91 (6): 1001–27.

Hall, Warren A., and J. A. Dracup. 1970. *Water Resource Systems Engineering.* New York: McGraw-Hill.

Hotelling, Harold. 1931. "The Economics of Exhaustible Resources." *Journal of Political Economy* 39:137–75.

Knapp, Keith, and H. J. Vaux, Jr. 1982. "Barriers to Effective Ground-Water Management: The California Case." *Ground Water* 20 (1): 61–66.

Maass, Arthur, and Raymond L. Anderson. 1978. *And the Desert Shall Rejoice: Conflict, Growth, and Justice in Arid Environments.* Cambridge, Mass.: MIT Press.

Maass, Arthur, N. M. Hufschmidt, Robert Dorfman, H. A. Thomas, Jr., S. A. Marglin, and G. M. Fair. 1962. *Design of Water-Resource Systems.* Cambridge: Harvard University Press.

Pinder, G. F., and J. D. Bredehoeft. 1968. "Application of a Digital Computer for Aquifer Evaluation." *Water Resources Research* 4 (5): 1069–93.

Reidinger, R. B. 1974. "Institutional Rationing of Canal Water in Northern India: Conflict between Traditional Patterns and Modern Needs." *Economic Development and Cultural Change* 23:79–104.

Remson, Irwin, G. M. Hornberger, and F. J. Molz. 1971. *Numerical Methods in Subsurface Hydrology.* New York: Wiley-Interscience.

Van der Velde, E. J. 1980. "Local Consequences of A Large-Scale Irrigation System in India." In E. W. Coward, Jr., ed. *Irrigation and Agricultural Development in Asia.* Ithaca, N.Y.: Cornell University Press.

Wade, Robert. 1982. "The System of Administrative and Political Corruption: Canal Irrigation in South India." *Journal of Development Studies* 18 (3): 287–328.

Part I

Theoretical Issues

2

Market Failure and the Efficiency of Irrigated Agriculture

Alan Randall

The efficiency of irrigated agriculture is inherently problematical. Economists have believed since Adam Smith (1776), and been able to prove since Arrow and Debreu (1954), that exchange between myriad independent buyers and sellers secured by well-specified property rights results in efficiency. There are other ways to attain efficiency, of course—a central manager with enormous capacity to assemble and process information conceivably could accomplish it—but market exchange under favorable conditions has a built-in tendency toward efficiency.

Economists are aware of a variety of circumstances in which the conditions are not favorable for attaining efficiency through unfettered exchange. Property rights may be incomplete, jointness in use may undermine the independence of individual buyers and sellers, and side effects may proliferate. Some kind of central management may seem essential. As though to reinforce their faith that in the usual scheme of things markets work for the best, economists have dubbed these unfavorable conditions "market failures."

This volume has been organized around the concept of market failures, or "externalities," in irrigated agriculture. My task is to discuss the conceptual basis for a diagnosis of market failure and the solutions that economists have suggested. In this chapter, market failure is the central theme and irrigated agriculture merely one specific area of application. Subsequent chapters focus directly on irrigation issues.

The concept of market failure has an interesting history, which is well worth exploring. As I see it, there have been three main stages: (1) a "market failure–government fix" stage, associated with Pigou (1932) and Bator (1958); (2) an "exclusive private property or disaster" stage, rooted in the writings of Coase (1960) and subsequently developed by a slew of neo-Austrian resource economists; and (3) an emerging thrust in which game theory formulations and experimental evidence seem to be slowly but surely elucidating a vast and varied mosaic of possibilities that lie between the neo-Austrian dichotomy of private property and perdition. I consider it important to follow the argument through all three stages, because so many who would presume to offer advice on market failure issues seem to be mired in stage 1 or stage 2. Although I have made no detailed study of applications to the irrigation economy, I suspect that concepts emerging from stage 3 will be helpful in rethinking the problem of market failure in irrigated agriculture and suggesting novel solutions.

The Market Failure–Government Fix Stage

In the conventional wisdom of the market failure–government fix paradigm, there are four kinds of circumstance in which even a fundamentally competitive economy would experience market failure. These phenomena are externality, public goods, common property resources, and natural monopoly. For three of these phenomena, the conventional solutions call unambiguously for government action: to tax or regulate externalities, to raise revenue for public provision of public goods, and to regulate the pricing policies of natural monopolies. For common property resources, the range of endorsed solutions is broader. Regulation and taxation may be suggested, but it is also frequently suggested that the government specify private property rights and then stand aside as emerging markets restore efficiency.

I will argue that on close scrutiny all four concepts are wanting. All of the valid analytical content of these four terms is contained in two alternative concepts, nonexclusiveness and nonrivalry. Further, these two concepts eliminate much that is confusing and misleading in the four concepts they replace.

In 1954 Gordon introduced the notion of common property resources to the current generation of economists. Gordon's analysis and those of most subsequent authors have focused on the open access resource, that which is unowned. The analytics are basically correct in that context. The problem is that rights to the resource are *nonexclusive* and leave nobody in a position to collect the user costs that reflect increasing scarcity. If properly charged, these user costs would serve to confront current users with the costs their activities impose on future users. With nonexclusiveness, there are no incentives to ration current consumption, conserve for the future, and invest in enhancing the productivity of the resource.

A difficulty arises because the now standard analysis of common property resources is not applicable to property held in common, the *res communis* of ancient Roman law. Ownership is vested in some kind of collective, and rules of access (to some degree exclusive and enforceable, and often transferable under stated conditions) are instituted to adjudicate conflicts among the common owners. Common property organizations may be voluntary associations (corporations and clubs) or government agencies established to manage resources and provide services in the public trust. Ciriacy-Wantrup and Bishop (1975) have drawn attention to the myriad common property institutions that have been developed to handle resource management and exploitation conflicts in traditional and modern societies. They argue correctly that the conventional wisdom (which uses "common property" as a misnomer for "nonexclusiveness") is misleading in very important ways.

There is no question that persistent nonexclusiveness is a recipe for disaster. One may ask why economies based on free enterprise so often handle the nonexclusiveness problem by establishing some form of common property institutions rather than nonattenuated property rights.[1] The answer may lie in the traditional belief that private ownership is inappropriate for certain kinds of resources. More often, I suspect, the answer may be found in the high cost of establishing and enforcing private property arrangements. Where many users share a common fishery or a large oil or groundwater pool, the costs of specifying and enforcing traditional property rights may exceed any gains that might arise from market transactions thus permitted. High transactions costs are often the impediment to private property relationships. Further, it makes no sense to put the blame for high transactions costs on the large-numbers problem, as is often done. Rather, high exclusion and transactions costs are usually attributable to peculiarities in the physical nature of the resource itself.[2] For example, fencing the open sea or large underground pools of liquid resources is technologically more demanding and therefore vastly more expensive than fencing the open range.

The concept of public goods also generates confusion. At least one, and often both, of two quite separate phenomena are involved: nonexclusiveness and nonrivalry. Nonrivalry refers to Samuelson's (1954) notion of a good which may be enjoyed (consumed) by some without diminution of the amount effectively available for others.

The literature has paid much attention to whether both of these phenomena are necessary to make a good "public," or, if one is enough, which one? The question, however, turns out to be quite pointless. Nonexclusiveness and nonrivalry have different economic interpretations and analyses, and they may occur together or separately. Accordingly, I would abandon the term "public goods," along with "common property resources," and focus instead on "nonrivalry" and "nonexclusiveness."

Nonrivalry results from some material circumstances concerning the particular good or the conditions under which the good is provided and distributed. Rationing is not a problem with nonrival goods: once the good has been produced, additional users may be added without imposing any additional costs on the system. But determining the efficient quantity to provide is a special problem for nonrival goods. If marginal willingness to pay (WTP) is determined for each individual and aggregated (vertically) across all potential users, the efficient level of provision can be identified: the level at which aggregate marginal WTP just equals the marginal cost of provision, given that aggregate total WTP exceeds total cost. So far, so good. Without exclusion, a government could use general revenues to provide the efficient quantity of nonrival goods, as long as it had good information about aggregate marginal WTP. But when aggregate WTP data must necessarily be derived at the outset from self-reported individual WTP, there is concern that individuals may indulge in false reporting for strategic reasons.

If it were possible to exclude all who did not pay the going price, revenue could be generated directly from users. Private sector provision of nonrival goods would be possible, as would public provision financed by user charges. These kinds of arrangements have some appeal and may even be second-best solutions.[3] They would not, however, be efficient solutions. Some individuals with low but positive valuations would be excluded, and this would be inefficient since their use of the nonrival good would impose no additional costs.

For complete efficiency, perfect price discrimination is necessary. Each user would have to pay his own WTP. Obviously, this would require a very special and demanding kind of exclusion. A turnstile or a tollbooth may be sufficient to exclude those who do not pay the going price, but such devices would be entirely ineffective at excluding those who did not pay their own individual marginal valuations. A technology for price-discriminatory exclusion may one day be developed. One thinks of truth serum or new developments based on the polygraph. Up to now, more progress has been made along a rather different line of

Table 2-1. *A Classification of Goods Based on Concepts of Rivalry and Exclusiveness*

	Level of exclusion		
Type of good	*Nonexclusive*	*Exclusive*	*Price-discriminatory exclusive*
Nonrival	Private provision, or public provision financed by user charges, is impossible. Pareto-efficiency is unattainable.	Private provision and public provision financed by user charges are feasible and may permit second-best solution. Pareto-efficiency is unattainable.	Private provision and public provision financed by user charges are feasible. With perfect price discrimination, Pareto-efficiency is feasible.
Congestible	Private provision, or public provision financed by user charges, is impossible. Pareto-efficiency is unattainable.	Private provision and public provision financed by user charges are feasible. Time-variable charges may permit second-best solution. Pareto-efficiency is unattainable.	Private provision and public provision financed by user charges are feasible. With time-variable user charges and perfect price discrimination, Pareto-efficiency may be feasible.
Rival	Private provision, or public provision financed by user charges, is impossible. Pareto-efficiency is unattainable.	The ordinary private goods case. Competitive market equilibrium is Pareto-efficient.	Price discrimination leads to excess profits and violates conditions for Pareto-efficiency.

attack: incentive-compatible mechanisms. Typically, these are rather complex systems of multipart taxes carefully structured so that an individual who volunteers his true valuation will emerge better-off than someone who attempts to beat the system by strategic reporting of false values. Currently, incentive-compatible devices may be found in the realm of economic theory and experimental economics, but real-world applications are in their infancy.

While pure nonrival goods are relatively rare, so-called congestible goods characterize substantial sectors of the economy. Congestible goods have high initial capital costs and capacity constraints. When use is much less than capacity, nonrivalry is the order of the day: additional users impose only trivial costs on the system. As the capacity constraint is approached, congestion sets in and additional users impose rapidly increasing costs on the system. At full capacity it is literally impossible to add a user without simultaneously removing another. For congestible goods, the economic analysis is similar to that for nonrival goods, but an additional complication enters. Where the level of demand varies with the time of demand, appropriate prices may well be different at different times. In general, ordinary exclusion may permit second-best solutions in which revenues collected from users cover the costs of provision. Perfect price discrimination and the especially demanding form of exclusion it implies are required for complete efficiency in the provision of congestible goods.

It is possible to devise a system for categorizing goods according to three levels of exclusion (nonexclusiveness, ordinary exclusion, and price-discriminatory exclusion) and three kinds of goods (nonrival, congestible, and rival). It is possible to conceive of goods in all of the nine categories thus created. In Table 2-1 the results of economic analysis are summarized for all nine categories.

Let me review the argument to this point. The conventional concepts of common property resources and public goods are fraught with difficulty and confusion. It is more helpful to focus upon nonexclusiveness and nonrivalry, phenomena which may occur separately or together. Inefficiencies induced by nonexclusiveness may, given an adequate exclusion technology and the political-institutional will to implement it, be resolved by privatization. Where nonrivalry is the problem, ordinary exclusion is not enough; efficiency requires price-discriminatory exclusion. Finally, the existence of nonexclusiveness or nonrivalry presents a prima facie case for market failure. A conclusive case, however, requires a demonstration not only that market performance is imperfect but also that alternative institutions would do better.

Although there is nothing wrong with the economic analyses usually associated with natural monopoly, that phenomenon is entirely captured by the construct of congestible goods. Given the analytical possibilities emanating from a classification based on concepts of rivalry and exclusion, natural monopoly becomes redundant.

An externality is defined as a situation in which the welfare of one is influenced by activities under the control of someone else. In the conventional wisdom of stage 1, externalities were seen as market failures, that is, inefficient situations inviting governmental attempts at mitigation. Where the externality was harmful (that is, where the welfare of the affected party was diminished by the externality), the typical prescription was to regulate or, better yet, tax the externality into submission.

Since the writings of Coase (1960) and Buchanan and Stubblebine (1962), most authors have focused on the subset of externalities that causes inefficiency (the Pareto-relevant externalities, in the jargon). Many categories of interactions which satisfy the general definition of externality are resolved efficiently in markets; for them, no possibility of Pareto-relevance exists when markets function well. For other kinds of interactions—air pollution and water pollution are commonly cited examples—it is not immediately clear that, in the ordinary course of events, markets take care of the inefficiencies.

A Pareto-relevant externality is so defined that unrealized potential gains from trade are inherent therein. The Coase theorem states that, given nonattenuated property rights, market transactions will realize the gains from trade and thus eliminate the inefficiency. Some of the nuisance (for example, air pollution) will almost surely remain, but it will be Pareto-irrelevant: to reduce the nuisance still further would cost more than the benefits.

A special case of the Coase theorem, one of more interest to economics teachers than to policymakers, asserts that the equilibrium amount of the nuisance will be invariant with the initial specification of rights. Regardless of whether the law protects polluters or receptors at the outset, after all trading opportunities have been exploited the remaining pollution will be the same. To get this result it is necessary to assume that transactions costs are zero (specification, transfer, and enforcement of rights are costless activities) and there are no income effects. But transactions costs are always positive and income effects are significant in some important cases. Under these more realistic assumptions, initial assignment of rights *does* affect the equilibrium outcome. There will be more of the nuisance remaining at equilibrium when the law protects polluters than when it protects receptors. But the general result of the Coase theorem remains: each of these different equilibriums is efficient in its own terms. What is efficient depends on the initial distribution of endowments and rights.

The impact of the Coase theorem is that inefficient, or Pareto-relevant, externality cannot persist. The imperatives of trade make for an inherently unstable disequilibrium situation.

Pollution, for example, may persist in excessive quantities. If the initial assignment of rights favors polluters and transactions costs are so large as to preclude any trade, all of the pollution will remain at equilibrium. Excessive pollution is a persistent problem, although we have a theorem proclaiming that Pareto-relevant externality is not. There is no inconsistency here, it turns out. The high transactions costs cannot, as we have already seen, be attributed to the large-numbers problem. Rather, they must be due to other aspects of the situation: nonexclusiveness or nonrivalry or both.

In economies that maintain institutions conducive to trade and efficiency, those things called externalities cannot persist in excessive quantities unless accompanied by nonexclusiveness or nonrivalry or both. Inefficient externality is not, by itself, persistent. Further, the effects and analysis of, and recommended solutions for, externality and nonexclusiveness or externality and nonrivalry are the same as for nonexclusiveness or nonrivalry alone. The inescapable conclusion is that externality adds nothing to the lexicon of market failure.[4]

Whereas the conventional wisdom of stage 1 offers the concepts of externality, common property resources, public goods, and natural monopoly, closer analyses find content only in the concepts addressing exclusiveness or the lack thereof and the nonrival or congestible nature of certain goods. Further, perception of the policy significance of market failure has shifted in an important way. Whereas market failure was once treated as a universal rationale for a government fix, the current approach insists that policy imperatives follow only when the diagnosis of market failure is accompanied by a demonstration that some other arrangements would actually do better.

The Exclusive Private Property or Disaster Stage

In the process of developing and critiquing stage 1 of the intellectual history of market failure, I have laid much of the groundwork for discussing stage 2. This second stage took its cue from the mid-1950s analyses of Samuelson (1954) and Gordon (1954) and the voluminous literature that followed Coase's seminal paper (1960).

Samuelson and Gordon did not merely show that rivalry and nonexclusiveness, respectively, were substantial impediments to Pareto-efficiency in a decentralized economy. Their analyses predicted the total collapse of the nonrival and nonexclusive economic sectors unless government stepped in, coercively, to save the day. Further, as we have already seen, Gordon's analysis of the open access or nonexclusive resource problem was mislabeled: the implication was that it referred to common property arrangements as a broad class.

Coase's analysis focused on nonattenuated property rights as a sufficient condition for efficiency in an economy where externality and nonexclusiveness might otherwise cause problems. The burden of proof was switched to those who would claim market failure in any particular case (Demsetz 1964 and 1969). They could now be called upon to show that what appeared to be market failure was not actually an efficient market solution. The only escape route left open, it seemed, was to argue that property rights were attenuated in some important way. An obvious prescription was for the government to establish nonattenuated property rights and then stand aside as the market took care of things. Sustained government activism—the regulation, taxation, and public provision amelioratives typically prescribed in stage 1—was unnecessary.

As the Coasian tradition developed, it was argued with increasing generality that attenuation of rights was endemic in the public sector itself. That, of course, took the argument one rather large step further. A sustained posture of government activism in control of market failure was not merely unnecessary, it was undesirable.

From this foundation developed the conventional wisdom of stage 2. So-called market failures were caused mostly by attenuated property rights, and nonexclusiveness was far and away the greatest part of that problem. Privatization was the appropriate policy response to diagnosed inefficiencies. Thus, Anderson and Hill (1976) argued, essentially, that the economic history of the United States could be encapsulated as a triumphal march of private property institutions from east to west with the predictable result of prosperity unparalleled in other times and places. Schmid (1977) raised the argument (originated by Ciriacy-Wantrup and Bishop 1975) that Anderson and Hill had ignored a whole universe of institutional possibilities, some of them quite serviceable, between the extremes of exclusive private property and open access. The Anderson and Hill (1977) response was scathing: the possibilities to which Schmid referred were essentially uninteresting, since any efficiency properties these institutions possessed must surely be attributable to some degree of exclusiveness inherent in them. Further, incomplete exclusiveness implied incomplete efficiency; why not go all the way? In this, Anderson and Hill were faithfully reflecting the mindset of stage 2: most of the issues raised by the old-fashioned notion of market failure can be addressed with a simple dichotomy between exclusive private property, which promotes efficiency, and nonexclusiveness, which leads to the collapse of the economic sectors it afflicts.

For this simple analysis, nonrivalry poses a difficulty, since ordinary exclusion is not sufficient to restore a nonrival goods sector to efficiency. Some proponents of the conventional wisdom of stage 2 (Anderson and Hill, for example) tend to play down the issue of nonrivalry. Others (Buchanan 1977, for example) confront nonrivalry directly, favoring voluntary taxation schemes in the tradition of Lindahl (1958) and Wicksell (1958) and endorsing the modern search for incentive-compatible collective decision mechanisms.

One should credit the Coasian tradition with important accomplishments in exposing the fallacies of the stage 1 concept of market failure. Nevertheless, I would argue that the stage 2 alternative—private property or perdition—is itself quite unsatisfactory. Perhaps the neo-Austrian proponents of stage 2 have simply taken the analyses of Samuelson, Gordon, and Coase too far.

New Approaches

Perhaps no long-established prediction of economics has been so thoroughly refuted as Samuelson's and Gordon's prediction of total collapse in the nonrival and nonexclusive sectors. There is evidence all around us that these sectors are seldom efficient, which supports Samuelson's and Gordon's predictions with respect to efficiency. But there is also ample evidence that, despite their predictions, these sectors have not totally collapsed.

The Samuelson-Gordon tradition left an escape route: government could coercively regulate or tax and thereby provide what citizens will not provide through markets or other endogenous institutions. But this escape route is unsatisfactory. More contemporary analyses (reflecting the Coasian tradition and a variety of other influences) treat government itself as endogenous. From this perspective, government is not a wise external force capable of disciplining an unruly society. Rather, government emerges, warts and all, from society. How, then, can government (which is endogenous to society) impose upon society that which society cannot agree to impose on itself? Once the endogenicity of government is conceded, it is impossible to reconcile Samuelson's and Gordon's prediction of collapse with the observation that the nonrival and nonexclusive sectors seem to do no worse than limp along and often perform passably well. Clearly, Samuelson's and Gordon's theory of market failure is inadequate and misleading.

In the past two decades several novel and related approaches have emerged to shed new light on the possibilities for collective action. These approaches include game theory formulations of the nonrivalry and nonexclusiveness problems (Sen 1967; Runge 1981), resource allocation mechanisms (Hurwicz 1973), the theory of teams (Marshak and Radner 1971), incentive-compatible mechanisms (Groves and Ledyard 1980), and principal-agent models (Arrow 1986).

An early and influential game theory problem is the prisoner's dilemma. A two-person single-period prisoner's dilemma may be expressed as follows. First, we establish a minimal notation. Define:

S_i: the strategy of i ($i = 1, 2$)
$S_i = 1$: i contributes (or cooperates, and so on)
$= 0$: i defects (or plays a selfish strategy, and so on)
(S_1, S_2): the strategies of 1 and 2 which together determine the outcome of the game
$F_i(S_1, S_2)$: the payoff to i, which is some function of (S_1, S_2).

Two prisoners are interrogated separately by the police, who, having little independent evidence of their guilt, are willing to bargain for confessions. The police confront each with the following individual payoffs from the various possible outcomes:

$$\text{for 1, } F_1(0,1) > F_1(1,1) > F_1(0,0) > F_1(1,0)$$
$$\text{for 2, } F_2(1,0) > F_2(1,1) > F_2(0,0) > F_2(0,1).$$

Each is best-off if he confesses and the other does not and worst-off if he steadfastly denies the guilt of both and the other confesses. But both would prefer the "both deny" outcome to the "both confess" outcome.

Perhaps a different example may help clarify what is at stake. The disarmament game has the same structure. For country 1, the best of all worlds occurs when country 2 disarms unilaterally. Both countries are better-off when both disarm than when neither disarms.

It is clear that the outcome (0,0) is Pareto-inferior; that is, for both it is inferior to (1,1). But if the prisoners cannot communicate, each will choose his maximin strategy of 0. Thus the equilibrium outcome of the game is (0,0), a Pareto-inferior outcome.

The countries can communicate in disarmament negotiations. Nevertheless, although they may agree to the cooperative outcome (1,1), each has strong incentives to cheat subsequently in order to gain an advantage and to make sure the other does not. The noncooperative outcome (0,0) is likely to be what actually emerges.

By the 1960s it was widely held that the Samuelson-Gordon analyses of market failure could be reformulated as single-period *n*-person prisoner's dilemmas. Such reformulation would, of course, reconfirm Samuelson's and Gordon's prediction of total collapse in the nonrival nonexclusive sectors.

The single-period prisoner's dilemma was only the beginning, however. It was soon realized that the prisoner's dilemma is not necessarily the proper specification for nonrivalry and nonexclusiveness problems (Sen 1967; Dasgupta and Heal 1977). As Shubik (1981) observed, games of pure opposition have many uses in, for example, military tactics but relatively few applications in economics. Consider the following alternative formulations.

The assurance game:

$$F_1(1,1) > F_1(0,0) > F_1(0,1) = F_1(1,0)$$
$$F_2(1,1) > F_2(0,0) > F_2(1,0) = F_2(0,1).$$

In this game (1,1) is Pareto-optimal, but both parties prefer (0,0) to outcomes in which they play different strategies. The problem is assurance: each will play $S_i = 1$ if assured the other will, too. Once agreement is reached, there is no incentive to defect subsequently.

Runge (1981) argues for the assurance game on two grounds. First, it is the appropriate formulation for nonexclusiveness problems characterized by nonseparability among users (for example, where the cost function of one user is influenced by decisions made by others). Second, it appeals to fair-mindedness ("I will if you will"), arguably a rather common human instinct, rather than to, say, unilateral benevolence.

The unanimity game:

$$F_1(1,1) > F_1(0,0) = F_1(0,1) > F_1(1,0)$$
$$F_2(1,1) > F_2(0,0) = F_2(1,0) > F_2(0,1).$$

In this game, nothing of value is produced unless both parties contribute; successful action has to be unanimous. Thus, $F_1(0,0) = F_1(0,1)$, and $F_1(1,0)$ is the worst solution of all for 1 because 1 contributes but nothing is produced. Fairmindedness is not involved.

The cooperative solution (1,1) is Pareto-superior, but independent maximin strategies for each player will generate the (0,0) outcome. Coordination would result in the cooperative solution, and there would be no incentive for subsequent defection.

The unanimity game is characteristic of certain committee and legislative environments. It also applies to nonrival goods with high fixed costs, so that none is produced unless all players contribute.

The congestible goods game:

$$F_1(1,1) > F_1(0,1) > F_1(0,0) > F_1(1,0)$$
$$F_2(1,1) > F_2(1,0) > F_2(0,0) > F_2(0,1).$$

Benefits are nonrival, and the marginal cost of providing the good to an additional user declines as the number of users increases, as is characteristic of congestible goods operating well within the capacity constraint. The benefits from a single contribution are positive but are valued less than the cost of contributing. If both parties contribute, the nonrival benefits exceed the cost of either individual contribution, and the increment in nonrival benefits from the last-received contribution exceeds its cost. Free riding—for example, (0,1) allows 1 to free ride—is individually preferred to nonproduction (0,0). For either party, contributing while the other does not yields the worst outcome.

Again, the cooperative solution (1,1) is Pareto-superior, but independent maximin strategies by each player will generate the (0,0) outcome. With coordination, (1,1) will be chosen and there is no incentive for subsequent defection.

Although the assurance, unanimity, and congestible goods games differ a little, in all three coordinated strategies would permit stable, Pareto-optimal cooperative solutions. Further, it is argued that each of these games is a more appropriate specification than the prisoner's dilemma for a particular class of problems emanating from nonexclusiveness or nonrivalry.[5] For these problems, a more careful specification of the game theory formulation suggests that the collapse predicted by Samuelson and Gordon is by no means inevitable.

The story does not end here, but there is a fork in the road: one path of further inquiry concerns mechanisms for coordination, and the other concerns the outcomes when games, including the prisoner's dilemma, are repeated.

The demonstration that, for several relevant classes of games, coordinated strategies permit stable, Pareto-efficient cooperative solutions is not entirely comforting. Coordination is likely to be a costly activity, and complete coordination, if it requires consultation among all participants, may be prohibitively costly. Private (that is, rival and exclusive) goods markets work well because prices convey, in simple signals, sufficient information and incentives to accomplish coordination and neither centralized management nor direct consultation among all market participants is necessary. Perhaps signaling devices can be developed for adequate and cost-effective coordination so that cooperative arrangements in large organizations dealing with nonrival and nonexclusive goods are reasonably stable and efficient. This is the working hypothesis that motivates research on principal-agent models, team theory, resource allocation mechanisms, and incentive-compatible mechanisms.

For principal-agent models, the following situations are typical. Total costs of loss and damage may be reduced if insured parties have some incentives for loss-avoiding behavior; can insurance policies with appropriate incentives be designed? If the work effort of individual agents cannot be monitored directly, what incentives can the manager devise to encourage agent efficiency without incurring excessive turnover of agents? If emissions of individual polluters cannot be monitored fully, can the pollution control authority devise incentives for reasonably efficient pollution control? Given that bidders for federal oil leases have some knowledge that is unavailable to other bidders and the government, can auction rules be devised to maximize some objective, such as government receipts or some broader measure of social welfare?

Each of these problems is characterized by hidden action (the agent can take some actions unobserved by the principal) or hidden information (the agent knows some things the principal does not). An interesting variant is the problem of a single principal and many agents, where the principal can observe the combined output of all agents but not the individual output of any one of them. The relevance of this kind of thinking to nonexclusiveness and nonrivalry problems is obvious. This particular principal-agent problem is a team problem and must be formulated as a game; the example serves to illustrate the close relationships among the various approaches.

The literature on principal-agent problems is substantial and often highly mathematical. No attempt at careful review and evaluation is offered here, but some impressions can be conveyed. Considerable progress has been made in modeling information requirements and group performance, given various combinations of problems and incentives. Results about information requirements provide indirect evidence about the transactions costs associated with various arrangements. While principal-agent models reconfirm the efficiency of price signals in a neoclassical competitive economy, they offer no support for the "private property or total collapse" thesis of the neo-Austrians. A wide variety of workable arrangements, with outcomes falling between Pareto-efficiency and collapse, can be identified for diverse problems exhibiting aspects of nonexclusiveness or nonrivalry.

The literature on incentive-compatible mechanisms (Groves and Ledyard 1980) has identified the general form of a tax rule for which truthful revelation of willingness to pay for nonrival goods is the dominant strategy. The basic idea is that the individual's tax is independent of his announced WTP but depends on his message's effect on the total amount of nonrival goods collectively provided. Current theoretical results are restricted to cases in which utility functions are additively separable.

A growing literature reports the performance of experimental nonexclusive or nonrival goods economies (Smith 1980). A frequent result is that free riding is much less than universal even when incentives encourage it. Further, voluntary taxation produces near-optimal amounts of collective goods under tax rules that fall short of incentive compatibility.

Now, we take the second path and consider repeated games. Return to the prisoner's dilemma. For individual 1, the payoffs from the various outcomes are ranked:

$$F_1\,(0,1) > F_1\,(1,1) > F_1\,(0,0) > F_1\,(1,0).$$

For any *i*, the preferred outcome occurs when *i* defects while all others contribute. If the game is repeated, however, each player would learn rather quickly that playing the $S_i = 0$ strategy leads to a Pareto-inferior outcome (0,0) because others would surely defect, too. For 1, (0,1) is the preferred solution, but it is unstable. Since (0,1) quickly degenerates to (0,0), perhaps all players have an incentive to attempt to achieve the (1,1) outcome. This is the basic motivation for research with repeated prisoner's dilemmas.

The first result is not helpful. A prisoner's dilemma repeated many times quickly degenerates to (0,0). The reason is easy to see. For any *i* involved in a *t*-times repeated prisoner's dilemma, the preferred outcome arises when all contribute on the first $t - 1$ rounds and all but *i* contribute on the tth round. Each player is motivated to defect before the others. By infinite regress, the game degenerates to "all players defect" in the initial round.

Some favorable results are also evident, however. Cooperative equilibriums may be stable if the game is repeated indefinitely or stochastically many times (that is, players do not know when it will end). These results depend on various assumptions about what information is available to players and what strategies they use. Obviously, it is favorable for stable cooperative solutions if players can observe the previous-period strategy of each player. Radner (1981) and Klepper (1983) have, however, obtained favorable results with principal-agent games, indefinitely repeated, where only previous-period total group contributions are observable.

This kind of game calls for each player to announce at the outset that he will use a trigger strategy. For example, if the previous-period group output falls below some specified amount, the player will defect in the present period. The player's problem is to figure out the appropriate trigger amount. Klepper has identified a test statistic that players can use and has shown that, perhaps surprisingly, the test statistic becomes more precisely defined as the number of players increases. This result is favorable for stable cooperative solutions in repeated prisoner's dilemmas.

All of the results discussed to this point are derived with models that assume the players are unknown to one another. Results become stronger if reputation effects are considered (Akerlof 1983). If the players know one another and develop predictions about their behavior, and if the best way to earn a reputation for cooperative behavior is to cooperate repeatedly, Kreps and others (1982) have shown that cooperative solutions may be stable even for finitely repeated prisoner's dilemmas.

One rather obvious trigger strategy is the tit-for-tat: each player contributes on each round, until he believes that significant defection has occurred on the prior round, and then he defects. Axelrod (1982) has conducted computer simulations of repeated prisoner's dilemmas and reports that players using the tit-for-tat strategy regularly obtained more favorable outcomes over the long haul than players using other strategies. When all players but one use the tit-for-tat, cooperation may be restored in a repeated game that had degenerated. The defector, observing that his action was met with a tit-for-tat reaction on the following round, has a strong motivation to contribute on the next play. The tit-for-tat players would then contribute on the next round, restoring the cooperative solution, and the defector, having learned a lesson, would likely contribute thereafter.

Two final observations are in order. First, most of the favorable results cited were obtained with repeated prisoner's dilemmas. Clearly, the prospects for favorable results are more promising if the game that is repeated is not a prisoner's dilemma but one of the alternative games—the assurance, unanimity, or congestible goods game. Second, it occurs to me that a stochastically repeated game with reputation effects and opportunity for partial monitoring of individual contributions may be a sound metaphor for real life. The prospects for stable cooperative solutions for such a game are by no means trivial.

The general impression that emerges from this diverse literature is that the Samuelson and Gordon collapse thesis is refuted as convincingly by superior theoretical models as it is by observation of the real world. Again, none of this shows that the nonexclusive and nonrival sectors will achieve efficiency. Samuelson and Gordon were right about that. But total collapse is not inevitable. Theoretical and experimental results show that a wide variety of signals and incentives may result in reasonably stable, passably well-performing collective goods economies. These results mesh well with the observation by Ciriacy-Wantrup and Bishop that a wide variety of common property arrangements can be observed to operate, stably and effectively, in both traditional and modern societies.

Given the understanding that emerges from the stage 3 literature, the stage 2 "private property or perdition" argument of the neo-Austrians is seen as materially false. It may faithfully represent a particular ideological position, but the theoretical and observational basis to convert that ideological position to a valid policy prescription is lacking.

Let there be no argument: private property arrangements have some demonstrable merits for handling rival and exclusive goods. Further, where exclusion is effective and inexpensive, it is a meritorious remedy for problems arising from nonexclusiveness. But private property arrangements are at a severe disadvantage when nonrivalry is the problem or when private property exclusion is technologically difficult. The mind-set that emerges from stage 3—let us explore the myriad of diverse possibilities that lie between private property and simple open access—is more helpful, and more defensible, than the stage 2 "private property or perdition" mind-set.

Many of the analyses I have grouped together and labeled stage 3 are institutionally agnostic. In a principal-agent model the principal could be a corporate boss, a Mafia capo, the head of a bureaucracy, a Soviet commandant, an elected chief executive, or society at large. A stable cooperative solution could take the form of a voluntary association, a standing committee, or a constitutional government.

That makes these models adaptable to a wide variety of collective action problems. But they all exclude at least one institutional form: a stable solution imposed on an unruly society by an all-wise exogenous government. Thus, the thinking of stage 3 rejects the precepts of stage 1 as surely as it does those of stage 2.

Market Failures and Efficiency in Irrigation Economies

Irrigation economies are typically characterized by interdependencies. Irrigators draw water from a common river, channel, or groundwater pool, and the tailwaters return to the groundwater or the river downstream. Channels are congestible, and during times of peak water demand users impose costs on one another. Groundwater may be nonexclusive, as a result of overexploitation. In the absence of effective metering of individual withdrawals, this problem would also apply to water in rivers and channels. Polluted tailwaters may be a nonexclusive discommodity, leading irrigators to pay little attention to controlling damage from that source. Principal-agent problems may bedevil any administrative attempt to resolve these problems.

Typically, these difficulties are called market failures. How would an economist diagnose these problems, and what solutions would be prescribed? As we have seen, it would depend on the economist's concept of market failure.

An economist with a stage 1 mind-set would classify these problems according to whether they are attributed to externalities, public goods, common property resources, or natural monopolies. He would recommend that the government eliminate these problems by regulation or, preferably, taxation. An exception might be made in the case of common property resources, if the economist were convinced that private property institutions could be implemented at less than prohibitive cost. In that case, he might recommend that the government establish private property rights and then stand aside as decentralized markets take care of the problem.

An economist with a stage 2 mind-set would be deeply troubled by these diagnoses and recommendations. He would maintain that externality is not a viable concept, that the absence of an observable market may itself be an efficient market solution, and that the fundamental problem is the attenuation of property rights, in which case government agencies are even more susceptible to the problem than market economies. Ever skeptical of government fixes, the economist would insist that the case for market failure be established not merely by showing the inefficiency of markets but also by demonstrating that some other arrangement would do better.

All of that is perfectly acceptable. The stage 2 economist, however, would tend to ignore nonrivalry problems and concentrate on those caused by nonexclusiveness and would tend to misspecify the latter by positing a simple dichotomy between idealized private property and nonexclusiveness. Confronted with the above-mentioned market failures in the irrigation economy, his inclination might be to suggest that all of these problems could be solved once and for all if only a single individual were empowered to collect all of the rents generated by the irrigation project. Then the rent collector would have an unambiguous incentive to maximize the efficiency of the whole project. Alternatively, the economist might seek to maximize the scope of markets within the project, so that efficient prices would coordinate withdrawals and return flows.

Strong objections would likely be raised to both proposals. The "single rent collector" proposal would violate other valid policy objectives. Although internal markets (in withdrawals and return flows, for example) may have considerable merit, it is unlikely that they would resolve all of the market failures; in some cases effective exclusion is just too expensive. The stage 2 economist, schooled in the "private property or total collapse" tradition, would be unequipped to deal with these remaining problems.

In the emerging understanding that I have labeled stage 3, the stage 1 concepts and definitions of market failure are rejected, and the focus is instead on nonexclusiveness and nonrivalry. The stage 3 economist is skeptical of the ability of a paternal government to effectively impose efficiency on an aberrant economy via regulation and taxation. At the same time, however, the stage 3 mind-set rejects the stage 2 "private property or total collapse" diagnosis for the nonexclusive and nonrival sectors. The Samuelson and Gordon predictions of total collapse are seen to be refuted by observations of the real world and by newer, more realistic theoretical models that explore the possibilities for stable cooperative action. There is no denying that exclusive private property arrangements have desirable built-in incentives and thus are appropriate when conditions are favorable for them—that is, when exclusion costs are reasonably low and sociocultural traditions are amenable. There is, however, a willingness to consider the vast mosaic of possibilities that are now perceived to lie between the extremes of exclusive private property and uncontrolled access.

I do not bring with me prepackaged stage 3 solutions for market failure in the irrigation economy. Nevertheless, it seems clear that recent thinking allows some reformulation of more traditional analyses of market failure in irrigation and suggests some new possibilities.

Uncontrolled individual pumping of groundwater leads to excessive extraction, which lowers the water table and increases pumping costs for all irrigators. This problem is often diagnosed as a physical externality, whereas I would suggest that it is a simple case of nonexclusiveness with rivalry. The array of stage 1 and stage 2 solutions has been discussed, with pessimistic conclusions. Private control of the aquifer reduces to a prisoner's dilemma, and the stable cooperative solution invariably fails. Public ownership fails because the low-level adminstrators will cheat and the costs of monitoring them are prohibitive. Tax-subsidy solutions are viewed as promising in concept, but in practice they require a central government that is strong enough to impose them on a community of recalcitrant farmers.

Stage 3 thinking may be open to more alternatives and less pessimistic about those considered. It may be possible to structure private pumping as a repeated game. With reputation effects, monitoring of water tables, and partial monitoring of individual withdrawals, the chances for stable cooperative solution increase. Principal-agent models may suggest ingenious systems of incentives and signals to induce low-level public employees to better serve the objectives of the principal, the irrigation community.

Stage 3 thinking counsels us that the nonexclusive and nonrival sectors may be less vulnerable to economic collapse than economists have previously thought. It invites us to consider the myriad possibilities that exist between the extremes of exclusive private property, free-for-all, and a central government fix via manipulated prices or regulations.

Notes

1. Cheung (1970) defines nonattenuated property rights as exclusive, transferable, enforced, and in no way inconsistent with the marginal conditions for Pareto-optimality.

2. The large-numbers problem alone will never cause prohibitive transactions costs. The market for bread, with myriad buyers and sellers, works as well as any other and much better than the market for clean air, which has a similar number of potential participants.

3. See, for example, the discussion in Davis and Whinston (1967).

4. See Cheung (1970), Dahlman (1979), and Randall (1983) for elaborations of this argument.

5. Nevertheless, not all nonexclusiveness and nonrivalry problems escape the prisoner's dilemma. The *n*-person prisoner's dilemma seems to be the appropriate formulation for the problem concerning a nonrival and nonexclusive good with separable individual cost functions, increasing marginal costs, and decreasing marginal benefits. In that case, *i* would prefer an outcome in which he defects while all others contribute.

References

Akerlof, George. 1983. "Loyalty Filters." *American Economic Review* 73:54–63.

Anderson, T. L., and P. J. Hill. 1976. "The Role of Private Property in the History of American Agriculture, 1776–1976." *American Journal of Agricultural Economics* 58:937–45.

———. 1977. "The Role of Private Property in the History of American Agriculture, 1776–1976: Reply." *American Journal of Agricultural Economics* 59:590–91.

Arrow, K. J. 1986. "Agency and the Market." In K. J. Arrow and M. D. Intrilligator, eds. *Handbook of Mathematical Economics.* Amsterdam: North-Holland.

———, and Gerard Debreu. 1954. "Existence of an Equilibrium for a Competitive Economy." *Econometrica* 22:265–90.

Axelrod, Robert. 1982. "The Emergence of Cooperation among Egoists." *American Political Science Review* 75:306–18.

Bator, F. M. 1958. "The Anatomy of Market Failure." *Quarterly Journal of Economics* 72:351–79.

Buchanan, J. M. 1977. *Freedom in Constitutional Contract.* College Station, Tex.: Texas A & M University Press.

———, and W. C. Stubblebine. 1962. "Externality." *Economica* 29:371–84.

Cheung, S. N. S. 1970. "The Structure of a Contract and the Theory of a Non-Exclusive Resource." *Journal of Law and Economics* 13:49–70.

Ciriacy-Wantrup, S. von, and R. C. Bishop. 1975. "'Common Property,' as a Concept in Natural Resources Policy." *Natural Resources Journal* 15:713–27.

Coase, Ronald H. 1960. "The Problem of Social Cost." *Journal of Law and Economics* 3:1–44.

Dahlman, Carl. 1979. "The Problem of Externality." *Journal of Law and Economics* 22:141–62.

Dasgupta, P. S., and G. M. Heal. 1977. *Economic Theory and Exhaustible Resources.* Cambridge: Cambridge University Press.

Davis, O. A., and A. C. Whinston. 1967. "On the Distinction between Private and Public Goods." *American Economic Review* 57:366–73.

Demsetz, Harold. 1964. "The Exchange and Enforcement of Property Rights." *Journal of Law and Economics* 7:11–26.

———, 1969. "Information and Efficiency: Another Viewpoint." *Journal of Law and Economics* 12:1–22.

Gordon, H. S. 1954. "The Economic Theory of a Common Property Resource: The Fishery." *Journal of Political Economy* 62:124–42.

Groves, Theodore, and John Ledyard. 1980. "The Existence of Efficient and Incentive Compatible Equilibria with Public Goods." *Econometrica* 48:1487–506.

Hurwicz, Leonid. 1973. "The Design of Mechanisms for Resource Allocation." *American Economic Review* 63:1–30.

Klepper, Gernot. 1983. "Incentives for Allocating Public Goods under Incomplete Information." Ph.D. dissertation, University of Kentucky, Lexington.

Kreps, David, Paul Milgrom, John Roberts, and Robert Wilson. 1982. "Rational Cooperation in the Finitely-Repeated Prisoner's Dilemma." *Journal of Economic Theory* 27:245–52.

Lindahl, Erik. 1958. "Just Taxation: A Positive Solution." In R. A. Musgrave and A. T. Peacock, eds. *Classics in the Theory of Public Finance.* New York: St. Martin's Press.

Marshak, Jacob, and Roy Radner. 1971. *The Economic Theory of Teams.* New Haven, Conn.: Yale University Press.

Pigou, A. C. 1932. *The Economics of Welfare.* New York: Macmillan.

Radner, Roy. 1981. "Monitoring Cooperative Agreements in a Repeated Principal-Agent Relationship." *Econometrica* 49:1127–48.

Randall, Alan. 1983. "The Problem of Market Failure." *Natural Resources Journal* 23 (Jan.): 131–48.

Runge, C. F. 1981. "Common Property Externalities in Traditional Grazing." *American Journal of Agricultural Economics* 63:595–606.

Samuelson, P. A. 1954. "The Pure Theory of Public Expenditure." *Review of Economics and Statistics* 36:387–89.

Schmid, A. A. 1977. "The Role of Private Property in the History of American Agriculture, 1776–1976: Comment." *American Journal of Agricultural Economics* 59:587–89.

Sen, A. K. 1967. "Isolation, Assurance, and the Social Rate of Discount." *Quarterly Journal of Economics* 81:112–24.

Shubik, Martin. 1981. "Game Theory Models and Methods in Political Economy." In K. J. Arrow and M. D. Intrilligator, eds. *Handbook of Mathematical Economics.* Amsterdam: North-Holland.

Smith, Adam. [1776] 1977. *The Wealth of Nations.* London: J. M. Dent and Sons.

Smith, V. L. 1980. "Experiments with a Decentralized Mechanism for Public Goods Decisions." *American Economic Review* 70:584–600.

Wicksell, Knut. 1958. "New Principle of Just Taxation." In R. Musgrave and A. T. Peacock, eds. *Classics in the Theory of Public Finance.* New York: St. Martin's Press.

Comment

Kenneth D. Frederick

Randall (1983, p. 132) has suggested that externality is "a vacuous and entirely unhelpful term, and can be replaced by the more general term inefficiency with no loss of content." If we do replace "externality" with "inefficiency," the topic under discussion takes on a rather circular or tautological ring—the effects of inefficiencies on the efficiency of irrigated agriculture. Randall would also have us do away with several other terms that are well entrenched in the lexicon of economists. He suggests replacing the terms "common property resources," "public goods," "natural monopoly," as well as "externalities" with two alternative concepts: nonexclusiveness and nonrivalry. Personally, I find it difficult to purge these terms from my vocabulary, especially when the replacements seem so cumbersome. Nevertheless, in questioning the conventional terminology, Randall sharpens our understanding of what really underlies market failures and of the range of solutions offered by economists.

In focusing on the causes of market failure, Randall directs attention to a number of issues important to irrigation efficiency. Among these are the proper role of government in countering market failures; the importance of secure, nonattenuated property rights for the operation of markets; and the need for institutions to facilitate trade by reducing transactions costs. The discussion of the relative advantages and disadvantages of government intervention when market failures exist (or, in Randall's terminology, when there is nonexclusiveness or nonrivalry) is particularly insightful. I suspect, however, that the objectives of this volume will not be best served by dwelling further on theoretical issues that are likely to be more relevant to fine-tuning a relatively efficient system and for which ideology is central to the preferred approach. Although the existence of natural market failures may provide theoretical justification for government intervention, in practice government actions often aggravate rather than mitigate the source of inefficiency. Thus, some of the greatest improvements in irrigation efficiency could, at least potentially, come from narrowing the enormous gap that frequently exists between actual government policies and those that theory suggests would improve efficiency.

To illustrate the above point, consider the case of groundwater use in the Cuyo region of Argentina, where uncontrolled individual pumping from a common pool provides a classic example of inefficient water use: excessive extraction has lowered the water table and increased the pumping costs for all users. A study by Zapata (1969) suggests that the discrepancy between the private and social costs of additional withdrawals in Mendoza, Argentina, was about 30 percent of private costs when pumping had only a transitory or seasonal impact on the level of the water table. When allowance was also made for the future scarcity value or user cost of water being depleted over time, the gap between private and social costs was even higher because of externalities arising from the exploitation of a common pool. Yet, the principal discrepancy between private and social costs was probably the result of misguided government policies. As of about 1970, government credit, tax, and power pricing policies reduced private groundwater costs by 50 percent or more for many farmers (Frederick 1975). The combination of the natural market failures associated with use of a common property resource and inappropriate government policies led to the rapid depletion and inefficient use of the area's groundwater resources.

Salinity in the southwestern United States provides another example of the contribution of government policies and laws to inefficiencies, which are often assumed to result solely from natural market failures. Irrigation runoff contributes nearly half of the salts in the lower reaches of the Colorado River, and irrigation is the principal source of salts resulting from human activities. An individual irrigator has no incentive to reduce his contribution to the problem since the costs are borne by those downstream. In the jargon of economists, externalities (or nonexclusiveness) characterize the irrigator's use of the river. Unfortunately, both the federal and state governments tend to contribute to, rather than compensate for, the inefficiencies and salt loadings attributable to irrigated agriculture. Federally subsidized irrigation projects are among the largest contributors to the salts, and water pricing policies and restrictions on use of federally supplied water discourage conservation. State water laws also stifle incentives for farmers to adopt more efficient irrigation practices, which, by reducing both withdrawals and return flows, would result in lower salt contributions from irrigation. For instance, under the beneficial use doctrine of state water law a farmer risks losing rights to unused water. Not only are individual farmers encouraged to use rather than conserve water, but also the Colorado River Compact, which allocates water between the upper and lower basins, encourages the upstream states to put their full allotment to immediate use. Just as individual farmers are not permitted to sell unused water, the states in the upper basin cannot sell any of their allotment to potential users in the lower basin. Although the introduction of nonattenuated water rights would not solve the salinity problems in the Colorado basin, it would contribute to more efficient water use and thereby help reduce the salinity caused by irrigation return flows.

My remarks are not intended to imply that externalities and natural market failures are unimportant in the water area. I do believe, however, that a narrow focus on natural market failures may overlook some of the principal causes of existing inefficiencies in water use. And, all too often, it leads to misguided government intervention, which aggravates the problems.

Furthermore, where water is scarce, it is not enough to look only at irrigation efficiency. The shortcomings of focusing exclusively on irrigation are readily apparent in the western United States. Irrigators account for 90 percent of the region's consumptive water use, hold most of the senior water rights, and enjoy preferred access to streamflows. A large part of the problem of the growing scarcity of water in the region stems from the fact that irrigators have been able to withdraw water without considering the opportunity costs their use imposes on others. These costs can be very high. In the Columbia River, for example, an acre-foot of water at the head of the river can generate power valued in excess of $60 as it passes through generators already in place on its way to the ocean. Irrigators are not accountable for such costs or the forgone benefits associated with their withdrawals, nor are they able to reap any of the benefits if they use less water. Despite these substantial in-stream values, government agencies continue to push for new irrigation developments which appear to be marginal investments at best, even when water is assumed to be a free resource.

The efficient use of water in any region requires reducing the obstacles to, and the costs of, transferring water among various users. But even if these obstacles and costs are eliminated, another challenge to achieving an efficient allocation of water among all users remains: to provide for an efficient level of production of the nonmarketable outputs (or, in Randall's terms, the nonrival goods) such as fish and wildlife habitat and amenity values associated with freely flowing streams. Finding both conceptual and practical solutions for efficient allocation may well be the greatest difficulty in achieving efficiency in overall water use. These are not, however, the most pressing issues associated with improving water use or irrigation efficiency in the developing countries.

References

Frederick, Kenneth D. 1975. *Water Management and Agricultural Development: A Case Study of the Cuyo Region of Argentina.* Baltimore, Md.: Johns Hopkins University Press.

Randall, Alan. 1983. "The Problem of Market Failure." *Natural Resources Journal* 23 (Jan.): 131–48.

Zapata, Juan Antonio. 1969. "The Economics of Pump Irrigation: The Case of Mendoza, Argentina." Ph.D. dissertation, University of Chicago.

Comment

David M. Newbery

Randall first wishes to replace the familiar terms "externality," "public good," and "free access resource" with the more diagnostic concepts of nonexclusiveness and nonrivalry. I am somewhat in sympathy with this view, because while the conventional terms direct attention to the possibility of market failure, the concepts of nonexclusiveness and nonrivalry focus on the source or sources of the potential market failure and may thereby suggest appropriate remedies. If groundwater is nonexclusive and nonrival, could it be made exclusive by creating nonattenuated property rights?

This appears to be a step in the right direction, but it is only a small step, for many other characteristics are relevant to the issue of mitigating market failure. Ideally, we need a description of the set of feasible institutional arrangements which might handle resource allocation—feasible in the sense of being sustainable and accessible from the status quo. One would expect other aspects besides degree of exclusiveness and rivalry to be relevant: for example, the number of agents involved and possibly their relative power; the extent to which information about resource availability, distribution, production possibilities, valuation, and so on is available and the cost of acquiring further information; and whether the resource allocation game is repeated essentially unchanged or whether time enters in an essential way (as with depletable resources and some renewable resources), to name just a few. One might further wish to distinguish between unilateral and multilateral externalities, stock and flow externalities, production and consumption externalities, and the like, because the type of problem raised may differ significantly from case to case and may favor some institutional alternatives over others. One could almost argue that the old, familiar terms were less misleading, because they merely directed attention to the possibility of market failure without prejudging the appropriate institutional remedy.

Randall goes on to argue that we are now wiser and more cautious about the possibility and nature of remedies available. He traces an evolution from the simple view that market failure provided the rationale for government intervention, through its obverse—the view that equilibrium implied the attainment of the best feasible outcome and that therefore any disturbance of the equilibrium would make matters worse—to our present agnosticism.

The early writers argued that public goods would be catastrophically undersupplied (and public "bads" oversupplied) without government intervention, management, control, or finance. The later view was that the problem would be avoided, either by creating or defining nonattenuated property rights or directly by bargaining among the concerned parties. Randall helpfully reviews recent developments which model the resource allocation problem as a game with many players and concludes that neither extreme is plausible. Typically, bargaining will avoid catastrophic collapse but will fail to achieve full efficiency. Of course, the property rights school might then argue that the outcome is "constrained efficient" (that is, the best attainable outcome given transactions costs and the available set of institutions), but there is growing theoretical support for the view that bargaining may lead to equilibriums which are not "constrained Pareto-efficient." Intervention may be able to make everyone better-off (Newbery and Stiglitz 1981, chap. 15).

Randall's conclusion with respect to irrigation externalities is that neither having the water supply managed by the central government via taxes, regulations, and controlled prices nor having it remain exclusive private property is likely to provide the best solution, and instead we should consider the myriad possibilities that exist between these extremes. It is hard to dispute this conclusion, although one might argue that it is not very constructive. My own conclusion would not be very different. First, I would argue against the view that equilibrium implies that all feasible alternatives have been rejected in favor of the status quo. Just as technical progress in production is not only possible but to be expected, so comparable improvements in productivity through institutional change are also possible. Second, the economist's approach of trying to identify discrepancies between private and social benefit and cost still seems to be a useful first step. The next step is to identify and compare various ways in which these discrepancies can be reduced. In formulating their proposals, economists have often paid too little heed to the Pareto principle, perhaps for the following reason. Economists who are impressed with the concept of Pareto optimality tend to be impressed with the efficiency of the market economy and belong to the property rights school that argues against intervention. Economists who are skeptical of the moral validity of the status quo and who endorse redistributive taxation and intervention to correct market failure would see little merit in respecting the existing distribution of income; they might well argue for the desirability of pursuing the two goals of equity and efficiency by different means. "Get the price right and use income taxes to redistribute income!" is a useful maxim that has much to commend it as a long-run goal.

Interventions which involve adjusting private costs and returns toward social costs and returns typically lead to large transfers of income, and it should come as no surprise that those who lose will oppose the intervention. Given sufficient coercive power, this may not be a problem, but experience suggests that in practice it often is. Here, economics offers a solution. If an equilibrium is Pareto-inefficient, it should be possible to devise a remedy which makes no one worse-off and some people better-off. We may not like the existing or resulting distribution of income and wealth, but, if it cannot be readily improved, it seems perverse to forgo the improvement that is feasible in *levels* of welfare. If water is to be sold, rather than made available, how might this principle be put into practice? One obvious suggestion is to define the existing allocation as an entitlement available on existing terms, with additional water available at the new price (and shortfalls compensated at the same rate). This effectively defines property rights in the existing allocation, but leaves open the question of how these rights might be marketed or transferred. Such an approach is Pareto-safe, and O'Mara, in his introduction to this volume (chapter 1), attaches prime importance to this principle. It would seem preferable as a solution to the problem of excessive depletion of groundwater than centrally imposed taxes, public ownership, or private control of the aquifer, all of which Randall suggests are unsatisfactory. To me, Randall's solution seems less promising, because it is likely to lead away from price-guided signals. He argues for restructuring the problem of private pumping as a repeated game in which each farmer's depletion and the level of the aquifer are monitored. Excessive depletion will be observed and the main culprits identified, and all might be persuaded to reduce their depletion to a rate deemed efficient. Presumably, this would be done by establishing norms, but there would be no guarantee that the marginal product of water would be equated across farms by this procedure.

I conclude by agreeing with Randall that we should now explore the whole range of options available, rather than presuppose that taxes or market solutions will automatically cure the problem. I would, however, argue that quantifying the discrepancies between private and efficient incentives is a useful first step and that trying to devise Pareto-safe methods of confronting agents with efficient incentives is a good way to organize the search among alternative options.

Reference

Newbery, D. M., and J. E. Stiglitz. 1981. *The Theory of Commodity Price Stabilization*. New York: Oxford University Press.

3

Legal Considerations for Coping with Externalities in Irrigated Agriculture

George Radosevich

This chapter examines the role of institutions in the efficient use and development of water resources for irrigated agriculture. In the past two decades developed and developing nations alike have tried to find out why irrigation systems are not as efficient or effective as they are designed to be. The problem has become more complicated with the advent of new technologies and improved delivery and applications systems.

Partly as a result of the technological advances made in irrigated farming and partly as a result of nations' efforts to increase their food supply, improve the quality of life, and attain favorable international trade balances, virtually all of a nation's irrigation water will be characterized by "physical interdependence." That is, the economic agents in the system, such as farmers, are linked by physical mechanisms that affect the enjoyment (utility) of real income and the cost of generating real income. The level of efficiency or, conversely, the extent of market failure in an irrigation system depends largely on the ability of the social institutions to cope with the demand for available resources. When the private market system attempts to resolve the issue of physical interdependence, some of the costs (or benefits) produced spill over or are incurred (or enjoyed) by persons other than the user. This gives rise to inefficiencies, or costs or benefits that are not accounted for within the system. These spillover effects are referred to as externalities and may be external economies (benefits) or diseconomies (costs). If the individuals who produce the externalities are not willing to account for them, it is up to the social institutions to examine the situation, evaluate their relevance, and select a solution to make the system more efficient. To the economist, efficiency would be the achievement of Pareto optimality, the condition in which one can become better-off without someone else or society becoming worse-off.

This chapter has three objectives: to briefly describe the externalities that arise in irrigated agriculture, to discuss the national institutional frameworks for water development and management in a global context, and to discuss the legal and institutional causes of externalities and propose ways to increase efficiency in water utilization.

The Economics of Externalities

The literature is replete with accounts of man's early attempts at irrigation. Evidence can be found in the ancient tanks and canal systems of Sri Lanka and the early irrigation systems along the Nile, Tigris and Euphrates, Indus, and Yellow rivers. In the most fascinating examples of primitive technology—the *qanats* of Iran and the *krazes* of Baluchistan—water was directed to a small irrigation system at the base of a mountain through a series of underground conduits dug into the side of the mountain (Cantor 1970). All these early efforts required both ingenuity and greed—ingenuity to devise solutions and greed or ruthlessness to carry them out regardless of the consequences to others or to the environment. Most of the solutions were technical (see Daumas 1969), and, although they provided a temporary advantage over existing practices, they were not intended to address the interests of others who may have been adversely affected.

In this chapter we are concerned with the use of water resources in a social context. Originally, water, like fire and air, was considered common property because of its fugitive nature, in contrast to land and minerals, which are fixed in location. As long as the uses were marginal and the supply abundant, this free good remained unrestricted. Only when the use by one party interfered with that by another did the process of institutionalization begin.[1] The use of water as common property continued even as the number of users proliferated, until owing to either the location of the users along the water supply or their time of use the resource base could no longer accommodate any newcomers. Eventually, if no adjustments were made in the system, the resource would be misused, its quality would be diminished, and the surrounding environment would deteriorate—the very situation described by Hardin in his famous "The Tragedy of the Commons" (1968) and elaborated on under contemporary conditions by Hardin and Baden (1977).

When action was taken it began with the assertion of power. By natural extension of the exercise of power over land as property, the exercise of dominion and control over water resources created a property right to water. Cantor (1970, p. 18) asserts that in some arid areas where irrigation was essential, property rights to the use of water antedated property rights to land. In some situations, individuals in society exerted this power; in others, hydraulic societies emerged (Wittfogel 1957). Empires have even been built upon dams, canals, and irrigated agriculture (Worster 1985). As the rules of property use developed for water resources, water was classified as either a private good or a public good. The nature of this property right will be discussed later, but the important point here is the institutionalization of an interest in water from which individuals in society began to exercise control over use of the resource.

In most societies water is still a free good in the sense that users do not have to pay for a quantity of water. The charge, if any, is not for the commodity but rather for the resources needed to control and distribute it. According to the marketplace theory of economics, these direct costs should not exceed the benefit that can be gained by the users. But these are not the only costs or benefits involved in the use of water resources. There are others that are external to the user and that must be borne by society or other users outside the market system. These unaccounted-for spillover effects, or externalities, are of two types, pecuniary and physical, as O'Mara pointed out in chapter 1.[2] The distinction is that in the case of pecuniary externalities changes in input prices cause the interaction between industry output and firm costs, whereas with physical externalities some uncompensated costs are imposed on third parties as a result of a physical or technical relationship between those parties and other firms (or individuals). Such is the relationship among users of water from a common aquifer or stream. One user's action—for example, excessive pumping, diversion of water, or discharge of wastes—may cause others to deepen their wells, cease pumping altogether, improve their diversion works, cut production to adjust to a reduction in supply, or treat the water before using it. Or the effects could be beneficial: By lowering the water table the pumping or diversion could allow leaching of saline lands or mitigate the possibility of floods. The wastes discharged could contain useful nutrients. Whatever the effect, if the externalities are not accounted for, the market cannot achieve the most efficient allocation. This situation is termed market failure (Bator 1958).

During the past several decades, economists have devoted considerable attention to the issue of externalities and the effective use of resources. In chapter 1, O'Mara described three solutions widely discussed in the literature. The first is to adopt a system of corrective subsidies or taxes that would reward or penalize the producer of the externality and thus establish an efficiency equilibrium. The second is to develop a system of centralized control and management that would "internalize the external effects" among users or producers. The example provided is the unitizing of an oil field to distribute the total payment proportionately among all the well owners regardless of which well is pumping.

The third alternative is to assign private property rights. The justification for this, popularly referred to as Coase's theorem, has been simplified by O'Mara: given well-defined initial legal property rights and zero transactions costs, the market allocation will be efficient. Coase concluded that either market efficiency would be achieved by bribing the producer of the diseconomy to stop or the affected parties would resolve the situation through negotiation and without government intervention (Fisher and Peterson 1976).

Coase's theorem generated considerable debate among economists.[3] The most obvious concern has been the non-practicality of zero transactions costs; hence the criticism that true market equilibrium cannot be achieved as Coase would have it.[4] Although many economists prefer the concept of creating a system of legal property rights that would define the relations among parties, in actual practice this solution is inadequate to compel the producer of the external diseconomies to internalize the cost. Not only is it difficult to establish an all-inclusive system of well-defined property rights, but also, in the case of water resources, the physical interdependencies that exist in time and space make implementation impossible without government intervention and probably far too costly.

Externalities in Irrigated Agriculture

Externalities in irrigated agriculture can take many forms. For example, if all the users of a given stream take as much water as they need whenever they want it, one user's activities may interfere with another's. Depending upon their location along the stream, some users will be compelled to incur the cost of building a diversion structure to ensure their ready access to the water no matter what the others do. Upstream locations would be preferable for the diversions because they would offer superior control—that is, they would allow one to take what is needed before other users downstream. The downstream users may actually receive a benefit, or external economy, if flows are delayed until a time better suited to their needs. But they are more likely to incur an external diseconomy in the form of diminished flows and inferior water quality. This is because the return flows of water used in agriculture typically carry leached-out salts, agricultural chemicals from fertilizers, pesticides, and herbicides, and large quantities of sediments.[5]

By extension from this simple example, it is clear that externalities can exist not only among water users along a stream but also among all those people and groups of people throughout a river basin who use water—for whatever purpose (agriculture, domestic use, industry, power production, transportation, or recreation) and at whatever level of organization (local, regional or state, national, or international). Within a nation the users may be grouped together as a private irrigation company, an irrigation district with project facilities, or a confederation of user groups organized within a valley or river basin into associations, commissions, or authorities. The jurisdiction of these entities may conform to hydrologic or administrative boundaries.

Surface irrigation development has been responsible for various types of externalities. Irrigation projects have been reported to cause schistosomiasis in Rhodesia (Farvar and Milton 1972, pp. 102–8) and Egypt (pp. 116–36); changes in the insect population in Israel (p. 349); fishery and ecological effects along the Nile (pp. 159–78, 189–205) and the Mekong (pp. 236–44); and salinity problems in the Indus and Helmand Basins (pp. 257–75), in the Algerian Northwest Sahara (p. 276), and along the Upper Rio Grande (pp. 288–300). And sedimentation of lakes and reservoirs from irrigation runoff and erosion has created problems for hydropower production (pp. 318–42).

Groundwater use also creates situations in which externalities arise. Depending upon the characteristics of the aquifer, the effect of pumping one well may be to lower the water table to such an extent that the owners of adjacent wells will have to incur the cost of either deepening their wells or delaying their time of use until the aquifer is recharged. Indiscriminate placement of wells with varying pumping depths and discharge rates can impose excessive and unnecessary costs upon well owners and lead to misuse of the resource.

The situation can become even more critical if the groundwater is hydrologically connected to streamflows (Bredehoeft and Young 1983; and Daubert, Young, and Morel-Seytoux 1980). In such cases the pumping may not only interfere with other well users, but also diminish the surface flows and thereby interfere with surface water users. If the resource is treated as common property and rights and criteria for use are not defined, the total cost to society will be increased by the cost of constructing a more efficient surface diversion or a deeper groundwater system or of altering the water use practices. The result is a more costly commodity or a reduction in production. As will be discussed later, even with a system of private property rights and government management, market failure can occur.

In addition to quantity considerations, water quality and environmental pollution have been the subject of extensive analysis (Downing 1984; Seneca and Taussig 1974; and Thompson 1973). Irrigation return flows can have harmful effects on the downstream environment and on downstream water uses. In the United States, for example, irrigation return flows have been responsible for salinity along the Colorado River near the Mexican-U.S. border (Gardner and Young 1985), sedimentation in Idaho and Iowa, and agricultural chemical pollution in California. Downstream users bear the cost of cleaning or treating the degraded return flows or of any reduction in production occasioned by the use of degraded water. The presence of these additional costs indicates that the market cannot achieve its greatest efficiency.

If externalities are so extensively associated with the use of water resources, how can the level of market efficiency or extent of failure be brought within tolerable limits? Demsetz (1967) and others have concluded that at least part of the solution lies in the development of a concept of property rights. Demsetz states that from an economic perspective "a primary function of property rights is that of guarding incentives to achieve a greater internalization of externalities," and "property rights develop to internalize externalities when the gains of internalization become larger than the cost of internalization" (Demsetz 1967, pp. 348 and 350). Seneca and Taussig (1974, p. 87) take a slightly different perspective: "Externalities arise when property rights are not defined for certain scarce resources or when nominal rights are not enforced in practice. When private property rights can be enforced, externalities tend to disappear because profit opportunities for firms encourage them to internalize externalities . . . The market solution is feasible in simple cases of externalities, in which markets can be readily created among just a few easily identifiable buyers and sellers. The market solution is not feasible when externalities are pervasive and diffuse throughout society and the costs of organizing the market become prohibitive."

But what about environmental pollution and particularly the externalities associated with the return flows of water used for irrigation? Seneca and Taussig (1974, pp. 77) point out that these externalities not only arise from a lack of well-defined property rights, but also involve individuals and groups that are not easily identifiable.

Water Law Systems

Given the fact that the unrestricted use of water by any individual, firm, or group of users may cause an economic inefficiency in the system, what institutional mechanisms have evolved to counter this effect? The legal theorist's response would be a legal system to identify the rights and duties of the users and to plan, control, and manage use of the resource. Water law, the generic term for such a system, may be defined as the manner by which society decides how water will be allocated, distributed, managed, and regulated. The institutional framework for the development of water resources consists of policies, laws, organizations, and the process of implementation. In this context, water law will refer to the laws governing water within a nation or sovereign unit, as distinct from international water laws, which provide guidelines for conduct between nations sharing a common water resource.

Often there is a noticeable difference between the water law as written and the law that is implemented. It has been observed in some situations that the written law can be traced to an external influence, whereas the implementation reflects the extent of control necessary to satisfy national and local needs or the implementing agency's awareness of the extent of inefficiency.

The scope and complexity of national water law systems vary from country to country, because every system is the product of its environment. Three factors in particular help shape a country's water law system. The first is the general legal or political system of the country. David and Brierley (1978) have identified four major legal systems with distinct historical, national, or political roots. They point out that each political society has its own set of laws and that several of these systems may coexist in one nation. For this reason David and Brierley prefer the term "legal families." The four families are Romano-Germanic law, common law, socialist law, and philosophical, religious, and traditional law.[6] The presence of a specific family sets the legal climate of a nation and permeates all legal enactments and interpretations.

The law differs from other disciplines in that its jurisdiction, or venue, is limited to the territorial or administrative boundary of its enacting body. Thus water laws generally do not follow hydrologic boundaries, but rather conform to the administrative boundaries of the nation and its states or provinces. Primary jurisdiction over water may be at the national or subnational level or a combination of both. For example, the national government may exercise primary jurisdiction over navigation and commerce, and each state or provincial government may have primary jurisdiction over the allocation, distribution, and administration of water within its boundaries. Regardless of the system used, certain social costs and benefits are incurred which must be balanced against regional and national interests in order to determine their overall effect on the use of the resource.

The legal or political system also has an impact on a nation's socioeconomic conditions, particularly with respect to the classification of public and private rights and the extent to which one infringes on the other. In the case of water resources, many nations recognize a distinction between public and private waters and between public and private rights to public waters. The extent to which these rights are defined and enforced greatly determines the presence of externalities in the system.

The second factor that significantly influences the type of water law system that develops is the geoclimatic environment. Until recent times countries in humid areas tended to develop laws that emphasized the importance of drainage and flood control and addressed problems of water quality. In contrast, countries in arid areas tended to focus more on the allocation, distribution, and management of their scarce resource and to develop laws to mediate between the many and often conflicting uses and users of water. In the past fifty years, however, many nations, even those in humid areas, have adopted stringent provisions regarding the allocation and management of water resources. This is primarily because improved technologies have led to increased demand for water for large-scale irrigation projects and power production, and domestic and commercial needs have increased as populations have expanded. Institutional changes and particularly the adoption of more stringent legal interventions are responses to this situation that are consistent with the economist's theory of scarcity and man's conscious or subconscious willingness to achieve stability through regulation and management.

The hydrogeological environment is the third factor that helps shape a country's water law system. Nature's water system, the hydrologic cycle, encompasses the precipitation in the atmosphere; the surface water that accumulates in lakes, streams, and ponds; the subsurface waters of artesian, tributary, unconfined, and nontributary confined aquifers; subterranean streams and deep or fossil water. Corresponding to this natural system are the water quantity laws that structure the use of the resource. These laws range from a segmented to an integrated treatment that is based more on legal precepts than the laws of science (see Figure 3-1). A similar spectrum of approaches could exist for control of water quality, ranging from laws regarding discharges into surface water, contamination of groundwater, or injection or percolation of wastes, to the integrated management of surface and groundwater quality. The ultimate integrated treatment would synthesize the management of both water quantity and quality in recognition of the synergistic effect from any use of water that may result in an external improvement (benefit) or degradation (cost) affecting subsequent users and the environment. Adoption of this level of legal sophistication in a comprehensive system of water resource management would enable the system to gain the highest level of economic efficiency.

Water laws may pursue their historical role or undertake a modern role. Historically, water law was used primarily

Figure 3-1. *Interface of the Natural and Legal Systems*

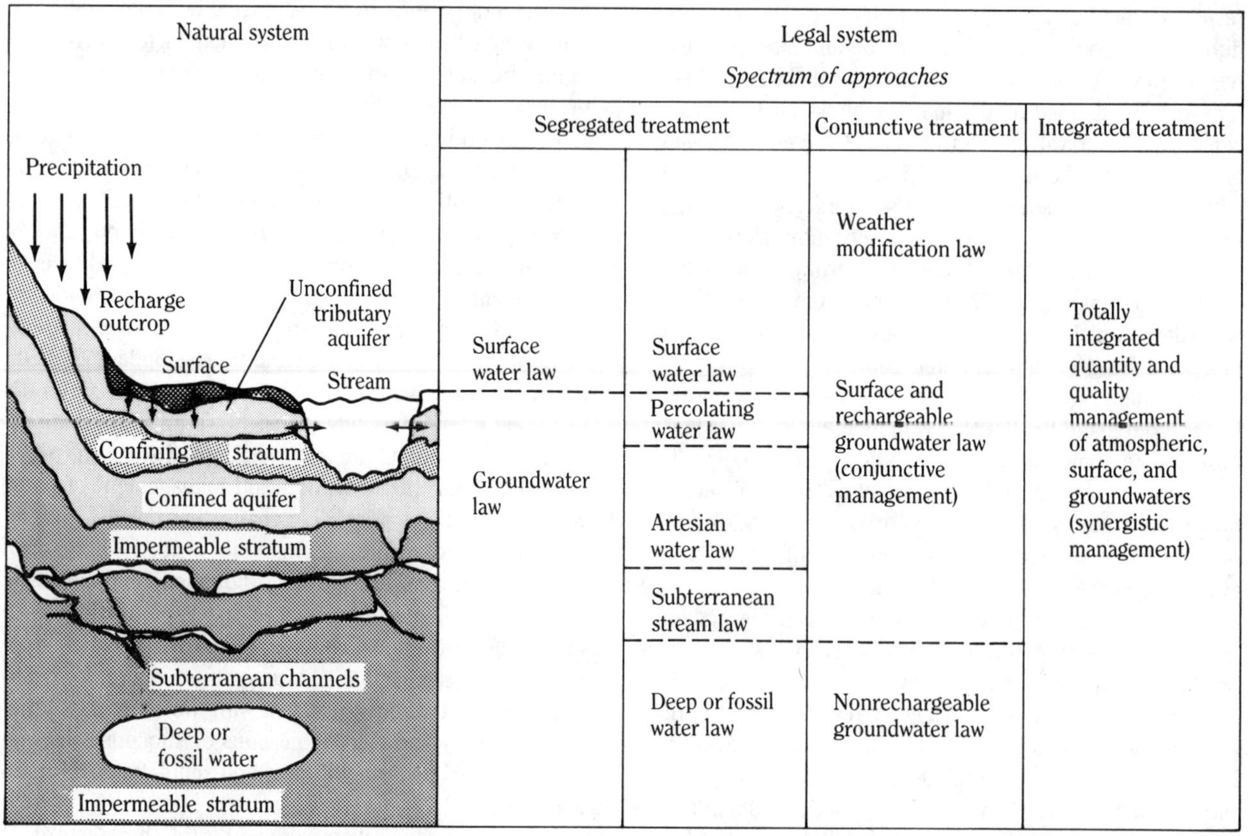

Natural system

Precipitation

Recharge outcrop

Unconfined tributary aquifer

Surface

Stream

Confining stratum

Confined aquifer

Impermeable stratum

Subterranean channels

Deep or fossil water

Impermeable stratum

Legal system: *Spectrum of approaches*			
Segregated treatment		Conjunctive treatment	Integrated treatment
		Weather modification law	Totally integrated quantity and quality management of atmospheric, surface, and groundwaters (synergistic management)
Surface water law	Surface water law	Surface and rechargeable groundwater law (conjunctive management)	
Groundwater law	Percolating water law		
	Artesian water law		
	Subterranean stream law		
	Deep or fossil water law	Nonrechargeable groundwater law	

Source: Adapted from John Muir Institute (1980), figure 1, p. 36.

to control the use of the resource and resolve conflicts among the users. Consequently, the law was often viewed as an agent for regulation, and an adversarial relationship existed between agencies and water users and among water users. Typically, resource use was regulated through separate acts for different segments of use such as surface water, groundwater, and water for irrigation. Prior to the 1950s most nations had water law systems following this historical role.

In its modern role water law is a tool for resource development and management. Modern laws, typically forming an integrated code, emphasize conservation, comprehensive planning and development, water quality control, and conjunctive use of surface and groundwater. The laws evince an awareness of the interdependencies that exist between water and other natural resources and among the various public and private uses of the resource. In modern systems agency action is directed more toward resource management than regulation, and policies are articulated to guide the implementation of the law.

Water laws can be classified as either national or federal systems. In a national system, the central government exercises primary jurisdiction over the allocation, distribution, administration, and development of the resource, generally through a national water code such as that of Egypt, Mexico, the Philippines, and Spain. In a federal system, allocation, distribution, and administration are primarily state or provincial matters, whereas interstate allocation, interstate commerce, navigational uses of interstate waters, and water quality control are handled by the central government. Federal systems are found in Argentina, India, Pakistan and the United States.

The form and style of water law system that emerges within each country greatly depends upon these three factors and their interaction. But inefficiencies (to use Randall's preferred term for externalities) are created in both systems. In a national system the policies and laws are often flexible to allow for implementation in different parts of the country. Consequently, at the grassroots, where government officials must deal directly with the water users, there are frequently problems of coordination between government agencies and within departments. On the positive side a national system ensures a measure of uniformity nationwide and makes it possible for water law to be implemented along hydrologic boundaries. Inefficiencies often created under federal systems include the lack of uniformity between national and state goals, policies, and laws; problems of inter- and intradepartmental cooperation and coordination; and implementation of the law along administrative instead of hydrologic boundaries. Federal systems do offer one advantage, however, especially if both geoclimatic and economic conditions vary widely across the country: the national and subnational governments can fine-tune their laws to the needs of local and regional water users.

Water law systems can also be classified as customary, traditional, or modern (Radosevich and others 1976, summary volume). Customary systems are generally based upon religious or customary law. Examples include the adat law of Indonesia, Moslem water law as practiced at the community level in many Islamic countries, and the Hindu-Bali system of water user organizations in Bali, Indonesia.

Traditional systems adhere closely to the historical role of water law. Here the emphasis is on water allocation and distribution, resolution of disputes, and the distinctions between private and public waters. The present water law of Spain and of many countries in South and Central America subject to the Iberian influence is traditional in nature, as are a number of state laws in India, Pakistan, and the United States.

Modern water law systems are distinguished from traditional ones by their emphasis upon the conjunctive use of surface and groundwater, basin planning and development, integrated control of water quality and quantity, and the articulation of policy guidelines. Examples are the new codes of the Philippines and Venezuela and the laws of England and of many states in the United States.

In 1975 an effort was made to identify the principal water law systems, their attributes, and the role they play in developing a nation's water resources.[7] The legal systems identified and their paths of influence in various parts of the world are illustrated in Map 3-1. The legal systems shown are the Soviet, the Chinese, the Roman, the Islamic, the British, the Spanish, and the combination of systems found in the United States.[8]

Property Rights to Water

The first step in any attempt to plan, manage, and control the development, allocation, and use of water resources is to determine the extent of the government's authority over the resource. Most countries have constitutional, legislative, or judicial declarations that specify the sovereign's control of all water resources—regardless of their origin or location—within the boundaries of the political entity. Some declarations, however, assert government jurisdiction over just water in watercourses, lakes, and streams and certain types of groundwaters, such as tributary groundwater or water in underground streams. In such cases it is necessary to determine the extent to

Map 3-1. *The World's Principal Water Law Systems and Their Paths of Influence*

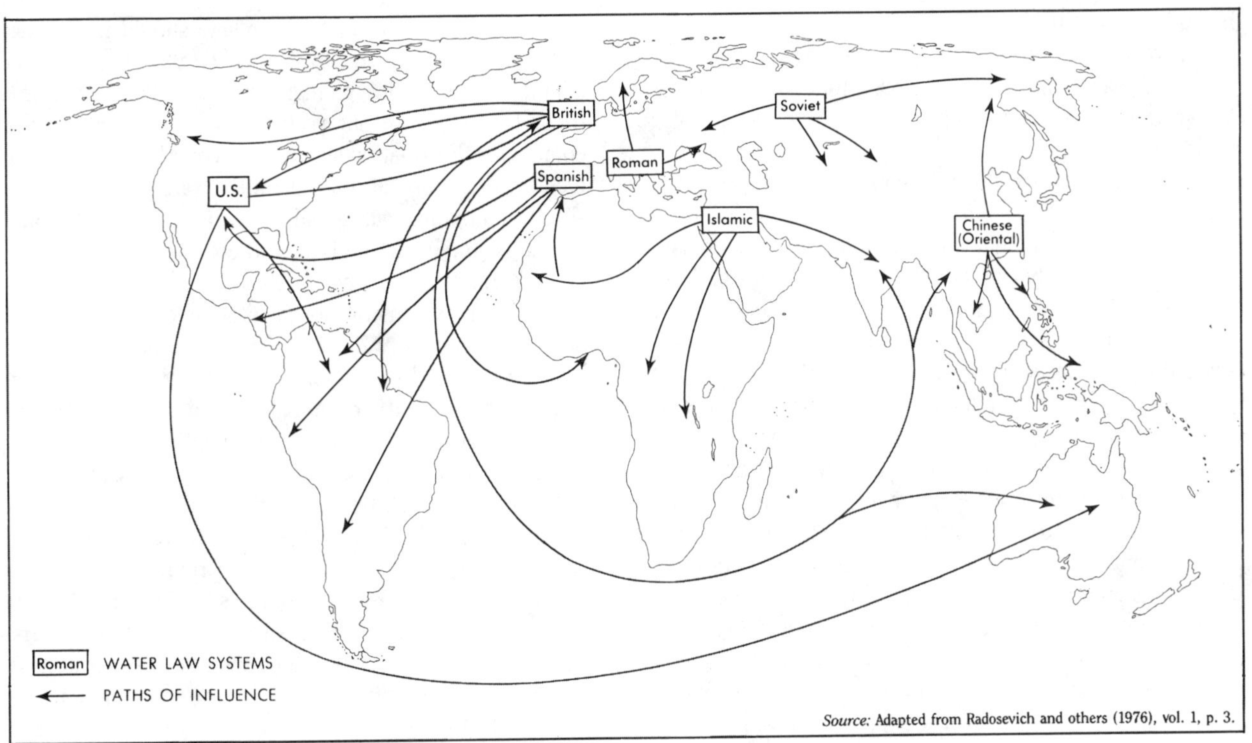

Source: Adapted from Radosevich and others (1976), vol. 1, p. 3.

which private rights exist, such as in diffused surface waters, nontributary groundwater, and, in some cases, percolating groundwaters. If privately owned water is a part of the landownership interests, the issue of externalities begins only after the water is used by the initial owner or leaves the property.

In general, then, waters referred to as common property are those that are declared to be public or state-owned. The right to use such waters, as well as those declared to be private, is generically referred to as a property right to water. This property right defines the relationship between society, the user, and the resource and is generally a constitutionally protected right or interest. Because water resources are fugitive in nature, in contrast to land and minerals, the right to water is generally a usufructuary right, or a right to use, and not a right in the corpus or body of water itself. In the United States and many other countries, the right to use water is a real property right. It is generally a salable and transferable right, similar to a title to land. When the right is exercised and the water put to use, the water changes its status from public or state property to personal property and thereby takes on all the benefits and liabilities associated with personal property. In countries with a definable property right to water, evidence of the right may take the form of a permit, decree, concession, or license that specifies the nature of the right, the source, the uses allowed, and the means by which the right could be lost. As a general rule a water right identifies the resource supply, the point of diversion, a specific quantity or a percentage of the available flow, the type and place of use, and sometimes, by implication, the time of use. For example, agricultural water rights can be exercised only during the growing season, whereas commercial, domestic, and municipal water rights may be exercised year-round. One of the basic functions of a water law system is to specify who has what rights to what water under what conditions.

In countries in which well-defined, legally protected rights do not exist, the small-scale development of water resources is generally guided by traditional or customary law, such as the tribal laws of Pakistan's Northwest Frontier Province and the adat law of Indonesia. Medium and large-scale development and the improvement of small-scale diversion structures are generally the province of government, which commits water to projects and is thereby responsible for the apportionment of quantities and the timing of delivery. Thus, it is difficult to talk about the creation of externalities in many Asian and African countries because to do so often amounts to an accusation of government inefficiency. Fortunately, the tendency in all nations is to identify at least an "interest" of the water users, if not a well-defined, legally protected water right.

Legal scholars would attribute the origin of property rights, and specifically those to water, to the articulation and social acceptance of man's expectations with respect to the resource and the attempt to prevent others from infringing upon those expectations. Demsetz (1967, p. 353), an economist, however, argues that "property arises when it becomes economic for those affected by externalities to internalize benefits and costs." He identifies three forms of ownership that are relevant to externalities arising from water use: communal ownership, private ownership, and state ownership. Concluding that the absence of property rights can generate market failure, Anderson (1983, pp. 4–7) wrote:

> The economics of property rights shows how the rules of the game affect the benefits and costs received and borne by the private decision makers. Individuals, not large groups or societies, make decisions, but they do so in an institutional framework . . . The economics of property rights tells us that decisions will be affected by benefits or costs faced by decision makers. These benefits and costs, in turn, will depend on the extent to which property rights are defined, enforced, and transferable, elements that are crucial to a well-functioning market . . . When private property rights are attenuated, costs can be externalized so that one person's gain becomes another's loss.

An example of this externalizing of costs is the situation in which water development is subsidized or the competition for water regulated in such a way that the highest economic value is not achieved.

Institutional Causes of Externalities

In theory, if not in fact, water laws should not remain static. These laws need to be dynamic, progressive, and adaptable to societies' changing needs and desires. Unfortunately, this is not always the case. The problems that come up with irrigation systems are complex and multifaceted. Outside of natural calamities, rarely is a problem strictly technological, social, economic, or institutional. But it is important to isolate the principal factors and analyze their significance.

The law may constrain or facilitate effective water management. Efficient water use depends upon how responsive the laws and regulations—and the government agencies charged with their implementation—are to the needs of the water users. In many instances the law tends to overregulate or otherwise discourage water users from using creative means to improve the use of the resource and hence their production. For example, there may be restrictions on transferring or trading water rights or turns to the use of water. Conversely, there are situations in which the law tends to underregulate, with the result that external costs are passed on to subsequent users. These conditions create serious problems which unfortunately may not be initially diagnosed as institutional constraints. And even when the law is recognized as the limitation, the agency charged with carrying it out may feel that it is easier to deal with the situation at the local level than to get the law changed.

In general, the law contributes to institutional failure in three cases. The first is the absence of law. As history has shown, the law cannot be all things to all people at all times; it is a product of society and evolves out of society's needs. But not all needs can be anticipated. Water users or society in general are often saddled with unaccounted-for costs, or externalities, when provisions of the following kinds are lacking: water policy statements, provisions pertaining to groundwater management and control, provisions on the planning and management of resource use and development, and provisions that stress the interdependency of water quantity and quality and establish guidelines for comprehensive and coordinated efforts to keep the two physical attributes in balance. Bangladesh, which has no groundwater laws, provides an example of how the absence of laws can create a tremendous amount of inefficiency in the system. As a result of the indiscriminate siting of shallow and deep tubewells and unregulated withdrawals of groundwater, rural water supplies have been placed in jeopardy and the pumping operations of shallow and deep tubewells interfere with one another. Not only are government and private expenditures for the wells subject to waste, but also the benefits may be appreciably less in terms of production, effective utilization of the electrical power supply, and social assistance to the landless laborers.

The second case is the inadequacy of the law. Although many nations have adopted surface and groundwater laws, they do not always clearly define the right of the parties to use the resource or adequately provide for government control of the development and utilization of water. For example, the groundwater laws of Indonesia and Thailand are inadequate to deal with the problems of groundwater mining and the need to designate affected areas for specific regulation. In the United States interstate mining of the Ogallala Aquifer, which extends from Nebraska through Colorado, Kansas, and Oklahoma and into Texas, has pointed out similar shortcomings in the law. Groundwater mining often creates problems in coastal regions, where the level of extraction causes saltwater to intrude into fresh groundwater bodies. The water laws of a country may also be inadequate to prevent the overapplication of water, with the result that return flows may not only leach the salt out of the root zone, but also carry the dissolved solids back into the river system, damaging downstream users. This problem occurred along the Colorado River, with adverse effects in Mexicali Valley, Mexico. The inadequacy of a law often becomes apparent when there are interprovincial, interagency, or interbasin conflicts.

In the third case, the law may be inappropriate to the present problems or may not provide the proper guidance. A notable example is the imposition of water user fees or charges in an effort to encourage a more efficient use of water. In few countries can a blanket law address the issue of water user fees and categorically charge all users the same rate. In Indonesia and Thailand, provisions for charging water fees exist in the law, but to date no fees have been levied (ESCAP 1981). Part of the reason is that users are reluctant to pay a water rate when they cannot be guaranteed a dependable supply of water. A more appropriate law would base water charges on the cost to the government of constructing and operating the system and would provide for a portion of the fee to be used for operation, maintenance, and rehabilitation of the system.

Implementation constraints often occur as a result of the difference between responsibility and authority. Sometimes canal officers are authorized but not required to perform certain tasks. Sometimes they are responsible for certain tasks but lack the authority to carry them out. This type of institutional inadequacy leads to mediocre conduct by government officers and personnel at all levels and to disrespect for, or mistrust of, the legal and administrative system by the water users.

Another organizational constraint has to do with the focus and orientation of the implementing agency. In most countries the agency responsible for providing water to irrigated agriculture is required to perform the functions of construction, operation, and maintenance of the storage, delivery, and distribution system. The Royal Irrigation Department in Thailand, for example, has these responsibilities. In other situations one organization may be responsible for construction (such as the Water Power Development Authority [WAPDA] in Pakistan and the Bureau of Reclamation in the United States) and another entity, such as an irrigation department or a department of water resources, may be charged with the allocation and distribution of water and with operations and maintenance activities. In any case, greater emphasis is usually placed upon design and construction of projects than on operations and maintenance. Project design is often a creative, highly visible activity, whereas operations and maintenance tasks are diffused over a wide area, perpetual in nature, and require endurance more than ingenuity to perform.

Examination of the operations and maintenance activities in several countries clearly indicates that few agency field personnel consider their role significant for meeting national or regional objectives. Most of their activities are routine chores and are therefore regarded as unimportant. It is easy to see how this type of attitude could lead to poor job performance and poor agency-farmer relations (what farmer would be willing to pay water charges in an irrigation system that failed to function properly or to deliver adequate quantities of water?). At the same time, it is also easy to see how canal and field personnel could develop such an attitude if they are not properly informed of the national, regional, local, and agency goals and policies and thus have little appreciation of the importance of the water users' production to the national economy.

Specific areas within an institutional framework that foster externalities include the following:

- The historical view of water laws. Problems arise when laws pertaining to the use of water are treated as completely separate from laws dealing with the use of other resources (land, oil, gas, minerals, air) or user laws (power, transportation, fisheries, and mining). For example, irrigation systems have been disrupted by the operation of hydropower structures, fisheries have suffered when other uses have lowered the level of water in lakes and streams, and mining operations have discharged waste and contaminated water into systems serving municipal, industrial, and irrigation users.
- The organizational system of jurisdiction over water. In most countries at least two government agencies vie for control of the nation's resources: the irrigation department and the agriculture department. In some developing countries the irrigation department delivers water through a series of canals to laterals serving farmers, and the jurisdiction shifts to the agriculture department at a headgate on the lateral. Misuse or waste of water by the farmers cannot be controlled by the agriculture department because it lacks jurisdiction over the source of supply.
- The division of ownership or control of water between government and private users. The resulting conflicts tend to ensure inefficiency.
- The allocation of water according to type of user and the spatial and temporal conditions of the water source.
- The basis—individual, subproject, project, or regional—on which water is allocated. This often explains the failure to account for return flow characteristics throughout a river system.
- The delineation of government and water user duties and responsibilities.
- The distribution process within a command area. Problems arise when the law or local custom fails to ensure equitable distribution to the head, middle, and tail of the lateral network.
- The administration of the water law and the exercise of property rights to water. Restrictions on the transfer of water rights and the lack of communication between the government and the water users often cause significant waste of water.
- The procedures for dispute resolution or the lack thereof. Dispute resolution can occur among water users and at both administrative and judicial levels. Within a command area or common unit, dispute resolution among users is appropriate and effective only if peer pressure will ensure compliance. Administrative resolution is often timely and essential to effective management and control of the system, but frequently reflects agency bias. If misused it may make the relationship between government and water user one of master and servant rather than agency and client. Judicial resolution may often be more just and fair, but may prove more costly and time-consuming. Unfortunately, it is not free from exogenous factors that can influence the outcome. For example, in a dispute between a wealthy farmer and a poor farmer, the former can hire a popular and flamboyant lawyer while the latter may have to seek voluntary legal aid.

Institutional Essentials for Coping with Externalities

Before proposing any institutional change in a country, it is necessary to have a thorough understanding of the existing legal and organizational system and the extent to which the law is implemented. This requires an analysis of existing law and related legislation, decrees, and orders; of documents setting forth national objectives, targets, and policies (development plans and programs adopted by the government); and of policies, rules, and regulations adopted by the irrigation, agriculture, and other relevant agencies. In the effort to improve the efficiency of water use within a country two distinct approaches, as well as a combination of these approaches, can be used (Prang and Tomb 1983). The first is to examine the *supply* to determine to what extent surface and groundwaters are available and the methods used to harness, store, and deliver them. The second approach is to examine the *demand* conditions (type, time, and place) and alternatives (conservation, reuse, and improved management). A combination of supply and demand solutions is obviously the most workable approach. The purpose of this analysis is to construct a model of the system and to determine the direction in which the nation or region (state) desires to go—that is, to identify national or regional (state) *objectives.* Regardless of the extent of supply or demand, a primary national objective should be conservation of the nation's water resources, in the sense of striving for the highest and best use. Other likely objectives are improving the standard of living and quality of life and maintaining a balance of payments equilibrium. In addition, externalities should be identified and taken into account.

After the goals and objectives have been identified, the institutional framework should be designed or modified as necessary to achieve those ends. The foundation of this framework is policy. The relationship of policies to the process of implementation and the achievement of goals and objectives is illustrated in Figure 3-2. The key institutional elements and processes are identified, and an approach or model is provided for analyzing institutional status, capability, and effectiveness. It is suggested that a sound and successful water development and management program can be achieved and maintained through a process known as management by results (MBR).

Policy is defined as "intelligently directed actions toward consciously determined goals, as distinct from aimless drift and blind faith" (U.S. Library of Congress 1976). It is the key ingredient in a successful national program of water resource development. Information flows to and from policy, and policy is what should be accomplished in the process of managing the resource. Water policy sets the direction for development and serves as guidelines for the agencies that implement the laws. A basic distinction between policies and laws is that policies may change periodically as different administrations come into power or as new directions are adopted by the government. In contrast, laws often remain the written and rendered authority for actions, interventions, or sanctions. The policies therefore provide the impetus for implementation of the laws.

Problems of inefficient water use are common (1) when there are no official policy statements, (2) when official policies are not transmitted through the hierarchy to the local level, (3) when official policies are too ambiguous to be implemented, and (4) when official policies cannot be implemented because of interagency constraints.

Policies, as shown in Figure 3-2, include national and regional policies as well as those of a general and specific nature. This allows for coordination of government activities when other interests and numerous implementing agencies are involved. For example, under a general policy to increase food production, a specific water policy could be directed toward increasing the efficiency of irrigated agriculture, developing inland waterways for transport, or promoting conjunctive use of surface and groundwater. Based upon these policies, strategies (approaches) and tactics (means) for implementation should be formulated to identify programs, their priorities, and the support to be provided.

These are the sociopolitical elements of the institutional framework. To achieve the intended results, these elements are thrust or pulled through an "implementation filter" (consisting of laws, organizations, and data) which produces action upon and by the water users.

Within this implementation filter, the water law system should encompass all of the following:

- Declaration of *ownership* of water. Generally, all water resources, regardless of their location, belong to the people, the public, or the state.

- A process for *allocation* and *reallocation* of water to allow for the creation of identifiable rights or interests of individuals or groups of individuals such as water user associations or districts, municipalities, and industries. A "well-defined nature of property rights" should be established to minimize externalities and provide for their internalizing.

Figure 3-2. *Implementation of Water Policy Based upon Management by Results (MBR)*

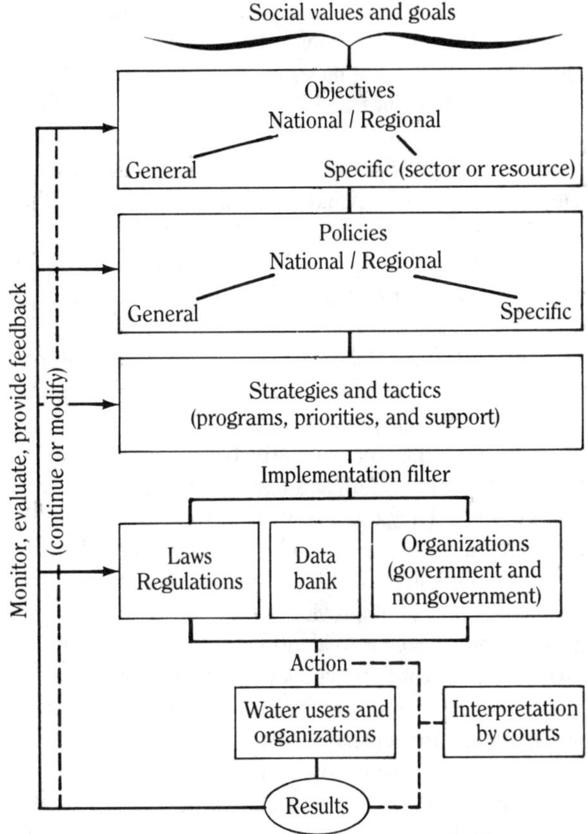

- A process for the *distribution* of water for all types of use and to all types of users. It should take account of the spatial and temporal nature of both water and the demand for it, the return flow characteristics within the basin, and possible transbasin diversions and their effects.

- Provisions for the conjunctive use of surface and groundwater. Although most of the law should pertain to water in general, one section of the law should address issues related to surface and groundwater and specifically the interdependence of surface water and tributary groundwater. As a practical matter, this section should address the efficiency of water use and the externalities that arise from the supply of and demand for water.

- Provisions for the integration of control over the quantity and quality of water. A specific law or section of a code should address the control and management of water quality. This section may set stream and discharge standards or establish a system to regulate the discharge of wastes and other pollutants. More specifically, this section should stress the importance of relating water quantity laws to considerations of water quality.

- Provisions for the planning, development, and conservation of water resources.

• Provisions for enforcement of the law and dispute resolution.

The second component of the implementation filter comprises the organizations related to irrigated agriculture. This includes the government agencies oriented toward achievement of the national objectives and policies. Often, agencies such as budget and personnel departments play a significant role in the effectiveness of the line agency (irrigation, agriculture, and development authority). Their role and operation should be examined.

This component also includes nongovernmental organizations designed or created to facilitate the effective use of water by the users. Usually called water user associations, these nongovernmental organizations range from informal to formal entities with simple to complex structures (Radosevich 1977). Their primary purposes are the distribution of water within the command area beyond the government outlet and the operation and maintenance of the distribution network. They often serve as a communications link between the government and the water users. In some cases they are instrumental in resolving disputes. From the standpoint of improving water use efficiency, their most important role is to identify a common body of interests. The physical boundaries of a water user association (whether it be a public or private company, an irrigation district at the project level, or a river basin authority at the multiple-project level) should be delimited in such a way that all members are hydrologically connected—that is, so that all members receive their water from the same lateral, canal, stream, or subsurface system. This will allow costs and benefits to be "unitized," or spread among the water users within a particular supply network, and will make each association accountable for externalities within a catchment area.

Water user associations also provide a mechanism for introducing incentives or imposing penalties. Members who adopt efficient application practices, who attempt to prevent the waste of water, and who are aware of their responsibility to use water in such a manner as to allow reuse by others can be rewarded by lower water charges or organizational assessments or by the distribution of surplus waters. Similarly, members who are indifferent to the rights of others or who abuse their rights through misuse or waste of water can be assessed a higher water charge or organizational fee or have their allocation of water reduced. The so-called free-rider problem is greatest in areas where all the water users act as individuals within a system. In the context of a water user association, the diseconomies caused by a free rider can more easily be identified and internalized to the free rider, rather than externalized to other members or to the system of users in the river basin.

The third component of the implementation filter is a hydrologic data base. The data base should include all relevant information on water supply and use, as well as projections of demands. It should be accessible to administrators and water users. Conventional data-gathering methods, such as stream gauging systems, should be employed, and measuring devices should be used on all major diversion structures. In addition, sophisticated methods of data collection should be explored. The government might consider the use of remote sensing devices, satellite imagery, and data collection platforms (DCPs) to validate and verify data on supply and water use as well as to facilitate data gathering. Several countries have begun using such sophisticated techniques so that decisionmakers at all levels can have access to current data. From the standpoint of water use efficiency, this data base is critical to identifying water misuse, seepage, and deep percolation losses resulting from overapplication of water, drainage problems, and water quality degradation. It can also be used to detect unaccounted-for benefits within the system. In theory, if not fact, the primary objective of an irrigation delivery network is to provide an adequate and timely supply of water to the users. Therefore, the desired result should come from the implementation process (filter operation and action upon and by the users). The process should be monitored and evaluated. If adjustments are deemed necessary, they can be made directly or indirectly on any component shown in Figure 3-2. The system is kept on track by determining whether the desired results are being achieved. Hence the suggestion that irrigation systems can achieve greatest efficiency and effectiveness when subjected to management by results (MBR).

Management by results is proposed in place of the management by objectives (MBO) system made popular by Drucker (1974) and McConley (1983). Under MBO it is the objectives that drive the analysis in evaluating the success of the program; under MBR it is the results that are evaluated. MBR may lead not only to changes in the policies, laws, and programs but also to the reformulation of objectives.

Conclusion

This chapter identified externalities that arise in irrigated agriculture and described how the institutional framework for water control and management can help to mitigate or eliminate them. Externalities are unavoidable in the context of water development and utilization, and particularly in the large-scale use of water by irrigated agriculture. But although externalities are unavoidable, the problems they bring are approachable, and numerous institutional alternatives are available. A dynamic economy requires that water administrators be alert to the occurrence of externalities in the system and know how to account for the benefits and the costs in order to achieve the highest degree of efficiency possible.

When evaluating the institutional framework, it is important to be aware of the role externalities play in water development. Obviously, dams, reservoirs, canals, and other physical structures within the system are extremely important. It is, however, equally important to provide proper institutions that allow the physical system to work. One cannot take a simplistic view of the role of institutions and the intricacies and difficulties in their appropriate development. Ciriacy-Wantrup (Bishop and Andersen 1985, p. 300) has said "a laissez-faire attitude toward institutions is no less inappropriate than regarding them as constraints."

Where serious questions of water use inefficiency exist, where levels of production are below expectations, and where projects fail to achieve their objectives, the search for solutions should begin not with technology but rather with the failure of the institutional system to employ properly the technologies already introduced. After an analysis of the policies, laws, and organizations and of the way they are implemented, it may be found that, in addition to institutional changes, new technologies are needed or the old ones need to be modified. The only constraint in designing and implementing water laws and their institutional components is the lack of human ingenuity to incorporate all that we know about the topic into the analysis. A policy, law, or organization will not make the system more efficient; it serves only as a mechanism for introducing advanced technologies and appropriate techniques which are economically feasible, socially acceptable, politically defensible, and easy to implement.

Notes

1. The term "institution" here is used as defined by Ciriacy-Wantrup (Bishop and Andersen 1985, p. 27) as a "social decision system that provides decision rules for adjusting and accommodating, over time, conflicting demands (using the word in its more general sense) from different interest groups in a society."

2. For an excellent discussion of externalities, see Seneca and Taussig (1974) and Staff and Tannian (1972). Young and Haveman (1985) recently provided a superb report on the economics of created resources. They describe the characteristics of water and the rationale for intervention, taking into account the external and secondary effects and the institutional arrangements that are available for coping with economic (externality) issues.

3. In 1960, Coase's paper "The Problem of Social Cost" was published in the *Journal of Law and Economics* (vol. 3, pp. 1–44). In 1973 and 1974, the *Natural Resource Journal* published the "Coase Theorem Symposium" in volumes 13 and 14. Randall (1983, pp. 131–48) takes particular exception to the Coase theorem. See Mishan (1971) and Buchanan (1973) for a review of the subject.

4. Demsetz (1982) argues that Coase did take into account the possibility of transactions costs but that in a laissez-faire economy these costs would also be internalized by the parties through negotiations or through a final determination of relative rights by the court.

5. An unusual case of not having a right to a natural external economy occurred in A-B Cattle Company v. U.S., 58 P.2d 57 (Colorado 1978). Downstream irrigators who used the muddy waters of the Arkansas River to silt and seal their ditches complained when the U.S. Bureau of Reclamation replaced this water with clean, clear reservoir water. The court ruled that they held a water right, not a silt right. Also, in Metropolitan Denver Sewage Disposal v. Farmers Reservoir and Irrigation Company, 499 P.2d 190 (Colorado 1970), the court held that irrigators downstream from the city's old sewage treatment plant were not entitled to the point of discharge of sewage containing nutrients. A right to this external benefit was denied downstream users as not being subject to the "no harm rule," when the City of Denver began discharging wastes from a new treatment plant below the irrigators' headgate.

6. Easterly (1977) has made a similar grouping.

7. The International Conference on Global Water Systems was held in Valencia, Spain in 1975, and the proceedings are recorded in the four volumes and summary report of Radosevich and others 1976.

8. Several unique systems were also analyzed, such as the permit system of water control in Israel and local water user systems found in Bali, Indonesia, and in Argentina (Radosevich and others 1976). The United Nations has sponsored studies of other water law systems as follows: in Africa (Caponera 1979), in Asia and the Far East (ECAFE 1967 and 1968), in Europe (Caponera 1979 and 1983), in South America (Valles 1983), and in selected Moslem countries (Caponera 1973 and 1978). The U.N. has also conducted a comparison of legal regimes for water use and abstraction (Teclaff 1972) and for the administration of water (United Nations 1974). Teclaff's *Water Law in Historical Perspective* (1985) is a synthesis of his earlier works.

References

Anderson, Terry L., ed. 1983. *Water Rights: Scarce Resource Allocation, Bureaucracy, and the Environment.* Cambridge, Mass.: Ballinger.

Bator, F. M. 1958. "The Anatomy of Market Failure." *Quarterly Journal of Economics* 72:351–79.

Bishop, R. C., and S. O. Andersen. 1985. *Natural Resources Economics: Selected Papers of S. V. Ciriacy-Wantrup.* Boulder, Colo.: Westview Press.

Bredehoeft, John D., and R. A. Young. 1983. "Conjunctive Use of Ground and Surface Water for Irrigated Agriculture: Risk Aversion." *Water Resources Research* 19, 5 (Oct.): 1111–21.

Buchanan, James M. 1973. "The Coase Theorem and the Theory of the State." *Natural Resource Journal* 13, 4 (Oct.): 579–94.

Cantor, L. M. 1970. *A World Geography of Irrigation.* New York: Praeger.

Caponera, Dante A. 1973 and 1978. *Water Laws in Moslem Countries.* Irrigation and Drainage Papers 20/1 and 20/2. Rome: Food and Agriculture Organization.

———. 1979. *Water Laws in Selected African Countries.* Legislative Study 17. Rome: Food and Agriculture Organization.

———. 1979 and 1983. *Water Laws in Selected European Countries.* Vol. 1, Legislative Study 10; Vol. 2, Legislative Study 30. Rome: Food and Agriculture Organization.

Daubert, J. T., R. A. Young, and H. J. Morel-Seytoux. 1980. "Measuring External Diseconomies from Ground Water Use in Conjunctive Ground and Surface Water Systems." In Dan Yaron and Charles S. Tapiero, eds. *Operations Research in Agriculture and Water Resources.* Amsterdam: North Holland.

Daumas, Maurice. 1969. *A History of Technology and Invention: Progress through the Ages.* Vol. 1, *The Origins of Technical Civilization.* Vol. 2, *The First Stage of Mechanization.* New York: Crown.

David, Rainey, and J. E. C. Brierley. 1978. *Major Legal Systems in the World Today.* New York: Free Press.

Demsetz, Harold. 1967. "Towards a Theory of Property Rights." *American Economic Review* 57 (May): 347–59.

———. 1982. *Economic, Legal, and Political Dimensions of Competition.* Vol. 4. New York: Elsevier.

Downing, Paul B. 1984. *Environmental Economics and Policy.* Boston: Little, Brown.

Drucker, Peter F. 1974. *Management: Tasks, Responsibilities, Practices.* New York: Harper & Row.

Easterly, Ernest S., III. 1977. "Global Patterns of Legal Systems: Notes toward a New Geojurisprudence." *Geographical Review* 67, 2:209–20.

ECAFE. Economic Commission for Asia and Far East. 1967 and 1968. *Water Legislation in Asia and the Far East.* Part 1, Water Resources Series no. 31; Part 2, Water Resources Series no. 35. New York: United Nations.

ESCAP. Economic and Social Council for Asia and Pacific. 1981. *Proceedings of the Expert Group Meeting on Water Pricing.* Water Resources Series no. 55. New York: United Nations.

Farvar, M. T., and J. P. Milton, eds. 1972. *The Careless Technology.* Garden City, N.Y.: Natural History Press.

Fisher, Anthony C., and Frederick M. Peterson. 1976. "The Environment in Economics: A Survey." *Journal of Economic Literature* 14, 1 (March): 1–33.

Gardner, Richard L., and Robert A. Young. 1985. "An Economic Evaluation of the Colorado River Basin Salinity Control Program." *Western Journal of Agricultural Economics* 10 (1): 1–12.

Hardin, Garrett. 1968. "The Tragedy of the Commons." *Science* 162:1243–48.

Hardin, Garrett, and John Baden. 1977. *Managing the Commons.* San Francisco: W. H. Freeman.

John Muir Institute. 1980. *Institutional Constraints on Alternative Water for Energy: A Guidebook for Regional Assessments.* Washington, D.C.: U.S. Department of Energy, Office of Environmental Assessments.

McConley, Dale D. 1983. *How to Manage by Results.* 4th ed. New York: AMACOM Book Division.

Mishan, E. J. 1971. "The Post-War Literature on Externalities: An Interpretive Essay." *Journal of Economic Literature* 9, 1 (March): 1–28.

Prang, George W., and Karen A. Tomb. 1983. "License to Waste: Legal Barriers to Conservation and Efficient Water Use in the West." *Mineral Law Institute* 28: 25–1 to 25–67.

Radosevich, G. E. 1977. *Improved Agricultural Water Use: Organizational Alternatives.* Fort Collins, Colo.: Resources Administration and Development, Inc.

Radosevich, G. E., and others. 1976. *Global Water Law Systems.* Vols. 1–4 and Summary. Fort Collins, Colo.: Department of Agricultural and Natural Resource Economics, Colorado State University.

Randall, Alan. 1974. "Coase Externality Theory in a Policy Context." *Natural Resource Journal* 14, 1 (Jan.): 35–46.

———. 1983. "The Problem of Market Failure." *Natural Resource Journal* 23 (Jan.): 131–48.

Seneca, Joseph J., and Michael K. Taussig. 1974. *Environmental Economics.* Englewood Cliffs, N.J.: Prentice-Hall.

Staff, Robert, and Francis Tannian. 1972. *Externalities: Theoretical Dimensions of Political Economy.* New York: Dunellen Press.

Teclaff, Ludwik A. 1972. "Abstraction and Use of Water: A Comparison of Legal Regimes." New York: United Nations.

———. 1985. *Water Law in Historical Perspective.* Buffalo, N.Y.: William S. Hein.

Thompson, Donald N. 1973. *The Economics of Environmental Protection.* Cambridge, Mass.: Winthrop.

United Nations. 1974. *National Systems of Water Administration.* New York.

U.S. Library of Congress, Congressional Research Service, Science Policy Research Division. 1976. *Science Policy: A Working Glossary.* Washington, D.C.

Valles, Mario F. 1983. *Water Legislation in South American Countries.* Legislative Study 19. Rome: Food and Agriculture Organization.

Wittfogel, Karl L. 1957. *Oriental Despotism.* New Haven, Conn.: Yale University Press.

Worster, Donald. 1985. *Rivers of Empire: Water Aridity and Growth of the American West.* New York: Pantheon.

Young, R. A., and R. H. Haveman. 1985. "Economics of Water Resources: A Survey." In A. V. Kneese and J. L. Sweeney. *Handbook of Natural Resources and Energy Economics.* Vol. 2. Amsterdam: Elsevier.

Comment

Mark W. Rosegrant

Radosevich views a system of water law—including policies, laws, organizations, and the process of implementation—as the institutional framework for the development of water resources. Although this conception of the legal system is useful in distinguishing among national legal families and among the socioeconomic, agroclimatic, and hydrogeological conditions that help shape a country's legal system, it is so broad as to confuse the role of externalities in failures in irrigated agriculture. This confusion about the causes of failures leads in turn to proposals for legal solutions which appear to be too rigid for the dynamic process of irrigation development and management in developing countries.

For example, Radosevich attributes the efficiency of groundwater exploration and development in Bangladesh to "the absence of law." But the problems in Bangladesh stem not from the legal system, but from inappropriate public investment policies and inappropriate incentives provided to private investors.

The government's decision to invest in deep tubewells, despite their extremely high investment and operating costs, led to initial problems with groundwater development in the 1960s. These problems were compounded by the poor choice of sites: wells were constructed in areas with high seepage and percolation rates. Furthermore, farmers were not trained in irrigation techniques. The result was command areas with yields far below potential and prohibitive costs per unit of production (Johnson 1986).

Additional problems have emerged as a result of the recent shift to shallow tubewells. The rapid growth in the use of shallow tubewells has been encouraged in part by heavy subsidies on investment costs, which are channeled through the banking system to individual farmers. Because neither the incentives nor the system of loan distribution has taken into account the need to encourage appropriate siting of wells and restraint in groundwater withdrawals, development has not proceeded logically relative to the resource base (Pitman 1983). This, like the problems of inappropriate investment in deep tubewells, is a matter of policy priorities and incentive structure, rather than a legal problem.

Radosevich goes on to note that "the law may be inappropriate to the present problems or may not provide the proper guidance." He cites water pricing as an example of an area in which the law fails to give guidance. In the same paragraph, however, he argues on the one hand that the law should permit water user fees to be charged on a volumetric basis to encourage more efficient use of water and on the other hand that "a more appropriate law would base water charges on the cost . . . of constructing and operating the system."

This inconsistency in recommended laws on water pricing points out the danger of strictly codifying the method of financing irrigation systems. The appropriate method of charging for irrigation systems is a pragmatic, empirical decision which may vary by type of irrigation system and level of development.

The continuing debate on the use of volumetric pricing of water to achieve efficiency in water use is a good example of an empirical problem. Although the literature often argues the theoretical superiority of pricing over quantity allocation of water, this is in fact not the case. As Weitzman (1974) points out, there is nothing to recommend one allocation system over the other from a theoretical point of view.

The reason most often cited for the superiority of prices as allocational devices is that they economize on information and transactions costs. In the case of diversion irrigation systems with a large number of small farmers, however, the information and transactions costs of administering efficiency pricing appear to be much larger than the costs of improving control of supply, which could generate nearly the same efficiency gains of volumetric pricing. This is because the costs (in both manpower and hardware) of implementing and administering an improved system of quantity allocation which would achieve approximately optimal water allocations are likely to be much lower than the costs of monitoring use at all levels down to the individual farm use. Furthermore, it would be very difficult to find the "right price" in an efficiency pricing system with nonlinear and stochastic water supply. If prices are set too low, demand would be excessive, especially among upstream users, and if prices are set too high, water would be wasted to system drainage.

In smaller pump systems, however, efficiency pricing of water may be much more cost-effective than quantity allocations. The price of water can be determined from the marginal cost of water (including appropriate adjustment for the scarcity cost of groundwater as a common property resource), and monitoring of the pricing system is relatively cheap. The wide variation in appropriate solutions to pricing issues argues against strict codification of pricing regulations.

Radosevich does an excellent job of pointing out the importance of organizations in the development and management of irrigation. But again, implicit in some of his arguments is the assumption that certain organizational forms lead automatically to appropriate incentives. The discussion of water user associations (WUAs) in particular is too sanguine about the effects of these organizations.

Several countries, including Indonesia, Mexico, and the Philippines, have emphasized the development of water user associations as a method of organizing farmers to improve the irrigation system through better maintenance and more efficient and equitable water allocation. But documented instances of WUAs actually improving system performance are rare. This may be because the organizational structure and system of incentives of these associations are inappropriate for these tasks.

Although water user associations are intended to be modeled after the organizations in successful, small indigenous and traditional irrigation systems, most WUAs that have developed under government supervision lack certain salient features of the indigenous organizations. Hunt (1985, p. 30) notes that the analogy between irrigation communities (indigenous or traditional irrigation organizations) and WUAs is weak at best: "Irrigation Communities are based on rewards as well as rights and duties, WUAs on duties alone. Irrigation Communities are systems of rights, duties and roles with substantial local control, WUAs are duties alone, with no rights, and no local control, and no system among these things. Irrigation Communities are vertically integrated to the headgate, so that the rights to water are protected and institutionalized, and in WUAs there are no group rights to water, and no arenas to legitimately discuss problems with water delivery."

WUAs have not solved the problems of management and maintenance of irrigation systems because sufficient attention has not been paid to the incentive structures which are most crucial to the successful functioning of the indigenous organizations. This is, in general, the problem with a formal legal approach to the analysis of irrigation development and management. Solutions from this type of analysis will tend to be too rigid. The role of law and the legal system should be to empower the government to deal with externalities, but the law should be flexible enough to allow site-specific solutions to problems. The main analytical focus should be on policies and adaptive organizational structures and roles which foster the incentives necessary to implement the main tasks in developing and managing irrigated agriculture.

References

Hunt, Robert C. 1985. "Appropriate Social Organization? Water Users Associations in Bureaucratic Canal Irrigation Systems." Waltham, Mass.: Department of Anthropology, Brandeis University. Processed.

Johnson, Sam H., III. 1986. "Social and Economic Impacts of Investments in Groundwater: Lessons from Pakistan and Bangladesh." In K. C. Nobe and R. K. Sampath, eds. *Irrigation Management in Developing Countries.* Studies in Water Policy and Management no. 8. Boulder, Colo.: Westview Press.

Pitman, G. T. K. 1983. "Groundwater Planning in Bangladesh: Some of the Major Issues." Dhaka: Master Planning Organization, Ministry of Irrigation, Water Development, and Flood Control. Processed.

Weitzman, Martin L. 1974. "Prices vs. Quantities." *Review of Economic Studies* 41, 4 (Oct.): 477–91.

Part II

Case Studies of Conjunctive Use

4

CALIFORNIA

Responses to Some of the Adverse External Effects of Groundwater Withdrawals in California

Jack J. Coe

This chapter recounts some of the experiences California has had with groundwater overdraft. After a brief summary of water conditions in California and a general discussion of the adverse effects of excessive withdrawals of groundwater, the chapter describes what has happened in four areas of the state afflicted by overdraft. The chapter concludes with a review of recent statewide responses to the problem.

Water Conditions in California

California suffers from a maldistribution of water supply and demand with respect to both location and time. Most of California's supply is in the northern part of the state, where natural replenishment occurs during the winter. Most of the demand for water comes from the south and occurs in the summer. Seventy-three percent of the state's runoff occurs north of Sacramento, but 75 percent of the state's water use occurs south of this point.

The large metropolitan areas and the nearly 10 million acres of irrigated agriculture tend to obscure the fact that much of California is naturally desert. Average annual precipitation in Bakersfield is only 8 inches. It is 15 inches in Los Angeles and 10 inches in San Diego. Even Sacramento averages only 19 inches of rainfall a year. The southeastern part of the state averages less than 3 inches.

The effects of the maldistribution of water supply have been aggravated over the years by a concurrent increase in population, irrigated agriculture, and water demand. During the period 1950–80 the population of California increased by 116 percent, reaching 23.8 million in 1980. During the same period, irrigated acreage increased by 38 percent, reaching 9.5 million acres. Water use jumped by 121 percent, and irrigation accounted for most of the increase. To meet the escalating demand, many areas pumped groundwater in amounts exceeding basin replenishment. Groundwater accounted for half of California's total water supply in 1950. Although by 1980 it had dropped to 24 percent of the total, the decrease reflected an increase in the development and use of surface water supplies, rather than a decrease in annual groundwater extractions.

Looking for ways to redress the maldistribution of water supply in the state, California's Department of Water Resources developed a master plan for the future development and management of the state's water resources. The California Water Plan, as it was called, was formally presented in report form in 1957. The first phase of the plan, the State Water Project, instituted the transfer and distribution of state water to areas in need of supplemental supplies. The project began delivering water to San Joaquin Valley and the southern San Francisco Bay area in the 1960s and to Southern California in 1972.

Nevertheless, for some areas that had overtaxed their local groundwater supplies—such as Kern County in the southern portion of the San Joaquin Valley—the initial deliveries were not sufficient to eliminate the overdraft. In 1983 the State Water Project had a firm yield of little more than half the quantity planned for ultimate delivery and covered by contract commitments. Overdraft in the state in 1980 was estimated at 1.8 million acre-feet a year.

Three factors have fostered the extensive use of groundwater in California since World War II: the continuing need for additional supplies, improvements in well drilling and pumping techniques, and the absence of systematic controls on groundwater use. In California, permits and licenses are not required for the use of groundwater as they are for surface water. Landowners are free to drill wells on their properties and pump water unless the water

Figure 4-1. *Policy and Physical Responses to Overdraft*

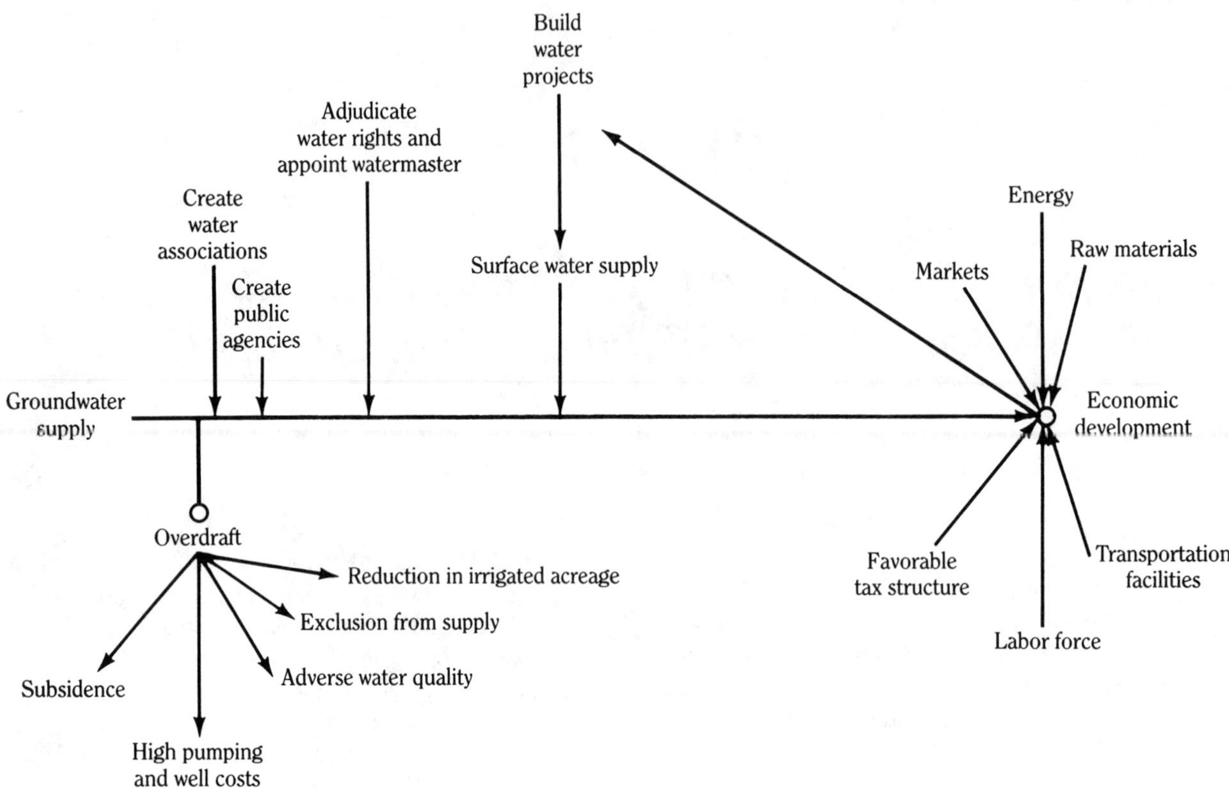

rights have been adjudicated by the courts. Groundwater is considered to be appurtenant to the land, and the right to its use is analogous to a riparian surface water right.

Adverse Effects of Overdraft

Overdraft has caused the following adverse effects, or externalities:

- Increased groundwater pumping lifts and costs
- Well deepening and pump lowering
- Land subsidence
- Degradation of water quality (including intrusion of seawater)
- Exclusion of some pumpers from the water supply
- A change in crop patterns to higher paying crops or, eventually, to nonirrigated crops.

Responses to Adverse Conditions of Overdraft

The adverse effects of overdraft and the policy and physical responses they have generated in many areas of California are depicted in Figure 4-1. Economic development is dependent on raw materials, a labor force, a market for the goods and services produced, and, among other things, an adequate supply of good water. In many locations where surface supplies were meager, groundwater resources were developed. When overdevelopment gave rise to detrimental effects, the typical response was for those experiencing problems to create a water association to provide a forum for discussion. Usually consultants were retained to provide advice. Once a plan had been agreed upon and it was necessary to levy taxes, condemn property, or contract to import water, a public agency was created. In some cases the groundwater rights were ajudicated to ensure equitable allocation of the scarce resource. But adjudication of water rights does not provide additional water, and the importation of supplemental supplies has been the solution to overdraft in California whenever feasible.

The importation of surface water over long distances is perceived differently by different groups. Originally, such facilities as the Los Angeles, Colorado River, and California aqueducts were considered engineering marvels. But some call them rescue projects and view the importation of supplemental water as an attempt to repair the damage done by past mismanagement of the economy and water resources. Others feel that a proper management plan permits overdraft to occur in order to provide an economic base for financing the water importation projects that will in turn eliminate the overdraft.

Case Histories

The experiences of four areas in California that confronted the problem of overdraft are described below. There are both similarities and differences in the manner in which the residents of the areas have responded to the overdraft, and solutions have been found in three of the areas.

Santa Clara Valley

Setting. Santa Clara Valley is situated just south of San Francisco Bay and covers about 500 square miles. Average annual precipitation is about 13 inches. The groundwater basin consists of a large pressure or confined zone surrounded on three sides by a forebay or unconfined zone. Water-bearing sediments extend to a depth of 1,000 feet.

History of Land and Water Use. Before World War II, Santa Clara Valley was largely an agricultural area, growing irrigated fruit trees, grapes, and truck and field crops. Urbanization occurred after the war, and in recent years the region has become the center of high technology industry earning the nickname Silicon Valley. The population in 1983 was about 1.4 million. Historically, groundwater was the principal source of water supply, and there were many flowing wells. In 1949 the Department of Water Resources estimated the overdraft at 52,000 acre-feet a year. Subsequent observations of groundwater levels have indicated that the amount of overdraft has increased since then. In this valley, as in other locations, use of, and responsibility for, groundwater resources have shifted from the private sector to public agencies. Initially, the overdraft was caused by agricultural water users. Eventually, however, the increase in land values and urban pressures led the farmers to sell their property to developers, and the agricultural valley became a small megalopolis. Public agencies serving urban users increased the overdraft and became responsible for financing corrective measures.

Adverse Effects of Overdraft. Groundwater pumping in excess of replenishment caused a loss of artesian well flow and lowered groundwater levels. Levels have dropped 120 feet since 1910, and most of this occurred before 1950. The decline in groundwater levels caused an increase in pumping lifts and in pumping costs. New wells had to be constructed deeper because of the overdraft, and in some instances existing wells had to be deepened.

Groundwater overdraft in the Santa Clara Valley also led to land subsidence and deterioration of water quality. The point of maximum land surface decline was near San Jose, where a drop of 13 feet occurred between 1910 and 1970. The quality of groundwater in the valley was degraded by the intrusion of seawater from the southern portion of San Francisco Bay and the invasion of connate brines from surrounding areas.

Responses to Overdraft. In most areas the first response to a water problem is the creation of a water association to provide a forum for discussion of possible solutions. The Santa Clara Valley Water Conservation District was created in 1929 for just such a purpose. By 1955 the conservation district had constructed ten local conservation dams to store floodwaters for subsequent release to groundwater recharge facilities. (In 1968 the conservation district was merged with the Santa Clara County Flood Control and Water District.) In 1961, after it had become evident that local water resources were fully developed and an import supply was required, the County Flood Control and Water District signed a contract with the State Department of Water Resources to get up to 100,000 acre-feet of water a year from the proposed State Water Project. The first shipment of water was received in 1965 and spread in the basin. In 1967 state water was first treated and delivered directly. The local district has also arranged to import water from the Federal Central Valley Project through Pacheco Pass Aqueduct. Delivery of this water is expected by the mid-1980s.

It also became obvious that in addition to the importation of water and the full use of local supplies some control of groundwater pumping would be necessary. Adjudication of groundwater rights was rejected as a solution. Instead, groundwater zones were created and a pump tax was levied which varied from one zone to another depending on the cost of alternative supplies. In 1983 the pump tax on groundwater extraction in the main portion of the Santa Clara Valley was $54.00 an acre-foot for municipal and industrial use and $13.50 an acre-foot for agricultural use. Under the current management plan, in which water is treated as a public utility, water from all sources is priced the same, but surcharges are assessed on the basis of benefits received. To halt seawater intrusion, an experimental barrier was constructed using injection wells and treated waste water.

Current Conditions. Groundwater levels in the Santa Clara Valley have stabilized at about 30 feet above historic lows, and land subsidence has essentially ceased. Because the problem of seawater intrusion was found to be minor and confined to the shallow, tight aquifer, the seawater barrier has been discontinued.

Outlook. Additional water will have to be imported into the area to supplement the deliveries from the state. Also, the full potential of the State Water Project of 4.23 million acre-feet a year needs to be developed. A better method of moving water across the delta would not only increase the yield, but also improve water quality.

Coastal Plain of Los Angeles County

Setting. The Coastal Plain of Los Angeles County is located southwest of downtown Los Angeles and encompasses about 420 square miles. Average annual precipitation is about 15 inches. Groundwater is obtained from several confined aquifers at various depths and from a forebay in the northeastern part of the plain. The plain is composed of the West Coast Basin and the Central Basin.

History of Land and Water Use. Before World War II, a mixture of agriculture, oil production and refining facilities, and isolated communities existed on the coastal plain. The chief crops were beans, truck crops, and cut flowers. Groundwater was the principal source of water. After World War II, the area urbanized rapidly. Groundwater from overdrafted aquifers and Colorado River water imported from the Metropolitan Water District of Southern California were used to meet the increasing demands for water. The population in 1983 exceeded 3 million. The accumulated overdraft in the Coastal Plain in the 1940s was estimated at 1 million acre-feet.

Adverse Effects of Overdraft. As a result of overpumping, groundwater levels dropped to more than 100 feet below sea level in the 1940s. Pumping costs increased, and some wells had to be deepened. Seawater intrusion was a problem along the coast of Santa Monica Bay, and many wells were abandoned.

Responses to Overdraft. The West Basin Water Association was created in 1946 to develop solutions to the problems of declining groundwater levels and seawater intrusion brought on by the overdraft. In 1950 the Central Basin Water Association was created. To levy an ad valorem tax to raise funds to purchase imported water for artificial recharge and to oversee direct deliveries of imported water to purveyors, two municipal water districts were created: one in the West Basin in 1947 and the other in the Central Basin in 1952. Local water interests in the Coastal Plain concluded that it was necessary to adjudicate groundwater rights to equitably allocate the supply among pumpers. Adjudication was completed in the West Basin in 1961 and a year later in the Central Basin. The court decree restricted groundwater extractions to 290,000 acre-feet a year and appointed the State Department of Water Resources as watermaster.

The Central and West Basin Water Replenishment District was created in 1959 to manage the Coastal Plain. Its responsibilities included the purchase of imported water for artificial recharge and the operation and maintenance of spreading basins and the seawater barriers. It was the first— and is still the only—replenishment district created under laws of the State of California. To obtain funds to purchase supplemental water the replenishment district levied a pump tax. This shifted costs from the property owners to the pumpers. In 1982–1983 the pump tax was $30.00 an acre-foot.

To prevent the intrusion of seawater from Santa Monica Bay, the Los Angeles County Flood Control District constructed an experimental barrier in the mid-1950s using injection wells and imported water. The barrier was designed, constructed, and operated with state funds. Subsequently, a longer, permanent barrier was completed at local expense.

The Coastal Plain receives about 300,000 acre-feet of water a year from the Metropolitan Water District. The imported water is a mix of Colorado River water and State Project water from northern California. It is used directly, spread in the forebay, and injected into wells at the seawater barrier. In addition to imported water, storm flows and reclaimed waste water are spread in the forebay.

Current Conditions. Seawater intrusion has been controlled through operation of the barrier. In addition, groundwater levels have been raised, which has reduced the costs of pumping and of operating the barrier.

Outlook. Maintaining stability within the Coastal Plain groundwater basin is dependent upon an assured supply of imported water. This in turn requires increasing the yield of the State Water Project.

Coastal Plain of Orange County

Setting. The Coastal Plain of Orange County covers most of the county and encompasses about 300 square miles. Average annual precipitation is about 13 inches. The Santa Ana River traverses the area and contributes about 50,000 acre-feet a year to the groundwater supply. Groundwater is obtained from several confined aquifers at various depths and from the forebay in the northeastern part of the plain.

History of Land and Water Use. The Coastal Plain consisted of irrigated agriculture and scattered communities before World War II. The principal crops were citrus fruits and truck crops. After 1945 the area urbanized rapidly, the population grew by more than 10 percent a year, and assessed valuation increased at the rate of $200 million a year. In 1983 the population was 1.9 million. The accumulated overdraft was estimated in 1956 to be 700,000 acre-feet.

Adverse Effects of Overdraft. Overdraft caused a decline in groundwater levels until they were 10 feet below sea level in the forebay and as much as 23 feet below sea level in the pressure area. This in turn caused well construction and pumping costs to increase. Seawater intrusion was a problem at two locations along the coast.

Responses to Overdraft. The Orange County Water District was created in 1933 to look into the problem of rapidly dropping groundwater levels. Since then, numerous amendments to the original district act have greatly expanded the district's role in managing the basin. Approximately 100,000 acre-feet of imported water are spread in the forebay each year. In 1983 the imported supply contained a fairly equal mix of Colorado River water and State Project water. Seawater intrusion barriers have been constructed, using both injection and extraction wells. Most of the injection water at the main barrier is waste water which has received tertiary treatment by the Orange County Water District.

A pump tax is levied to secure the funds for importing water. In 1983 the tax was $15.00 an acre-foot. In addition, a basin equity assessment is applied to control the pattern of groundwater pumping. The Municipal Water District of Orange County, a member agency of the Metropolitan Water District, has expanded its surface distribution of imported water. As in the Santa Clara Valley, groundwater rights in the Coastal Plain of Orange County have not been adjudicated. Instead, the local water interests have decided that the resource should be handled like a public utility. Thus, a water supply is guaranteed all users, and water from all sources is priced the same.

Current Conditions. As a result of the physical and policy responses discussed earlier, seawater intrusion has been controlled and groundwater levels have been stabilized.

Outlook. As in the Santa Clara Valley and the Coastal Plain of Los Angeles County, successful management of the Orange County Coastal Plain is dependent on increasing the yield of the State Water Project.

Kern County (within the San Joaquin Valley)

Setting. The portion of Kern County that is in the southern part of the San Joaquin Valley covers 6,840 square miles. Average annual precipitation is about 7 inches. The area is traversed by the Kern River, which is the groundwater basin's principal source of replenishment. The depth of water-bearing sediments is as much as 4,500 feet. There is an upper unconfined aquifer and a lower confined member, which is the principal aquifer. About 20 million acre-feet of water are stored within the top 200 feet of saturated sediments. Dewatered storage capacity is 11 million acre-feet.

History of Land and Water Use. Kern County remains primarily an agricultural area. Unlike the other areas discussed, it has not become an urban complex. Irrigation commenced with diversions from the Kern River in the early 1900s. After the development of deep well drilling and pumping techniques in the 1940s, groundwater became the main source of supply. In 1980 the population—most of it around Bakersfield—had reached 325,000. Irrigated acreage in that year was 944,000 acres, 782,000 acres of which overlay groundwater. The value of agricultural products was $1.27 billion in 1980. The area irrigated in Kern County has increased constantly since World War II despite the overdraft, which was estimated at 500,000 acre-feet a year in 1950. In 1980, because of expanded artificial recharge operations, the overdraft was reduced to 346,000 acre-feet a year, but by 1989 it will increase to 590,000 acre-feet a year, if no additional water supplies are made available.

Adverse Effects of Overdraft. The overdraft has caused groundwater levels to drop about 200 feet. This led to an increase in pumping and well construction costs. As in the Santa Clara Valley, the overdraft has caused land subsidence. Between 1925 and 1982 land surface dropped as much as 10 feet in places. The overdraft is also responsible for the degradation of water quality that has occurred as water of inferior quality moves from the western part of the county to the pumping trough.

Responses to Overdraft. When it became obvious in the 1930s that the agricultural economy could not be supported by local water resources alone, Kern County was included in the service area of the Central Valley Project. Water from the project was received through the Friant-Kern Canal starting in 1955. In 1961 the Kern County Water Agency (KCWA) was created for the purpose of contracting for water from the State Water Project. Subsequently, fifteen water districts were created as members of KCWA. The state signed a contract with KCWA in 1963 for the delivery of up to 1.1 million acre feet of state water a year plus any surplus water that might be available. State water was first delivered in 1968. To provide an institutional mechanism for the apportionment of costs and benefits, Improvement District number 4 (Bakersfield) was created in 1972.

For many years now water has been spread on the Kern River fan to help reduce the overdraft. Artificial recharge has also been tried in unlined canals. In 1981 the amounts of artificial recharge were as follows: 116,300 acre-feet in the Kern River, 255,600 acre-feet from the State Water Project, 7,400 acre-feet from the Central Valley Project, and 28,200 acre-feet from minor streams—for a total of

407,500 acre-feet. KCWA reports that since the drought of 1976–77, additions to the groundwater basins have exceeded extractions by 1.1 million acre-feet. An innovative "banking" concept, initiated in 1982, supplements the "overdraft correction" program. The advantage of the banking program is that water in storage belongs to KCWA or is sold to a member agency. The water in storage can be used during droughts when surface water is not available from the import projects.

To provide a source of revenue and to offset any special benefits received, a pump tax was initiated in 1975 in Improvement District number 4. In 1982–83 the tax was $10.00 an acre-foot for water used for agriculture and $20.00 an acre-foot for all other uses. The funds from the pump tax are used to offset the costs of local transport of state water, treatment of water delivered directly to the users, and artificial recharge. Other areas do not need a pump tax to discourage groundwater pumping because groundwater costs exceed the cost of surface water supplies. Another mechanism used to equate costs with benefits is the zone-of-benefit charge (ad valorem tax) that KCWA levies against those who do not receive or pay for state water but benefit from the higher groundwater levels caused by those who do.

To help protect groundwater quality, the Kern County Board of Supervisors has passed a well construction ordinance that prohibits perforating well casings in both the upper, unconfined (saltier) aquifer and the lower, confined (principal) aquifer.

Current Conditions. The accumulated overdraft in Kern County is still large. Subsidence is continuing, but at a lower rate since the State Project water became available. There are still about 73,000 acres of undeveloped irrigable land in Kern County. If this acreage is put under irrigation without additional surface supplies, the overdraft will be increased.

Groundwater conditions in Kern County are perceived differently by different groups. There are some that contend that large corporate farmers are getting rich by expanding their irrigated acreage and are exacerbating the overdraft. Some also contend that this will cause permanent damage as a result of land subsidence and the intrusion of water of inferior quality. Further allegations are that the long-term costs of solving the problem, if it can be solved at all, exceed the short-term benefits and that those who benefit now are probably different from those who will pay later. KCWA and its member agencies, however, believe that they have an excellent management program and that the only problem is the lack of imported supplies. They point out that areas in the state where overdraft has been eliminated have all received imported water.

Outlook. The debate over what should be done to eliminate the overdraft in Kern County will undoubtedly continue. The future magnitude of the overdraft is dependent on (a) the availability of surplus water from the State Water Project; (b) the increase in the yield of the State Water Project; (c) the amount of land to be irrigated (total water demands in KCWA will increase by 224,000 acre-feet a year by 1990 if the new lands are irrigated); and (d) the availability of additional supplies from the Central Valley Project.

Local water interests report that they will continue to avoid adjudication of water rights at all costs because they believe it will be expensive and time-consuming. In addition to overdraft, a problem that worsens each year is the adverse salt balance in the county. Importation of water brings not only water but salts. Because there is no natural outlet to the basin, the salts accumulate. Over the long term, the costs of solving this problem could be substantial. Although a master drain is an authorized element of the State Water Project, up to now, local interests have shown little inclination to pay their share of the capital and operating costs of a drainage program.

Statewide Responses to Groundwater Overdraft

In 1977, the Governor's Commission to Review California Water Rights Law recommended that state legislation be enacted to create local groundwater management districts in areas without existing effective management. It also recommended that the new districts or other groundwater authorities develop groundwater management plans and submit them to the State Water Resources Control Board for evaluation and comment. The board would be authorized to ask the state attorney general's office to seek judicial relief if no, or inadequate, management plans were submitted or if approved plans were not implemented properly.

Legislation was introduced to implement the foregoing recommendations, but was never passed because of strong opposition from water agencies in areas where groundwater management had already been implemented and in areas where overdraft existed. Primarily as a result of the lack of groundwater legislation, an initiative which included much of the Governor's Commission's recommendations on groundwater management was placed on the ballot in November 1982. Again because of strong opposition from water interests in the state, the initiative was defeated. Yet another effort which would have affected the overdraft failed: Legislation which provided for additional yield for the State Water Project was passed in 1980. The program (Senate Bill 200) included water development in northern California, a peripheral canal to convey water from north of the Sacramento–San Joaquin delta to an existing state pumping plant and canal south of the delta, and other measures. Areas to receive additional water included Kern County. But in a referendum held in June of 1982 the voters rejected the legislation.

The water problems in California will not go away as a result of the elections in June and November of 1982, and we can expect to see further responses to the adverse external effects of groundwater overdraft in California. These responses will assist in finding solutions to the problem.

Comment

F. L. Hotes

Coe mentions that the California Water Plan was formally presented in 1957. At the time it was probably the most comprehensive and massive plan ever developed for the transfer and distribution of large quantities of water (water usage of 63 billion cubic meters a year under ultimate development). Although possible projects for the first phase were identified, the California Water Plan was not a rigid plan of specific works. It was considered to be a general and coordinated master plan for the progressive and comprehensive future development of the water resources of the state by all agencies (including private projects), subject to: (a) more detailed investigation and study of component features of the plan to determine their need, engineering feasibility, economic justification, financial feasibility, and recommended priority of construction; and (b) continuing review, modification, and improvement in the light of changing conditions, advances in technology, additional data, and future experience.

Regulation of water supplies would depend upon conjunctive use and operation of more than 200 significant groundwater basins in the state, including the basin underlying the Central Valley, where an underground storage capacity of 38 billion cubic meters was visualized. Although the legislature and the people have as yet authorized conjunctive operation by a public agency in only a handful of these basins (generally those with critical overdraft and water quality problems), all of the basins provide natural storage and benefits under present unregulated usage. Ultimately, as water usage increases and groundwater levels drop and as water quality problems become more prominent, public regulation of basins will probably increase.

Seawater intrusion was a problem in three of the four examples of groundwater basin operations discussed by Coe. In the Los Angeles and Orange County cases, injection wells were part of the solution, but in the Santa Clara Valley such wells were not used at all. Since the costs of piping good water to the well sites and injecting it into underground aquifers are much higher than the cost of pumping water, the decision on injection wells should depend upon detailed economic analysis of the specific situation. It may be more economical to recharge the supplemental water in spreading grounds near the head of the aquifer, as was done in the Santa Clara Valley, or it may be more economical to deliver the supplemental water to users in the area of seawater intrusion through existing or new transmission systems. In all three cases the same amount of supplemental water is needed to offset the overdraft condition causing the intrusion, but the relative economics of the three methods of delivering the water (or a combination thereof) should be explored.

5

PAKISTAN

Large-Scale Irrigation and Drainage Schemes in Pakistan

Sam H. Johnson, III

The Indus Basin of Pakistan, with its deep alluvial soils and the abundant water resources of the Indus River and its tributaries, is potentially one of the world's most important food-producing areas. It contains the largest contiguous irrigation system in the world, and there is no inherent technological reason why the basin could not produce up to three times its present output. Pakistan's rapid population growth, as well as that of other countries in South and Southeast Asia, will result in a billion additional people in the region within the next four decades. To support this population there must be concern with more than food self-sufficiency. The Indus Basin should be envisioned as one of the great export suppliers of food for the future decades. The potential gains for Pakistan and for the world from better utilization of the Indus are extremely high (Falcon 1976).

Unfortunately, in spite of a favorable climate and vast resources of land and water, the Indus Basin agricultural system is responding very slowly to the challenge before it. The "Indus Food Machine" is struggling just to increase food production fast enough to keep up with population growth. A long list of reasons can be developed to explain why agricultural output is increasing so slowly. Variable weather patterns and rapidly increasing energy prices provide short-term explanations, but a more serious long-term impediment to increased crop productivity is soil salinity (Pakistan 1978). Because of poor drainage of the vast, nearly flat Indus Plain, waterlogging and resulting salt accumulation in the soil are slowly destroying the fertility of much of the irrigated land. Although it has recognized the problem, the government of Pakistan has not been very successful in finding a long-term solution. From the 1930s through the 1950s, the government experimented with various drainage schemes involving drilled wells (tubewells). Then in 1960, starting with Salinity Control and Reclamation Project I (SCARP-I), a full-fledged commitment was made to implement the most extensive, and expensive, vertical tubewell drainage scheme in the world.

The objective of this chapter is to record the historical development, implementation, and management of Pakistan's SCARPs, using field and management data from SCARP-I, SCARP-II, and Khairpur SCARP. The direct economic feasibility of the SCARPs is considered and compared with that of private tubewell development.

Historical Overview

More than 2,000 years ago the inhabitants of the Indus Plain had constructed inundation canals to irrigate areas along the banks of the river. But it was not until 1859, when British army engineers completed the Upper Bari Doab Canal, that large-scale perennial irrigation was introduced to the Indus Plain itself (see Map 5-1). The canal system was designed according to the same principles as other North Indian systems constructed by the British during this period. There were no storage reservoirs, and supply depended on seasonal variation of discharge of the river systems. Canal design was determined by the government's economic and social objectives, which were to spread the available water as extensively as possible and in such a way as to prevent famine, to "bring to maturity the largest area of crop with the minimum consumption of water," and to ensure equitable distribution throughout the command area (India 1873). Farmers and their families moved into newly watered lands by the hundreds of thousands. By 1920 the northern part of the Indus Plain, the Punjab, was called the breadbasket of India (Paustian 1930). In the south the pattern of land settlement was somewhat different, with large landowners holding most of the land and sharecroppers on small parcels, but even here the availability of water turned the barren land into a productive resource.[1]

Between the 1830s and the 1960s the Indus Plain,

Map 5-1. ***Indus Plain: Rivers, Dams, and Link Canals***

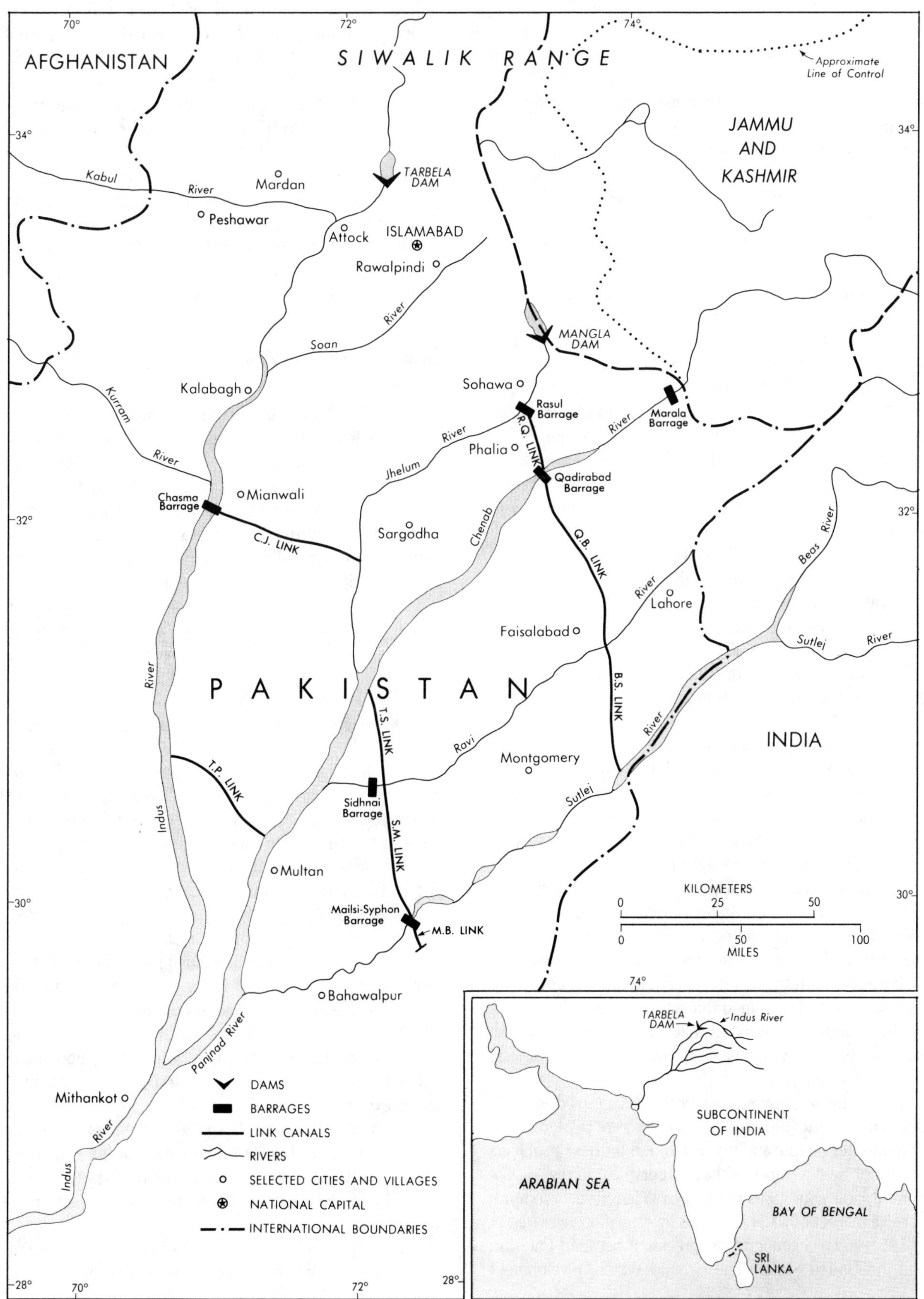

which encompasses more than 207,000 square kilometers and stretches 1,200 kilometers from the Himalayan foothills to the Arabian Sea, was covered with the world's largest contiguous block of irrigated land. Here the Indus and its tributaries drain a watershed of more than 943,000 square kilometers to serve an irrigated area of 13 million hectares (Taylor 1965). Yet the bounty of the irrigation system was not perfect. Given the gentle slope of the Indus Plain, 0.2 meter per kilometer, drainage soon became a problem in many areas. By the late 1800s waterlogging and soil salinity had already been recognized as serious problems in the Indus Plain, particularly in areas between the rivers known locally as *doabs.* In 1917 a drainage board was established in the Punjab to study the problem, and the Irrigation Research Institute at Lahore and the Directorate of Land Reclamation Punjab were created to develop remedial measures. Provincial irrigation authorities took steps to reduce canal seepage losses, and construction of a large network of surface drains was also undertaken. Steady-state conditions in the position of the water table were reached in the 1940s in some areas where evaporation from the water table at or near the land surface balanced recharge from canal leakage (Mundorff and others 1976).

Soon after the independence in 1947, Pakistan became increasingly concerned about the growing waterlogging and soil salinity problems in the Indus Plain. By 1950 more than 2 million hectares of irrigated land had gone out of production, and additional land was going out of production at the rate of 29,000 hectares a year. The government of Pakistan requested help from the Food and Agriculture Organization (FAO) of the United Nations, and in response the FAO in 1950 sent drainage and reclamation experts to study the problem. These experts provided some valuable suggestions, but they were frustrated in their study by the lack of technical baseline data. Their main recommendations were to carry out more surveys and investigations and to select a few acres for pilot schemes. In 1952, again at the request of the government of Pakistan, the United States Bureau of Land Reclamation sent a drainage engineer, E.R. Maierhofer, to study the damaged areas of the Punjab and the Khairpur-Shikarpur area in the south. He reviewed the proposals of the FAO experts, but opined that actual large-scale projects would be much more worthwhile than an occasional review of the achievements of a few small pilot projects (Maierhofer 1952).

As a result of these studies, it was recognized in 1954 that both waterlogging and soil salinity were related to the groundwater regime and that a solution to these problems could not be developed without a complete knowledge of groundwater conditions in the area. Therefore, a comprehensive project of water and soils investigations was begun in 1954 under a cooperative agreement between Pakistan and the United States. The Ground Water Development Organization, which later became the Water and Soils Investigation Division of Pakistan's Water and Power Development Authority (WAPDA), was created in August 1954 to serve as the base organization. Broad objectives of the research project were to inventory the water and soil resources of the Punjab, to determine aquifer characteristics, to delineate fresh and saline water zones, and to identify and describe relations among various components of the hydrologic system, including canal seepage, irrigation, waterlogging, and salinity. The insights gained from these studies, which are briefly summarized in the following paragraphs, have provided much of the scientific and technical base for the various reclamation and drainage projects that are now in operation in Pakistan.

Groundwater Hydrology

The Indus Plain was formed by alluvium deposited by the Indus River and its tributaries—the Jhelum, Chenab, Ravi, Beas, and Sutlej—in an extensive depression lying between the Himalayas and the Salt and Sulaiman mountain ranges. This vast plain is underlain by an extensive aquifer system in which groundwater may move both horizontally and vertically. Locally, however, lenses of fine-grain materials form horizontal boundaries or semiconfining layers that impede vertical movement of water and result in semiartesian conditions. Near the center of the plain fairly small bedrock hills protrude through the alluvial deposits to form local boundaries to the movement of groundwater; elsewhere the alluvial deposits are from 300 to 500 meters thick (Mundorff, Bennett, and Ahman 1972).

The climate of most of Pakistan, and particularly of the Indus Plain, is arid to semiarid (less than 200 to 500 millimeters annual precipitation). Only in the submontane zone of the Himalayan foothills is the precipitation more than moderate; there it may exceed 2,000 millimeters. Along the northern rim of the Upper Indus Plain the annual precipitation ranges from 750 to 1,000 millimeters. A marked decrease in annual precipitation occurs southward of the rim—about 475 millimeters at Lahore and 200 millimeters at Multan—and in the Lower Indus Plain rainfall of 100 millimeters or less annually is not uncommon. The bulk of the runoff in the Indus River and its major tributaries is generated in the montane and submontane zones of their headwaters and is derived both from melting snow and ice and from the rains of the southwest monsoon. The Indus River and its tributaries carry huge loads of sediment derived from erosion of the Himalayas, the Hindu Kush, and contiguous mountain ranges, as attested by the vast accumulation of stream deposits in the Indus Plain.

Before the development of canal irrigation in the nineteenth century, the groundwater hydraulic system was in a

Figure 5-1. *Water Table Profiles along Line B–B′, Chaj, Rechna, and Bari Doabs*

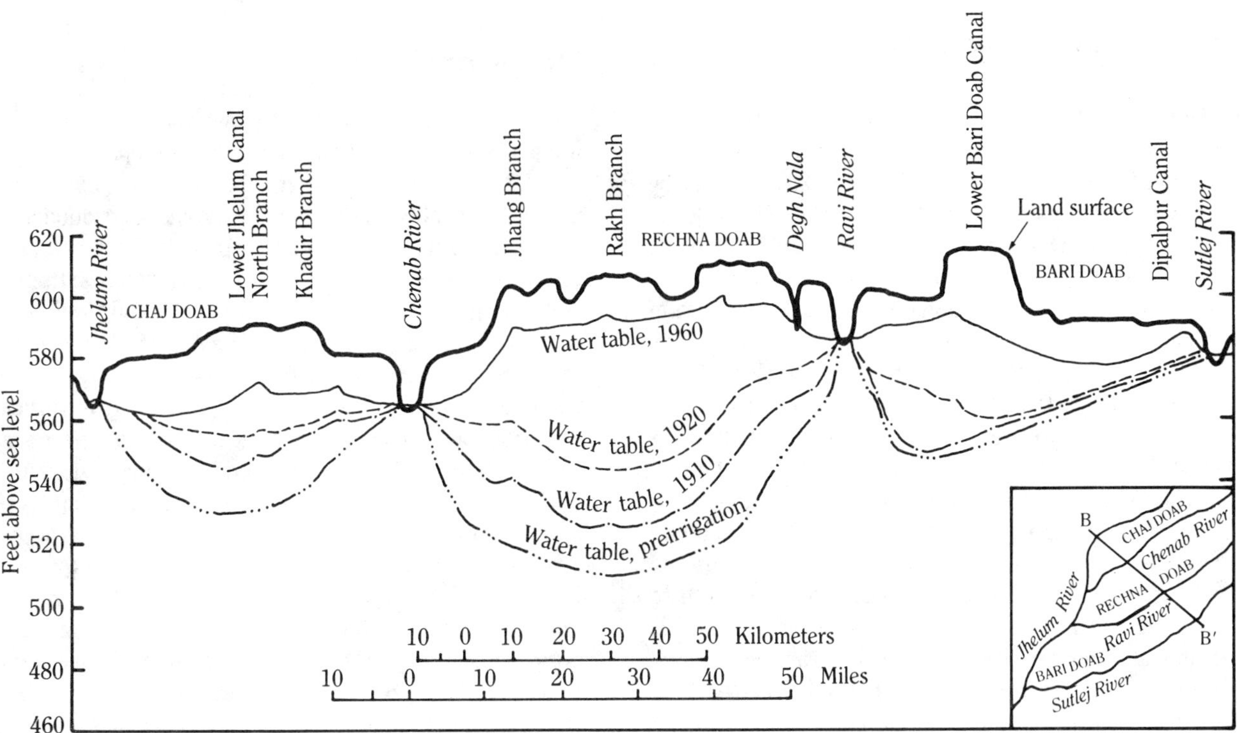

state of dynamic equilibrium. Over moderately long periods of time, recharge to the groundwater reservoir balanced discharge, and there were no long-term changes in groundwater levels. Equilibrium conditions of the hydraulic system before the introduction of irrigation reflect the compound effects of geology, climate, topography, and drainage. Of particular importance in the hydrologic budget is the distribution of precipitation. Where precipitation was below a threshold value, probably 250 to 350 millimeters under prevailing temperature conditions, recharge from precipitation was not a significant factor in the groundwater budget, and infiltration of river water was the dominant source of groundwater replenishment.

Irrigation changed the natural hydrologic environment of the Indus Plain. The canal system introduced additional sources of recharge and caused a rise of the water table in and around the irrigated areas. Seepage losses were greatest near the bifurcation points in the upper parts of the doabs because of the greater density of canals and less near the rivers because the water table was already close to the surface (Mundorff and others 1976).

Figure 5-1 illustrates the change in the depth to the water table from before the introduction of irrigation to the early 1960s. The water table in the middle of the doabs rose from 20 to 30 meters during this 80- to 100-year period. This rise initially proceeded at a constant rate until the water table approached the land surface (Greenman, Swarzenski, and Bennett 1967).

As it became evident that the rising water table and the resulting waterlogging and salinity threatened the future of irrigated agriculture in the Indus Plain, studies were made to determine the root of the problem. Carlston (1953) carefully examined previous research and identified three main factors: leakage from the irrigation-distribution system, infiltration of applied irrigation waters, and, to a lesser degree, increased recharge from rainfall runoff because of obstruction of natural drainage courses. Recent studies indicate that leakage and infiltration combined contribute about 80 percent of the recharge to the aquifer, with the remaining 20 percent coming from rainfall runoff, link canals, and the rivers themselves.

Studies of water quality show that the alluvium beneath about two-thirds of the Upper Indus Plain and less than one-third of the Lower Indus Plain is saturated to an average depth of 150 meters or more with waters of acceptable quality for irrigation supply. The average concentration of dissolved solids in the supplies in the Upper Indus is less than 800 milligrams a liter. In the Lower Indus, groundwater of acceptable quality is generally confined to a narrow strip along the rivers. Total annual recharge to groundwater within canal-commanded areas with a diversion of 123 billion cubic meters is estimated to be 59.2

billion cubic meters, of which 40.7 billion cubic meters occurs in usable groundwater zones (WAPDA 1979).

SCARP Studies and Development

After 1958, drainage and reclamation works were transferred to WAPDA. In 1961 a plan for eradicating waterlogging and salinity in the whole of Pakistan was prepared by WAPDA with the assistance of its consultants Harza Engineering Company International, Tipton and Kalmbach, and Hunting-MacDonald. Before the WAPDA plan was completed, however, preparations were made to reclaim 490,000 hectares of land in a project known as Salinity Control and Reclamation Project I (SCARP-I). Using $15.2 million made available to the government of Pakistan by the United States Development Loan Fund, SCARP-I was begun in 1960. The project area for construction of SCARP-I was in the center of the interfluvial area between the Ravi and Chenab rivers known as Rechna Doab (see Map 5-2). A main objective of SCARP-I was to demonstrate that vertical tubewells could be used effectively both to lower the water table over a large area and to provide additional water for intensified irrigation and leaching of salts from saline-affected soils (Malmberg 1975).

While SCARP-I was under construction, WAPDA and its consultants were completing their ambitious program to eliminate waterlogging and salinity throughout Pakistan. In this plan the Upper Indus Plain was divided into ten reclamation projects of from 0.4 to 1.6 million hectares, and the Lower Indus Plain was divided into sixteen projects of from 0.3 to 0.8 million hectares. In all, the program encompassed the construction of 31,500 tubewells, 12,500 kilometers of major drainage channels, and 42,000 kilometers of supplemental drains serving more than 12 million hectares in the Northern and Southern Zones (Ahmad 1974). A panel of experts headed by Roger Revelle (hereafter referred to as the Panel) was sent by the American president to study the problem of waterlogging and salinity in Pakistan. The Panel prepared a comprehensive report on agriculture, drainage, and reclamation in Pakistan which examined technical, institutional, and organizational solutions. Engineering aspects of the report, often called the Revelle Report or the White House Report, were generally along the lines of the WAPDA program, although the Panel used sophisticated computer models to demonstrate that the pumping of groundwater by public tubewells could provide an intermediate solution to waterlogging and salinity problems. These tubewells could also provide much-needed additional irrigation water (White House 1964).

In addition to preparing its own proposals, the Panel reviewed WAPDA's plan. Both plans were also reviewed by a group of three consulting firms, called the Irrigation and Agriculture Consultants Association (IACA), appointed by the World Bank in 1963 at the request of the president of Pakistan. Reports by the three firms were consolidated in a World Bank study (Lieftinck, Sadove, and Creyke 1968).

Study Recommendations

Recommendations prepared by WAPDA, the Panel, Tipton and Kalmbach, Harza Engineering Company International, IACA, and various other expert groups contain similar elements. Most agree that the groundwater aquifer of the Indus Plain is a vast, relatively untapped resource that can be used to supplement surface water supplies. In general, all agree that the aquifer is high-yielding with substantial storage capacity, but there is some disagreement about its magnitude. There has always been disagreement about the merits of horizontal drainage versus vertical drainage, but usually, after detailed study of the issue, vertical drainage is selected. Vertical drainage does not remove the salts from the system, however, and some type of horizontal removal system is usually stipulated as necessary for long-term stability. The Panel and Tipton and Kalmbach both suggested pumping highly saline groundwater into rivers during periods of heavy runoff and disposing of salt in desert salt flats or lagoons. While these alternatives may be technically feasible, they represent such explosive administrative and political issues that they have never gone beyond the discussion stage. The proposal to mine groundwater to a significant depth (30 meters was suggested by the Panel) was one of the most obvious areas of disagreement among the various groups. In some places this question was settled by the courts. In the Khairpur SCARP, for example, the court ruled in favor of farmers who claimed that their date palm orchards would be damaged if the water table was held below 4 meters. In most SCARPs, however, this question was settled by default since none of the systems came close to seriously mining the groundwater resource. Other matters of serious debate were the proper mixture of saline groundwater and fresh surface water and, an issue that continues to be discussed today, the relative merits of publicly installed deep tubewells and privately installed shallow tubewells. Public tubewells were recommended almost unanimously, largely because the original studies were completed before there were many private tubewells in the Indus Plain. Nevertheless, a few Pakistani and foreign consultants, most notably Ghulam Mohammad of the Pakistan Institute of Development Economics, argued that public tubewells should be installed only in areas where the groundwater was too saline to be applied to lands without dilution with canal water. In areas where the groundwater was of good quality, they said, development should be left to private users, with the government providing the electrical grid and credit schemes for purchase of pumps and motors (Mohammad 1964). More recent studies by such groups as the World Bank's Indus Basin Review Mission,

Map 5-2. *Location of SCARPs*

the Punjab Government Special Committee on the Working of SCARPs (Punjab 1971), Mundorff and others (1976), and the WAPDA Master Planning Division (1979) have the benefit of hindsight; their strong arguments for private tubewells rest on changes that occurred after the earlier recommendations were made.

Table 5-1. *Implementation of Public Tubewell Projects*

Project	Zone	Gross area (million hectares)	Number of tubewells	Period	Installed capacity (cubic meters per second)	Costs[a] (millions of 1977 dollars)
SCARP-I	Northern	0.49	2,069	1959–63[b]	180	25
SCARP-II	Northern	0.67	2,205	1963–73	298	90
SCARP-III	Northern	0.43	1,635	1966–73	203	40
SCARP-IV	Northern	0.23	935	1967–73	127	20
Khairpur	Southern	0.18	540	1969–70	48	10
Rohri North	Southern	0.32	1,192	1973–77	69	50
Larkana Sukkur Shikarpur	Southern	0.01	87	1973–75	8	n.a.
Total	—	2.33	8,663	—	933	235

—Not applicable.
n.a. Not available.
a. These figures do not reflect all associated costs.
b. 256 tubewells installed from 1954–58.
Source: WAPDA (1971).

SCARP Design

SCARP-I, which was completed in 1963, demonstrated that the water table could be successfully lowered by tubewells uniformly distributed over a large area, and additional public tubewell projects were accordingly undertaken in both the Northern and Southern Zones.

More than 8,000 public tubewells covering more than 2.3 million hectares were installed between 1959 and 1977 (Table 5-1). More than 12,000 tubewells covering 3 million hectares had been completed by 1979, and construction was still underway in 1982. Total costs are estimated to have exceeded $500 million. Originally the SCARPs were viewed as drainage projects whose by-product of supplemental irrigation water helped to justify the projects economically. In SCARP-I and some areas of SCARP-II (Lalian, Khadir, and Mona) capacities of the tubewells were set so that the combined water supply from surface and groundwater at the watercourse head was 1 cubic meter per second for 2,144 hectares. In subsequent SCARPs a cropping intensity for the area was projected and the tubewell capacities were determined on the basis of how much water would be needed to meet this requirement, with or without canal supplies, depending on the area. Table 5-2 illustrates the projected changes in cropping intensity expected after the SCARPs were in operation. The larger projected increases in the more recent SCARPs reflect the change in design criteria discussed above.

In general, the capacities of the tubewells ranged from 56 to 142 liters per second. The choice of tubewell capacity was made by considering the tubewell requirements of one or more adjoining watercourse commands, or *chaks.*[2] As chaks vary from 80 to 400 hectares, one tubewell often served up to three chaks. Distribution works for each tubewell required structures for proportional allocation of tubewell supplies to the watercourses to be served. At first it was thought that link watercourses which connected the tubewell to the main watercourse channel for each chak would be excavated by the farmers. In SCARP-I and parts of SCARP-II, however, farmers were unable or unwilling to dig these link watercourses. They were usually completed by the contractors in the more recent SCARPs. No provisions were made for enlarging the main watercourse channel and distribution system even though they were expected to carry two to three times the previous flow quantity.

Table 5-2. *Cropping Intensities in SCARP Projects*
(percent)

Project	Culturable cultivated area (million hectares)	Preproject intensity	Projected intensity	Actual intensity, 1975–76
SCARP-I	0.46	89	150	116
SCARP-II	0.60	83	130	102
SCARP-III	0.37	54	120	97
SCARP-IV	0.22	63	150	91
Khairpur	0.13	106[a]	135	109[a]
Rohri North	0.28	98	150	n.a.

n.a. Not available.
a. Questionable value.
Source: WAPDA (1979).

Table 5-3. *Water Duties for Different SCARP-II Sections*

Section	Chaks per section	Average hectares commanded per inlet	Average design (liters per second)	Hectares supplied with 1 cubic meter per second
Mona	11	266	128	2,078
Lower Hujjan	7	243	113	2,150
Phalia	10	175	95	1,844
Bhusal	19	202	115	1,757
Sohawa	25	179	99	1,811

Source: Data collected in SCARP-II by author and researchers.

SCARP Problems

In contrast to the elaborate models developed to determine if vertical tubewell drainage would be successful in the Indus Plain, little effort seems to have been spent on identifying how they should be operated and by whom. In particular, minimal attention was paid to issues of organization and human behavior. No effort was made to educate the water users about their role in the system, nor, as indicated by the size of the public tubewells, was there any real appreciation of the difficulties involved in organizing farmers across one or more watercourse areas. As a result, the performance of the SCARPs was quite different from what the original planners had anticipated. The following section, using data from SCARP-II, illustrates some of the project-level problems.

SCARP-II: Project Performance

SCARP-II is in the Upper Chaj Doab between the Chenab and Jhelum rivers. As mentioned earlier, some schemes within SCARP-II were designed with a fixed water duty of about 1 cubic meter per second for every 2,144 hectares. Others were designed to meet a projected cropping intensity, which in general meant that they had a higher water duty. Water duties for five SCARP-II sections are detailed in Table 5-3. Phalia, Bhusal, and Sohawa were designed to meet a projected cropping intensity rather than an arbitrary fixed water duty. For reasons explained below, however, actual flows were often much less than the designed supplies (see Table 5-4). The weighted average delivery into the channel from the sample in the table is about 78 percent of design capacity. This overestimates actual deliveries because power rationing limited pumping to 40 percent of an assumed utilization rate of 20 operating hours a day.

The measured reduction in flow was due in part to a decline over time in the output capacity of the tubewells (Table 5-5). The decline in tubewell capacity in a sample of eighty-one tubewells in SCARP-II/A was 21 percent. WAPDA (1979) records indicate that the overall decline of tubewell capacity in all of SCARP-II/A is more than 30 percent. Other reasons watercourses have not received their full design flow are submergence of watercourse channel inlets (either by tubewell flow or because the channel could not carry both canal water and water from the tubewell), low flow in the distributary, and the poor design and condition of link watercourses. In a sample of twenty-two tubewells in the Phalia section that were operating at designed pumping capacity, actual water flow (tubewell water plus canal water) in the main watercourse channel was only 67 percent of designed capacity. The rest of the flow was either lost in the tubewell link watercourses or restricted by submergence of the inlet.

Link watercourse channels. Connecting watercourse channels to the tubewell outlet in SCARP-II was more difficult than in SCARP-I because the SCARP-II wells in general had a higher output capacity and usually served two or more watercourses. The project plan assumed that farmers would construct the link connections between tubewells and watercourses. In practice this has not worked out. All of the tubewells in SCARP-II are connected to watercourses, but many of the connections, especially those in SCARP-II/A, are unsatisfactory. The high-capacity tubewells were designed so that their supplies would flow through a sophisti-

Table 5-4. *Actual Delivery Compared with Design Delivery: SCARP-II*

Section	Chaks per section	Average design (liters per second)	Actual delivery (liters per second)	Percentage delivered
Mona	11	128	98	77
Lower Hujjan	7	113	92	81
Phalia	10	95	63	67
Bhusal	19	115	89	78
Sohawa	25	99	82	82

Source: Data collected in SCARP-II by author and researchers in 1977.

Table 5-5. *Decline in Tubewell Output: SCARP-II/A*

Section	Number of tubewells sampled	Average design capacity (liters per second)	Measured output (liters per second)	Average percentage delivered
Phalia	31	105	91	86
Bhusal	16	116	82	71
Sohawa	27	99	79	80
Lower Hujjan	7	102	74	72
Weighted average	—	—	—	79

—Not applicable.
Source: Data collected in SCARP-II/A by author and researchers in 1977.

Table 5-6. *Watercourse Losses and Delivery Efficiency: SCARP-II*

Section	Number of measurements	Losses per 300 meters (liters per second)	Losses per 300 meters (percentage of initial flow)	Delivery efficiency
Mona	6	13.3	13.8	65.8
Mona	20	6.2	10.0	64.0
Lower Hujjan	7	9.9	11.7	n.a.
Phalia	30	6.8	12.8	n.a.
Sohawa	25	14.7	15.7	n.a.
Bhusal	19	13.9	15.6	n.a.
Weighted average	—	—	13.5	—

— Not applicable.
n.a. Not available.
Source: Data collected in SCARP-II by author and researchers in 1977.

Table 5-7. *Total Annual Pumpage as a Percentage of Acceptance Capacity: SCARPs*

Year	SCARP-I		SCARP-II		SCARP-III		Khairpur	
	Pumpage[a]	Percent	Pumpage[a]	Percent	Pumpage[a]	Percent	Pumpage[a]	Percent
1962–63	2,790	59	n.a.	n.a.	n.a.	n.a.	n.a.	n.a.
1963–64	3,095	66	n.a.	n.a.	n.a.	n.a.	n.a.	n.a.
1964–65	3,004	64	n.a.	n.a.	n.a.	n.a.	n.a.	n.a.
1965–66	3,073	65	503	7	n.a.	n.a.	n.a.	n.a.
1966–67	2,088	44	398	5	n.a.	n.a.	n.a.	n.a.
1967–68	2,287	48	442	6	n.a.	n.a.	196	16
1968–69	2,424	51	897	12	n.a.	n.a.	459	36
1969–70	2,401	51	1,781	23	n.a.	n.a.	407	32
1970–71	2,386	51	2,121	28	n.a.	n.a.	599	48
1971–72	2,293	49	1,989	26	759	14	645	51
1972–73	2,003	42	1,960	26	744	14	558	44
1973–74	1,781	38	2,482	33	2,632	31	567	45
1974–75	1,699	36	3,178	42	2,762	52	331	26
1975–76	1,949	41	2,611	34	1,779	33	447	35
1976–77	1,576	33	n.a.	n.a.	n.a.	n.a.	n.a.	n.a.

n.a. Not available.
a. In millions of cubic meters and assuming twenty pumping hours per day.
Source: WAPDA (1979).

cated diversion box which allocated the water to two or more watercourses. In actual practice many of the diversion boxes are being bypassed and the tubewell water is serving only one watercourse (USAID 1970).

In some instances the water has flooded an area around the tubewell, creating a watering hole from which the water is being diverted into the watercourse. Because of these large flooded areas it is difficult to measure accurately how much tubewell water actually enters the watercourse. Measurements taken in twenty-one link water-

Table 5-8. *Usable Pumping Capacity since Acceptance*
(percentage of capacity)

Year	*SCARP-I*	*SCARP-II/A*	*SCARP-III*	*Khairpur*
1969–70	66.8	83.9	—	—
1970–71	62.2	81.4	—	—
1971–72	61.9	78.2	—	—
1972–73	60.3	74.5	85.9	—
1973–74	57.6	71.3	80.8	—
1974–75	51.9	65.2	80.3	—
1975–76	52.1	65.4	74.4	73.7

—Not applicable.
Source: WAPDA (1979).

courses in the Phalia section during canal closure, when the only water entering the watercourse was tubewell water, showed an average loss of 19.5 percent of the tubewell discharge.

Once the water enters the watercourse channel, it comes under the control of farmers served by that channel who are supposed to maintain the channel and distribution system. Often this is not done because of neglect, ignorance, and village conflicts. Table 5-6 summarizes measurements of losses in 107 watercourse channels in SCARP-II. The average watercourse channel losses on unimproved water channels varied from 10 percent to 15.9 percent per 300 meters of length and averaged 13.5 percent. Given farmers' needs for water, these losses represent a critical shortage, especially at middle and tail sections of the chak. With losses of this magnitude, by the time the water reaches 1,500 meters from the head of the watercourse, users have lost half of the initial flow entering the system. Assuming an average delivery of 79 percent of the design flow entering the system and losses of 13.5 percent per 300 meters, users 1,500 meters down the watercourse channel are receiving only 40 percent of their design allocation. A sample of measurements in the Sohawa section indicates, at 1,000 meters, that farmers are receiving only 44 percent of the design flow. Similarly, a sample from the Bhusal section indicates that farmers 1,000 meters from the junction of the main channel and the link watercourse channel are receiving only 38 percent of the design flow.

Operating schedules. Although high water tables are not a serious problem in SCARP-II, tubewells are designed both to lower the water table and to provide supplemental irrigation water. In areas where the water table is very close to the surface, it is necessary to pump the tubewells more in order to lower the water table. In areas where the water table is more than 3 meters from the surface, the tubewells can be pumped more on demand. Almost all SCARPs followed a pattern of increased groundwater pumping during the initial years and declining groundwater pumping thereafter (Table 5-7). Pumping in SCARP-II rose to a peak in 1974–75 and then fell. Given current budget restrictions and declines in pumpage capacity, this trend is likely to continue (Table 5-8). With respect to changes in the depth to water table within the SCARPs, these pumping figures meant rapidly falling water tables in the initial years and then rising water tables as quantity pumped declined (Table 5-9). In the Punjab the depth to water table was less than 3.0 meters in June 1959 for about 3.8 million hectares and in June 1978 for about 3.9 million hectares (WAPDA 1979). The data indicate that despite the development of the SCARPs, there has been no real change in the amount of land with high water tables over the past twenty years. Similarly SCARP-II water tables have not changed appreciably over the first ten years of tubewell operation (Table 5-10).[3]

SCARP tubewells are supposed to be operated on schedules developed by the Irrigation Department. These vary

Table 5-9. *Depth to Water Table in SCARPs*

Project	*Year*	*Percentage of water table*			*Average depth (meters)*
		0–1.5 meters	*1.5–3.0 meters*	*3.1+ meters*	
SCARP-I	1961	13.5	61.2	25.3	2.5
	1971	0.0	8.1	91.9	5.1
	1977	2.2	30.4	67.4	3.8
SCARP-II	1965	9.1	23.4	67.5	3.6
	1971	0.2	11.2	88.6	4.9
	1977	5.3	26.6	68.1	3.7
SCARP-II/A	1968	14.9	57.5	27.6	2.6
	1971	1.1	34.4	64.5	3.8
	1977	13.4	53.1	33.5	2.7
SCARP-III	1964	41.2	42.5	16.3	1.9
	1975	6.2	19.2	74.6	3.6
	1977	12.0	52.4	35.6	2.8
SCARP Khairpur	1960	29.7	70.3	0.0	1.7
	1977	25.0	62.0	13.0	2.1

Source: WAPDA (1979).

Table 5-10. *Depth to Water Table, SCARP-II, 1966 and 1977*

Depth (meters)	Percentage of area 1966	Percentage of area 1977
Less than 1.5	15	14
1.5 to 3.0	57	53
More than 3.0	28	33

Source: WAPDA (1979).

among wells in perennial canal areas, nonperennial canal areas, and uncommanded areas. Schedules do not allow for rainfall, power failures, or personnel problems and therefore must be considered as no more than general guidelines. Poor performance of tubewell operators is one of the main complaints about SCARP tubewells raised by both farmers and Irrigation Department staff. This makes it difficult to determine how many hours each tubewell is operated. Operators are supposed to keep a daily log of tubewell operating hours, but because they are frequently absent from the tubewells for long periods and the farmers have to operate the tubewells themselves, the logbook often can provide only a rough estimate of actual operating hours. Nor can monthly pumping hours be determined by reading the electric meter connected to each tubewell, for most meters do not function properly and are useless in determining power consumption and pumping hours. Data available on actual operating hours are therefore sketchy and must be used with caution. Some general operating characteristics can be described, but exact operating schedules cannot be determined from existing data.

The Irrigation Department has two guidelines for the interagency scheduling committees which meet biannually to schedule tubewell operations in SCARP-II: over the year, pumps should run at 40 percent of annual capacity; and on days when the pumps are operated, they should run continuously from 12:01 A.M. until 12:00 P.M., with scheduled rest periods between 12:00 noon and 4:00 P.M. The exact rationale for these guidelines is not at all clear. The first appears to derive from power rationing instituted in 1972 as a result of the war between India and Pakistan. The second guideline may represent an attempt to pacify tubewell operators, whose working hours according to official labor legislation are only eight hours, or it may reflect a mistaken belief in the need to rest the tubewell motors. In Mona, where the tubewell rest period is from 5 P.M. to 9 P.M., the most common explanation is that these are peak hours for electricity consumption.

Given these guidelines, the main area of choice is the number of days a month that the tubewell should be operated. Schedules should take into account plant-water relationships, rainfall, and expected availability of canal water. In fact, the proposed Lalian pumping schedule varies little from month to month and bears little relation to that proposed by the Land Reclamation Department (LRD), which attempts to match expected water supplies with expected demand.[4] Nor do actual pumping schedules resemble either the proposed LRD schedule or that followed by private pump operators (Table 5-11). More flexible

Table 5-11. *Planned and Actual Tubewell Operation: SCARP-II*
(percentage of working hours in proportion to total available hours)

Month	Proposed schedule, Lalian, 1977–78: Days	Proposed schedule, Lalian, 1977–78: Percentage of month	Proposed schedule, LRD, 1967	Actual operation, all SCARP-II: 1974–75	Actual operation, all SCARP-II: 1975–76	Actual operation, all SCARP-II: 1976–77	Actual operation, Lalian: 1974–75	Actual operation, Lalian: 1975–76	Actual operation, Lalian: 1976–77	Private,[a] 1977
Wet season										
April	14	39	12	n.a.	56	38	n.a.	58	27	28
May	21	58	31	n.a.	53	40	n.a.	59	40	6
June	14	39	47	n.a.	45	41	n.a.	42	34	6
July	14	39	25	n.a.	35	46	n.a.	36	40	16
August	14	39	19	n.a.	43	24	n.a.	37	28	24
September	14	39	15	n.a.	33	27	n.a.	44	32	26
Dry season										
October	14	39	83	59	45	36	53	55	54	24
November	21	58	73	52	50	40	45	66	50	24
December	14	39	89	38	39	23	38	41	27	27
January	14	39	25	39	36	36	40	54	44	29
February	14	42	35	48	19	40	67	23	40	30
March	14	39	49	57	30	49	62	43	54	26
Annual total	182	42[b]	41	n.a.	40	37	n.a.	47	39	22

n.a. Not available.
a. Actual pumping of private tubewell in Mona, SCARP-II, 1977.
b. Percentage of year.
Source: SCARP-II Records and Land Reclamation Department (LRD).

Table 5-12. *SCARP-II: Lost Time Because of Various Component Failures, 1972–76*
(percentage of capacity)

Section	Year	Total utilization	Causes of shutdown: Nonavailability of electricity	Burnt motor	Transformer defect	Other failures[a]
Bhusal	1976	43.0	10.0	12.4	5.4	11.5
	1974	64.0	18.0	6.5	6.5	10.6
	1973	53.0	11.0	7.5	10.1	11.1
Sohawa	1976	44.0	10.0	15.9	10.6	12.8
	1975	51.0	22.0	5.7	11.9	17.2
Lower Hujjan	1976	33.0	11.0	5.9	n.a.	18.9
	1975	50.0	5.0	5.9	3.2	7.7
	1974	45.0	13.0	1.0	8.8	7.0
	1973	21.0	22.0	n.a.	n.a.	n.a.
	1972	30.0	36.0	n.a.	n.a.	n.a.

n.a. Not available.
a. Includes motor defect, starter defect, blown fuse, and cable problems.
Source: SCARP-II records.

groundwater pumping, closer to the schedule proposed by LRD, could prevent both over- and underpumping and potentially could support a higher cropping intensity.

Maintenance problems. According to WAPDA data, SCARP-II has seen a decline in the utilization rate during recent years, from an average of 49.7 percent of installed capacity in 1974–75 to 37.0 percent in 1976–77. As electricity charges have increased at a rate exceeding 12 percent a year (from 1978 to 1982) and budget allocations have not kept pace, the utilization rate is expected to continue to decline. Over the same period the allocation of funds for maintenance and repair work has decreased by 14–15 percent, with consequent impairment of operation (Table 5-12).

SCARP Program Performance

The entire SCARP program has been affected by the poor operating records of individual SCARPs. In addition, such factors as unforeseen increases in energy costs, shortened tubewell life, rapid development of private tubewells, and failure to achieve desired cropping intensities have combined to make SCARPs an economic and financial burden.

Economics. Depending upon the various consultants' assumptions and mandates, the estimated costs of relieving waterlogging and salinity problems throughout the Indus Plain ranged from $1.2 billion to $2.7 billion. Predicted benefit-cost ratios for these plans were as high as 7.5:1 and as low as 2.25:1. As vertical drainage programs of this magnitude had never before been tried, all ratios depended upon the underlying assumptions.

One assumption that was clearly incorrect in almost all of the proposed programs derived from significant underestimation of the number of private tubewells that would be developed, even with the implementation of the public tubewell schemes. Ghulam Mohammad's 1964 survey of 23,000 private tubewells in sixteen districts of the Northern Zone of West Pakistan established that private tubewells were very profitable and that the number installed would continue to increase (Mohammad 1965). His findings were confirmed: between 1965 and 1975 the number of private tubewells quadrupled (Table 5-13). Yet even the Lieftinck report (Lieftinck, Sadove, and Creyke 1968) failed to appreciate the fact that private tubewells had the potential to replace public tubewells in most of the nonsaline groundwater areas.

Another assumption that proved wrong concerned the length of life of the public tubewells. Most consultants originally predicted forty- or even fifty-year service lives. When it became apparent that the pumping capacity was quickly declining in almost all of SCARP-I and that a number of wells were facing critical problems with corrosion of the strainers and incrustation, the consultants first tried to replace the mild steel strainers with strainers made of stainless steel and fiberglass. It was soon obvious, however, that even these materials were seriously affected by minerals in the groundwater. As a result, the consultants reduced their estimates of tubewell life to twenty or twenty-five years. In 1971 the Special Committee on the Working of SCARPs set twelve years as the average life of a SCARP tubewell (Punjab 1971). Depending upon the acceptable degree of decline in pumping capacity and the amount public agencies are willing to pay for repairs, life is a relative term, but it seems likely that approximately fifteen years will be the practical life for most SCARP tubewells.

A third erroneous assumption concerned cropping intensity. Almost all of the early studies planned to double cropping intensities from 75 to 150 percent. This clearly has not happened; in a few areas cropping intensity rose to

Table 5-13. *Number of Private Tubewells in Pakistan*

Year	Punjab and Northwest Frontier provinces	Baluchistan and Sind provinces	Total	Annual increase
1965	29,007	3,447	32,524	—
1966	36,663	3,806	40,469	7,945
1967	45,103	4,250	49,353	8,884
1968	54,570	4,751	59,321	9,968
1969	63,000	5,267	68,267	8,946
1970	76,509	59,420	82,451	14,184
1971	83,337	6,665	90,002	7,551
1972	92,298	7,442	99,740	9,738
1973	101,425	8,050	109,475	9,735
1974	112,002	8,415	120,417	10,942
1975	122,702	9,694	132,396	11,979
1976	133,807	10,193	144,000	11,604
1977	143,355	10,675	154,030	10,030

— Not applicable.
Source: WAPDA (1979).

as high as 135 percent, but even in those areas cropping intensity later dropped back to 125 percent.[5]

Unfortunately, while changes to higher valued crops do increase revenue from water charges, higher yields do not. In SCARP areas, where the water supply theoretically has been doubled, double water charges are supposed to be assessed. But many farmers refuse to pay double charges because they claim that by increasing acreage they are already, in effect, paying double water charges. Water charges have not changed since 1969 and are not very significant (Table 5-14).

Government subsidy to SCARPs. Failure to increase cropping intensities as much as projected, increase water charges to reflect a general increase in prices, collect water charges effectively, and maintain a forty-year life for the tubewells has resulted in a massive government subsidy to the SCARPs. A number of groups using various assumptions have calculated the magnitude of this subsidy. The Committee for Financial Subsidy, Department of Land and Water Development, Government of the Punjab, in 1970 estimated the annual subsidy for SCARP-I to be $4.15 million excluding the power subsidy and $5.41 million including it. Another division of the Land and Water Development Department in 1970 estimated an annual subsidy of $2.56 million for SCARP-I assuming a life of forty years for the tubewells and $3.93 million assuming a life of twelve years, excluding the power subsidy (Punjab 1971). WAPDA estimates that the total annual subsidy for SCARP-I has varied from $2.5 million to $3.8 million, or from $5.15 to $7.82 per hectare (Table 5-15). The average annual subsidy over the period is estimated at more than $3.0 million, or $6.20 per hectare.

Table 5-14. *Water Charges, 1919 and 1978*
(dollars per hectare)

Crop	1919–20	1978	1919 in 1978 prices
Wheat	1.15	3.24	4.99
Rice	1.77	5.24	7.71
Sugarcane	2.25	10.23	9.78
Oilseeds	n.a.	3.74	n.a.
Cotton	1.00	4.49	4.34

n.a. Not available.
Source: WAPDA (1979).

Data for SCARP-II indicate a similar level of subsidy for that project. Operating expenditures for 1976–77 were $5.58 million, and recovered water charges were $2.91 million. If capital costs are amortized using the same assumptions as for SCARP-I, the annual net subsidy for SCARP-II exceeds $5.3 million, which is more than $8.50 per cultivable hectare. For SCARP-II in 1975–76, recovered water charges were 77.8 percent of 1975–76 demanded water charges and 62 percent of current year plus previously overdue charges. Even if SCARP-II personnel had been able to collect 100 percent of demanded water charges, the amount would have been almost $1 million short of annual operation, maintenance, and repair costs.

With approximately 3 million hectares under SCARP schemes and an annual subsidy of $7.50 to $8.50 per hectare, the annual national subsidy for SCARP schemes exceeds $22 million. Because the more recent SCARPs were more expensive and were purchased with funds provided at a significantly higher interest rate, these estimates are extremely conservative, but they do indicate the magnitude of government subsidy.

Public versus private tubewells. The rationale underlying the recommendation that the public sector plan a large role in groundwater development was that private development

Table 5-15. ***Costs and Revenues from SCARP-I***
(millions of dollars)

Year	Water supplied (billions of cubic meters)	Costs: Power only	Costs: Total operating and maintenance	Costs: Total including annualized capital[a] (A)	Water charge revenue collected[b] (B)	Net subsidy (A – B)
1964–65	3.01	1.21	2.08	3.85	0.41	3.44
1965–66	3.07	1.27	2.19	3.96	0.73	3.23
1968–69	2.42	1.06	1.97	3.74	1.20	2.54
1971–72	2.29	1.52	2.31	4.12	1.27	2.85
1973–74	2.01	1.44	2.27	4.04	1.17	2.87
1974–75	1.70	1.49	2.09	3.86	1.26	2.60
1975–76	1.95	2.07	3.31	5.10	1.27	3.83

a. Annualized capital costs at 8 percent over fifteen years. Includes an adjustment for 1972 devaluation of Pakistan rupee.
b. Attributed to SCARP water supply; that is, after deducting water charge revenue attributed to surface water supplies.
Source: WAPDA (1979).

- Would be inequitable and therefore not benefit most small farmers
- Would be haphazard and probably not accomplish the desired drainage function
- Could deteriorate the groundwater aquifer through uncontrolled pumpage
- Could not be expected to proceed at the rapid rate desired.

In the early 1960s, when Pakistan had limited experience with private or public development of groundwater, this seemed reasonable. By the mid–1960s there were more than 30,000 private tubewells, and some experts, both local and international, urged that private development in areas overlying fresh groundwater be stressed over public development (Eaton 1965). This advice was noted by the World Bank report (Lieftinck, Sadove, and Creyke 1968), but it was not strongly supported and therefore was, in effect, rejected. By 1978, Pakistan had acquired substantial experience in groundwater development in both sectors.

Results of private tubewell development have demonstrated that private tubewells can serve the needed drainage function and also improve cropping intensities. The public sector program has lagged far behind original and revised goals and has only partially performed its drainage function. Private tubewell investment has continued in SCARP areas because centralized management has been unable to meet the changing needs of the water users (Hussain, Ali, and Johnson 1976).

The operational status of public and private tubewells is strikingly different and goes far to explain why farmers have opted to install private tubewells even in areas served by public tubewells (Table 5-16). Only a few of the numerous postproject benefit-cost analyses of SCARP-I have attempted to compare SCARP-I with an equivalent private tubewell area; that is, with one where private tubewells supplement canal supplies. As part of the development of the Revised Action Programme for Irrigated Agriculture by WAPDA (1979), SCARP-I was compared with both the perennial commanded area in the Upper Rechna Doab (162,000 hectares), which borders SCARP-I, and the adjacent Lower Rechna Doab (Tandlianwala) area (110,000 hectares). In the Upper Rechna Doab in 1975 there was one private tubewell per 33 hectares, and in Tandlianwala there was one per 55 to 61 hectares (Table 5-17). There are private tubewells within SCARP-I.

Growth in the number of private tubewells and increases in cropping intensity have been faster in both areas than in SCARP-I. From this it can be inferred that the development of public tubewells slowed investment in private tubewells in the SCARP-I area. Assuming that, if SCARP-I had not been built, private tubewells in that area would have developed to a density of one tubewell for every 67 hectares, WAPDA calculated a rate of return for SCARP-I of 6 percent. When WAPDA data are used but tubewell density is increased to that of Tandlianwala (one tubewell for every 50 hectares), the rate of return on SCARP-I is less than 3 percent. Even with a density of one tubewell per 67 hectares, the predicted cropping intensity in 1976 would have been 122 percent rather than the actual intensity of 117 percent, and groundwater withdrawals would have increased by more than 22 percent.

Table 5-16. ***Operational Status of Public and Private Tubewells***

Division	Government tubewells: Total	Government tubewells: Percentage operational	Private tubewells: Total	Private tubewells: Percentage operational
Rawalpindi	868	81	2,300	96
Sargodha	1,527	67	10,700	93
Lahore	3,202	66	20,400	97
Multan	1,586	17	26,510	94
Bahawalpur	174	49	4,060	93
Punjab total	7,357	57	69,030	95

Source: Punjab (1971).

Table 5-17. *Tubewell Development and Cropping Intensities in and around SCARP-I*

Section	Area (hectares)	Number of private tubewells 1964	Number of private tubewells 1975	Cropping intensity (percent) 1964	Cropping intensity (percent) 1975
Upper Rechna	162,000	900	4,900	103	131
Tandlianwala	110,000	500	1,800–2,000	125	144
SCARP-I	468,000	570	3,000	105	117

Source: WAPDA (1979).

Management Problems

Many factors have contributed to the disappointing performance of the SCARPs—unprecedented increases in energy costs, technical design faults, foreign exchange constraints—but most of the underlying problems can be traced to poor organization and management. The planning process, especially planning for management and administration of the systems, did not address some of the most critical issues. Questions related to local participation in such activities as construction of link watercourses, organization of farmers' groups, location of tubewells, and choice of tubewell technology appear not to have received sufficient attention from the planners. The fact that operation and maintenance manuals were never prepared for the majority of the SCARPs and that no attempt was made to achieve optimal conjunctive use of canal and tubewell water illustrates that planning was deficient.

Management Structure

The Panel recommended that the SCARPs be managed as a project under a project management board with the authority to cut across line agencies at the field level. Of the northern SCARPs only SCARP-I was organized under a project director with administrative control over all services related to irrigation, agriculture, and cooperatives. Khairpur SCARP was also organized under a project director, but surface irrigation management was under a separate senior engineer. In SCARP-I the project approach did not succeed, and in 1970 the management structure was changed to a system of separate responsibility for irrigation, agriculture, and cooperatives by the respective government departments.

A problem that plagued SCARP-I and was certainly a serious problem for SCARP-II and Khairpur SCARP was that separate management circles were established for canal irrigation and for tubewell operation. In each case the boundaries of the two management circles were different. SCARP-II straddled part of the Upper Jhelum Circle as well as part of the Lower Jhelum Circle. Both circles were managed by senior engineers and there was no direct interaction of lines of authority. Operating SCARPs and canal systems as separate circles practically guarantees that there will be no attempt to integrate groundwater and surface water use.

Planning Faults

In all the SCARPs—including Mona, which is organized as a research area, and Khairpur, which is perhaps the best managed SCARP—no real attempt has been made to optimize the potential for conjunctive use of canal and tubewell water. Khairpur does monitor depth to groundwater in each well each month and adjusts pumping schedules as needed to maintain a desired groundwater level. Yet even there, as elsewhere, pumping schedules are not adjusted to reflect estimated canal supplies and crop water requirements. In many areas it would be possible to divert canal supplies to water-short commands and to make up for this deficit by pumping extra hours. Separate canal and tubewell management circles make this type of operating procedure almost impossible and thus severely restrict potential benefits of the conjunctive system.

Field staff management. Tubewell operators constitute the largest group of staff working in SCARP circles. SCARP rules stipulate that the operator must be a local person but cannot come from any of the villages served by the well. To reduce the danger of misallocation, the rules restrict tubewell operators from working within a radius of 24 kilometers from their place of origin. But since the operator is always supposed to be present when the tubewell is in operation, which is normally twenty hours a day, this rule is clearly counterproductive. It forces the operator either to cheat by leaving the system jammed open, thereby circumventing safety devices, or to turn off the tubewell even if water is scheduled. Usually farmers and operators work out some type of compromise which invariably costs the farmers money (Johnson, Early, and Lowdermilk 1977). An additional problem is that tubewell operators are highly unionized and therefore difficult to punish or dismiss. It has also been suggested that officials find it difficult to control their subordinates because a substantial portion of the operators' exactions are passed up the line. Whatever the truth of these allegations, it is clear that the day-to-day operations of tubewells are supervised loosely, if at all. The absence of any effective means of control over the activities of field staff—whether by senior officials, by standardized cross-checking procedures, or by the farmers through some ability to reward or penalize—demonstrates that planners did not think out how tubewells were to be operated and maintained in practice.

Watercourse-level organization. The decision to construct tubewells that could potentially serve more than 500 hectares and as many as 100 farmers reflects the fact that planners gave little thought to conditions at the local level. Even the most cursory investigation would have revealed that farmers along a single watercourse had difficulties organizing for operation and maintenance. Most of the court cases in rural areas involve conflicts over water and associated land. Provision of large, publicly owned and operated tubewells that were designed to serve two or more watercourses opened the door to all sorts of new conflicts. Investigation of farmers' organizational capacities, as well as their technical ability to deal with larger flows of water, would have indicated that smaller-capacity, more localized tubewells were better suited to existing conditions. The argument that larger public wells are more "economic" than smaller private wells rests on the unproved assumption that management would be the same under both systems. Planners failed to recognize, or ignored, farmers' limited capacity to cooperate at the watercourse level as well as the technical difficulties of redesigning watercourse channels to carry larger flows. This was plainly a gross error of planning and goes far toward explaining why the SCARPs have not been utilized properly at the local level.

Financial Problems

Within the irrigation system, relatively large sums are spent on special and emergency repair work, as opposed to ordinary system maintenance. This supports the view that in the past the Irrigation Department's allocation for recurrent expenditure has been inadequate for an effective program of preventive maintenance. Even with additional funds for a special five-year maintenance program aimed at strengthening canal banks, senior irrigation officials feel that the bulk of maintenance funds is spent on essential maintenance of main and branch canals. This leaves little for strengthening main canals, much less for work on distributaries and minor canals or drains which are in poor condition and vulnerable to breaching. Systemwide maintenance cycles which should be completed every three to five years will take more than twenty years at 1978–82 funding levels.

With current water fees and collection rates, total revenue from water is insufficient to cover even operation and maintenance charges. The addition of SCARP operation and maintenance expenses to the Irrigation Department's already overburdened operation and maintenance budget has further increased the department's deficit. In 1975–76 the Punjab Irrigation Department spent approximately $1.3 million on ordinary operation, maintenance, and staffing for 628,400 hectares served by the Lower Jhelum Canal. An additional $3.2 million was spent on operations and maintenance as well as staffing to provide tubewell drainage of 360,000 hectares served by SCARP-II within the Lower Jhelum Canal command. The combined operation, maintenance, and staff budget for 1975–76 in the Lower Jhelum Canal command was therefore $4.5 million. With recovered water charges of approximately $2.9 million, the deficit was $1.6 million. If emergency capital charges including emergency operation and maintenance costs are included, the deficit increases to about $2.2 million. This deficit does not take into account capital repayment costs for the SCARP system, and it assumes that all capital costs for the irrigation system are already sunk costs. For the entire Punjab the deficit in the Punjab Irrigation Department budget was $17 million in 1978–79 and was estimated to be more than $20 million in 1981. For all of Pakistan the annual deficit may exceed $40 million, again excluding past capital expenditures.

These deficits are the responsibility of individual provinces, but the provinces' ability to raise revenue has not increased enough to meet them. In the short run, the provinces have subsidized tubewell operation by underfunding required canal system maintenance and agricultural extension and crop and livestock research. They have also gone into debt to WAPDA for SCARP electric charges. The provinces must increase their revenue from water charges or reduce their costs of operating and maintaining tubewells, or both. They have already restricted funds for SCARP operation and maintenance, but this forces a reduction in the utilization rate and slows the rate at which tubewells are repaired. The result is a reduction in total pumpage and an increase in the per-unit costs of water and drainage.

Options

Although SCARPs have played an important role in demonstrating the potential for groundwater development, private tubewells have actually pumped more groundwater than have public tubewells. WAPDA has conceded that most of the reclamation of salt-affected land in the past fifteen to twenty years has been the result of improved water supply and individual farmer initiative rather than sustained public programs (WAPDA 1979). Given the abundance of information available about the SCARPs and the extent of the public subsidy required to keep the SCARPs operational, the government now needs to explore its options.

Freshwater Zones

More than 3 million hectares of land in Pakistan are served by SCARP tubewells with a sunk cost of more than $500 million. After fifteen years of SCARP operations, waterlogging and soil salinization within the SCARPs appear to have improved marginally, at least in the less salt-affected areas. Yet SCARP tubewells are becoming older and less efficient, and they must be pumped more hours each month even to hold their own, while the price of energy is increasing rapidly. WAPDA has recommended a phased replacement of existing public tubewells in freshwater zones with private tubewells as SCARP tubewells are exhausted (WAPDA 1979). Moreover, WAPDA has recommended increasing the projects' operating funds to permit higher utilization rates, using private workshops to reduce the duration of breakdowns, and launching distinct efforts to better integrate operation of surface water and groundwater supplies. WAPDA has also suggested using pilot studies to determine what would happen if public tubewell operators were replaced by farmer groups that have a stronger incentive to keep the well operating.

Unless the government can locate and invest vast sums of money to replace and rehabilitate the SCARP systems, these systems are inevitably going to deteriorate. Private tubewells will be built where the groundwater is of good quality and markets are available for increased output. There is no justification for continuing to subsidize SCARP systems in areas where private tubewells have already been installed and SCARP tubewells are in their final years. Farmer groups could be given the option of paying energy costs, establishing their own schedules, and operating the tubewells until the group decided this was no longer economic. But since SCARP tubewells are located at the head of the watercourse while private tubewells are located down the channel close to the owners' fields, distribution losses are considerably higher for SCARP tubewells. Therefore, only farmers in the head end of the watercourse command will normally be willing to pay to continue to operate SCARP tubewells. Farmers located away from the SCARP tubewells, given increasing maintenance costs and excessive energy costs per unit of water delivered to their fields, will quickly find that owning their own tubewell or sharing a tubewell with close neighbors is more economic.

In areas where tubewells are newer and there has been less private development, more effort could be made to form farmers' groups to operate SCARP tubewells until private tubewells become a better alternative. Giving farmers the freedom to operate the public tubewells on demand or to install their own tubewells should lead to a significant increase in total pumping of groundwater in the freshwater areas. This would accomplish the desired goals of the SCARP program at a mere fraction of the cost to the government, as has been demonstrated in non-SCARP areas of the Punjab and across the border in the Indian Punjab. Inevitably there will be areas where development is slow or where certain target groups appear to lag behind. WAPDA's action plan has recommended an intensified agricultural credit program to support private tubewell installations in areas not yet fully exploited. This would be administered through commercial banks with medium-term loans at commercial interest rates. Although this type of program can serve some select groups, perhaps a credit subsidy to manufacturers to encourage them to produce better and cheaper equipment, especially small diesel engines and high-efficiency electric motors, would have a better chance of accelerating the spread of private tubewells. Similarly, efforts to ensure that groundwater quality is taken into account when construction priorities are set for rural electrification schemes could speed tubewell development. Encouraging and legalizing the market for selling and trading water at the watercourse command level would also provide a strong incentive for small farmers or groups of small farmers to invest in tubewells.

Saline Water Zones

In saline water zones farmers find no incentive to install tubewells. Drainage, either by vertical tubewell or by surface drain, requires large-scale outlet channels to remove saline effluent, although freshwater skimming wells can serve a drainage role in some areas. The World Bank and WAPDA have estimated the rate of return for surface drainage in the Sukh Beas Scheme at 13 percent and 15 percent, respectively. But although drainage may be economic, most of the capital investment would have to come from the government. Based on past experience, WAPDA (1979) questions whether the government would allocate sufficient funds for operation and maintenance of either drainage tubewells or surface drains. Since drainage does not generate significantly higher incomes, but merely prevents degradation of existing incomes, farmers are reluctant to pay much to support public drainage schemes. Nevertheless, they might be willing to provide some support, and even if they gave only a few dollars per hectare (more than 8 million hectares may need drainage), they could contribute a large amount of the operating funds.

Although drainage needs are more immediately obvious in saline groundwater zones, they also exist in the freshwater zones. Eventually a drainage department fully as competent as the irrigation department is going to have to be created for all of the Indus Basin. Saline effluent disposal and provision of irrigation water must be given equal priority. Drainage activities need to reach a point where career bureaucrats perceive equal potential for advancement through drainage activities as they currently do through irrigation-related activities.

Conclusions

Because of its unique size and flat topography, Pakistan selected public vertical tubewell drainage as a medium-term solution to its problems of waterlogging and soil salinity. The decision to implement public vertical tubewell drainage rather than other alternatives was not made without careful study and considerable debate, but it did mean relying on a new technology never before attempted on such a large scale. With the information available at the time, original design parameters appeared realistic. But as it became apparent that many of the initial assumptions were proving incorrect—such as those about the willingness of farmers to invest in private tubewells, the economic life of tubewells, the availability of power, the costs of maintenance, and the level of cropping intensity—there was sufficient justification to question the original design. Recognized management problems with tubewell operators and jurisdictional conflicts between irrigation and tubewell circles should have led to needed internal administrative adjustments, but the bureaucracy did not respond.

The failure to change design and operational procedure was primarily a result of the administrative structure associated with large-scale water projects and its relationship to the public decisionmaking process. The decision to invest in SCARPs and the establishment of priorities for construction were in the hands of the central government, and construction was also under the control of a central government organization (WAPDA). Operational responsibility lay with provincial irrigation bureaucracies that had no control over project design and construction. Nor had they historically been actively involved at the level where the tubewells actually operate—the watercourse command areas. The situation was further complicated when tubewell operators formed unions and demanded rights that other irrigation employees had never been granted. Since the political system discouraged feedback from rural water users and provincial governments did not have the power to influence a rigid public decisionmaking process, it was extremely difficult to change a decision once it was made. International funding agencies must also bear a share of the blame because they were aware of many of these shortcomings, but continued to fund SCARPs without demanding revisions in either design or management.

Lessons for future development of additional SCARPs as well as large-scale irrigation or drainage schemes in other countries can be learned from Pakistan's experience. The most important are:

- In selecting a technology, particularly a new one, a system to monitor project implementation is a necessity. In conjunction with the monitoring system, a mechanism to alter design parameters and operation procedures as new information becomes available is also necessary.

- Administration of large projects also requires an internal organization to maintain constant project review. An external monitoring organization is rarely effective because it lacks power within the bureaucracy and is therefore unable to influence decisions related to sensitive administrative adjustments.

- Administrative jurisdiction must be clearly defined with no areas of ambiguity or overlap. Where questions arise, there needs to be a recognized decisionmaker who can quickly resolve the issue. New administrative organizations that attempt to take power from old, established bureaucracies can succeed only if the transfer of power and responsibilities is complete, if it is accepted by all concerned parties, and if career decisions for all personnel rest with the new organization.
- On projects that involve lengthy planning and construction periods, it should be recognized that farmers' expectations and behavior change over time. As economic circumstances change, farmers' reactions to perceived economic incentives will change. Project success depends upon the ability to adjust to changing farmer behavior.
- Water pricing and collection policies need to be tied to costs so that users who benefit from the system pay for the services. Subsidies will invariably lead to larger and larger requirements for public funds and therefore to inadequate maintenance and repair programs.

Even with improved management of the SCARPs, more private tubewell development, further expansion of surface water supplies, and increased intensification of agricultural production, salts have continued to accumulate in the soil and associated groundwater. Only when equality between salts flowing into the system and salts flowing out of the system is reached will long-term irrigated agriculture be possible.[6] This fact was known and plainly stated in the Revelle Report, but it has been ignored in the rush to develop SCARP areas. SCARPs are not a long-term solution; they only delay the eventual need to remove salts from the irrigated area.

Technologies to reduce salts in the system include flushing saline effluent down the rivers (Kemper and others 1978), diverting effluent to designated salt flats or lakes, and draining effluents by surface drains to the sea. This last alternative is the least destructive environmentally but the most expensive. All three methods will have to be used eventually to maintain a positive salt balance in the Indus Basin. Even with a least-cost mix of alternatives, drainage will require massive investments in human and physical capital, and users of the system will have to pay significantly more for water and for drainage.

The government will also have to provide additional funds and, even more important, commit additional administrative and technical personnel. Given a population which is expected to exceed 130 million by 2000 and a precarious food supply situation, Pakistan has no choice but to invest in protecting the long-term productivity of its most valuable natural resource, agricultural land in the Indus Basin. To make this investment both economically and financially feasible, however, users who benefit need to be persuaded to pay most of the costs. Provincial governments can no longer subsidize these services from other resources, as the SCARP experience has demonstrated.

Notes

Data were collected in Pakistan while the author was an assistant professor at Colorado State University (CSU). Research was supported in part by the United States Agency for International Development (USAID). B. Hasan from the Irrigation, Drainage and Flood Control Research Council (IDFCRC) jointly supported the study with CSU/USAID. M.A.R. Farooqi from IDFCRC and Zahid Saeed Khan and Peter Joseph from CSU/USAID provided assistance in collecting the field data. Additional information was obtained from the U.S. Geological Survey. Helpful comments were provided by Anthony Bottrall, Robert M. Hirsch, William O. Jones, Walter P. Falcon, Carl H. Gorsch, and Robert Dorfman. This chapter originally appeared in a somewhat different form in *Food Research Institute Studies*, vol. 18, no. 2 (1982), pp. 149–80.

1. Gotsch and others (1975) present an excellent review of traditional agricultural practices in the Indus Plain for those interested in microlevel agricultural and irrigation practices.

2. In Pakistan the canal water is distributed through minor canals and flows out of the turnout (*mogha*) to a village-level watercourse command (80 to 400 hectares). There are no headgates at the moghas and if a particular canal has water in it there is water in every watercourse command on that canal. There are more than 88,000 watercourse commands in the Indus Basin.

3. WAPDA (1979) records the average preproject depth to water table as 2.6 meters.

4. The LRD schedule ignores equitable distribution of water throughout the seven-day fixed irrigation water rotation schedule that is in operation for each chak, a significant constraint on the scheduling committee.

5. How much of the increase in crop yields is a function of new high-yielding varieties and how much is a function of additional groundwater supplies is unknown.

6. The salt-flow computer simulation model developed by the Panel predicted that, with no drainage and with canal water containing only 250 milligrams of salt per liter, salt concentrations after twenty-five years would severely impair crop production. The model predicted that surface drainage of approximately 10 percent of the quantity of tubewell water pumped over a fifty-year period would be needed to preclude excessive salt accumulation, although this drainage could be delayed ten or even twenty years, provided that the total drainage over fifty years equaled 10 percent of total pumpage (White House 1964).

References

Ahmad, Nazir. 1974. *Waterlogging and Salinity Problems in Pakistan.* Parts I and II. Lahore: Irrigation, Drainage, and Flood Control Research Council.

Carlston, C. W. 1953. *Report to Pakistan Government on the History and Causes of Rising Groundwater Levels in Rechna.* EPTA Report no. 90. Rome: Food and Agriculture Organization.

Eaton, Frank M. 1965. "Waterlogging and Salinity in the Indus Plain: Comment." *Pakistan Development Review* 5, 3.

Falcon, Walter P. 1976. "Agricultural Policy in Pakistan." Islamabad: Ford Foundation.

Gotsch, Carl H., Bashir Ahmad, Walter P. Falcon, Muhammad Naseem, and Shahid Yusuf. 1975. "Linear Programming and Agricultural Policy: Micro Studies of the Pakistan Punjab." *Food Research Institute Studies* 14, 1.

Greenman, D. W., V. W. Swarzenski, and G. D. Bennett. 1967. "Groundwater Hydrology of the Punjab, West Pakistan, with Emphasis on Problems Caused by Canal Irrigation." United States Geological Survey Water Supply Paper 1608-H. Washington, D.C.

Hussain, Muhammad, Barket Ali, and S. H. Johnson, III. 1976. "Cost of Water per Acre Foot and Utilization of Private Tubewells in Mona Project SCARP-II." Publication no. 62. Bhawal: Directorate of Mona Reclamation Project.

India. 1873. *Northern India Canal and Drainage Act.* Act no. VIII. New Delhi.

Johnson, Sam H., Alan C. Early, and Max K. Lowdermilk. 1977. "Water Problems in the Indus Food Machine." *Water Resources Bulletin* 13, 6.

Kemper, W. D., Mian M. Ashraf, Munir Chandkry, and S. H. Johnson. 1978. *Potential for Building and Utilizing Fresh Water Reservoirs in Saline Aquifers.* Annual Technical Report. Fort Collins, Colo.: Water Management Research Project, Colorado State University.

Lieftinck, Pieter, A. Robert Sadove, and Thomas C. Creyke. 1968. *Water and Power Resources of West Pakistan—A Study in Sector Planning.* 3 vols. Baltimore, Md.: Johns Hopkins Press.

Maierhofer, C. R. 1952. "Reconnaissance Report on the Drainage, Waterlogging, and Salinity Problems of West Pakistan." Denver, Colo.: U.S. Bureau of Reclamation.

Malmberg, Glenn T. 1975. "Reclamation of Tubewell Drainage in Rechna Doab and Adjacent Areas, Punjab Region Pakistan." United States Geological Survey Water Supply Paper 1608-O. Washington, D.C.

Mohammad, Ghulam. 1964. "Waterlogging and Salinity in the Indus Plain: A Critical Analysis of Some of the Major Conclusions of the Revelle Report." *Pakistan Development Review* 4, 3.

———. 1965. "Private Tubewell Development and Cropping Patterns in West Pakistan." *Pakistan Development Review* 5, 1.

Mundorff, M. J., G. D. Bennett, and Masood Ahman. 1972. "Electric Analog Studies of Flow to Wells in the Punjab Aquifer of West Pakistan." United States Geological Survey Water Supply Paper 1608-N. Washington, D.C.

Mundorff, M. J., P. H. Carrigan, T. D. Steele, and A. D. Randall. 1976. "Hydrologic Evaluation of Salinity Control and Reclamation Projects in the Indus Plain, Pakistan—A Summary." United States Geological Survey Water Paper 1608-Q. Washington, D.C.

Pakistan, Planning Commission. 1978. *The Report of the Indus Plain Research Assessment Group.* Islamabad: Government Printing Office.

Paustian, Paul W. 1930. *Canal Irrigation in the Punjab.* New York: AMS Press.

Punjab, Land and Water Development Board. 1971. *Report of the Special Committee on the Working of the SCARPs.* Lahore: WAPDA Press.

Taylor, George C. 1965. "Water, History, and the Indus Plain." *Natural History Magazine.* New York: American Museum of Natural History.

USAID (United States Agency for International Development). 1970. "Salinity Control and Reclamation Projects: Management, Operation, and Maintenance." Lahore: USAID Provincial Office.

WAPDA (Water and Power Development Authority). Central Monitoring Organization. 1971. *Review of Completed Salinity Control and Reclamation Projects.* Lahore: WAPDA Press.

———. Master Planning and Review Division. 1979. *Revised Action Programme for Irrigated Agriculture.* 3 vols. Lahore: WAPDA Press.

White House–Department of the Interior Panel on Waterlogging and Salinity in West Pakistan. 1964. *Report on Land and Water Development in the Indus Plain.* Washington, D.C.: U.S. Government Printing Office.

Comment

Max K. Lowdermilk

Johnson's account of Pakistan's SCARP program in three SCARP areas is significant because it

- Reviews the lessons of a public tubewell program designed to help supplement canal irrigation deliveries and help resolve waterlogging and salinity problems
- Describes a complex set of institutional constraints which continue to hamper the management of many large public groundwater schemes in South Asia
- Sets in perspective the unique role of private groundwater development and the need to build into the water resource planning process a formal conjunctive use program that allows for farmer involvement.

One might ask, What would have been the likely extent of externalities (waterlogging, salinity, social impacts, and so on) if the SCARPs and the private sector tubewell boom of the late 1960s and 1970s had not taken place? One could hypothesize, for example, that a larger area, especially in the Sargodha, Faisalabad, Gujranwala, and Lahore regions, would have been more adversely affected by waterlogging and salinity. Although it is not known exactly how much the water table in these areas was controlled by private tubewells, indeed it has been substantial. Private tubewells, as Johnson has shown, have also played a more significant role than public tubewells in increasing agricultural production in freshwater zones of the Punjab. What has been lacking is a workable comprehensive program to coordinate groundwater development with a sound conjunctive use policy. The top-down centralized model has not worked well.

The boom in private tubewells was fostered by a set of market incentives provided during the Ayub era. The terms of trade for agriculture were positive during a relatively long period of political stability. The rapid spread of private tubewells in the 1960s was documented by Ghulam Mohammad (1964), who strongly argued that public tubewells be located only in areas with saline groundwater. Owing to several exogenous factors such as increased support costs of equipment, rising energy costs, and less positive terms of trade, private tubewell development began to phase down some in the late 1970s, but nearly 190,000 tubewells had been constructed by 1978 (Lowdermilk, Early, and Freeman 1978).

One issue not raised by Johnson is the need for further study of the externalities associated with the practice of pumping saline groundwater in the Punjab SCARP areas and conveying it through the canal and river systems downstream. Salinity levels have been building up over the past twenty-five years or so in the Southern Province. To what degree has this increase been the result of the dumping of saline water in the north? Externalities with serious economic consequences may result from this practice and give rise to political conflicts between the two provinces. Horizontal drainage constitutes a complex technical problem because the two natural drainage routes are through the existing river systems and the Rajasthan Desert of India. The existing topographical constraints make technical solutions more difficult. For example, the land slope to the sea is only about 0.03 meters per kilometer.

Solutions that have been proposed other than horizontal drainage are improving the mixing of saline groundwater and freshwater in the north, shifting excess freshwater from Punjab SCARP areas to areas where canal water is scarce, and using skimming wells to utilize freshwater overlaying saline water. Skimming well technology was developed, tested, and demonstrated by the USAID–Colorado State University program in the 1970s. This technology may work in some places but will require much more refinement and careful monitoring.

Without a strong conjunctive use policy which can be carefully monitored, much of Pakistan's vast irrigated area will continue to be threatened. The new initiative to privatize many of the large SCARP public tubewells appears sound for areas with nonsaline groundwater, but in areas where the groundwater is highly saline, public tubewells for vertical drainage will remain a priority. The need for new laws regarding future groundwater development is now regarded as pressing.

As private groundwater development continues, research is needed to ascertain how the particular agrarian structure of Punjab and Sind provinces tends to skew benefits from such technology to large tubewell farmers. There is some evidence that private tubewell development in the past has been one of many factors creating problems of socioeconomic dualism (Gotsch 1972; Lowdermilk, Early, and Freeman 1978; Hiroshima 1978).

The lessons learned from Pakistan's experience since 1959 with both public and private tubewells provide several useful insights.

- With respect to cropping intensities, yields per unit, irrigation efficiencies, income per hectare, water control, and so on, private tubewells in freshwater zones have been more effective than public tubewells (Lowdermilk 1972; Lowdermilk, Clyma, and Early 1975; Lowdermilk, Early, and Freeman 1978, vol. 48C, pp. 63–66, 91–107, and vol. 48D, pp. 155–62; WAPDA's 1980 master planning chak studies). A large factor is better management by farmers and improved water control.
- Private tubewells have created a private water market in many freshwater areas of Pakistan. There is evidence that small farmers have benefited from this water market. Many private wells are jointly owned by groups of small farmers. Joint family farms pool and utilize their resources effectively. Most tubewell farmers, except those with more than twenty-five hectares, sell large quantities of water. Where the density of private wells is adequate, the price of water becomes competitive (Lowdermilk, Early, and Freeman 1978). More information is needed on the extent of this private water market where even small-scale entrepreneurs have invested in private tubewells primarily for the purpose of selling water.
- The issues of private tubewells and equity require more study. Small and marginal farmers can be given greater incentives to buy tubewells, as has been done in India. Also, there is a need to examine how small discharge wells with mobile pumpsets can be made more widely available to small farmers in Pakistan. A significant difference between the West (Pakistan) and East (India) Punjabs lies in the size of private tubewells.
- The principal problem with the large public tubewells has been poor management, a factor which was not anticipated when the designers estimated a forty- to fifty-year life for this technology. With present levels of management, ten years may be more realistic. Johnson has documented many management problems, such as operations and maintenance, lack of training and supervision of staff, lack of integrated management and farmer participation at the local level, and the absenteeism of tubewell operators. Other problems not documented by Johnson are the role of politics and corruption in the installation and operation of public tubewells. Those who advocate economies of scale for large public tubewells need to examine the potential management problems within each situation. The magnitude and mechanisms of political influence and its effects on public tubewell management and performance need more study. Such complex problems have undoubtedly contributed to unusual low performance and the prohibitive costs involved. As a result of such "management problems," the government of Pakistan and the World Bank have rightly moved to a policy of prioritizing the public tubewell systems in freshwater zones.
- Whatever conjunctive use policy is evolved, effective forms of farmer participation are a priority. An agenda of research and social experiments on how to do this in Pakistan would likely pay dividends in improving water management programs and relations with water authorities.

Johnson's paper provides a most useful contribution. Future research should build upon this study and test his basic hypotheses. Such studies can contribute to the policy dialogue taking place regarding Pakistan's irrigation sector. The technical inputs to improved water resource management are well known. What is not yet clear is how to develop practical programs on a large scale which allow for the proper fit and interface between the technical, social, organizational, legal, and political functions involved. Although farmer participation in these programs is now accepted as necessary for local-level management, too little is known about how to make this a reality.

References

Gotsch, Carl H. 1972. "Technical Change in the Distribution of Income in Rural Areas." *American Journal of Agricultural Economics* 54:326–41.

Hiroshima, Shigenochi. 1978. *The Structure of Disparity in Developing Agriculture.* Tokyo: Institute of Developing Economies.

Lowdermilk, Max K. 1972. "Diffusion of Dwarf Wheat Production Technology in Pakistan's Punjab." Ph.D. dissertation, Cornell University, Ithaca, New York.

Lowdermilk, Max K., Wayne Clyma, and Alan C. Early. 1975. "Physical and Socio-Economic Dynamics of a Watercourse in Pakistan's Punjab." Water Management Research Project Technical Report no. 42. Fort Collins, Colo: Colorado State University.

Lowdermilk, Max K., Alan C. Early, and David M. Freeman. 1978. "Farm Irrigation Constraints and Farmers' Responses: Comprehensive Field Survey in Pakistan." Water Management Research Project Technical Reports 48A–48F. Fort Collins, Colo.: Colorado State University.

Mohammad, Ghulam. 1964. "Waterlogging and Salinity in the Indus Plain: A Critical Analysis of Some of the Major Conclusions of the Revelle Report." *Pakistan Development Review* 4, 3.

6

THE NORTH CHINA PLAIN

Development of Groundwater for Agriculture in the Lower Yellow River Alluvial Basin

Huang Ronghan

The Yellow River, the largest river in the northern part of China, flows through Qinghai, Gansu, Shanxi, Shaanxi, Henan, and Shandong provinces and the Ningxia and Inner Mongolia autonomous regions. It is 5,460 kilometers long and has a catchment area of 752,000 square kilometers. The river carries an enormous amount of sediment, an average of 1.6 billion tons a year, and has an annual runoff of 56 billion cubic meters. Downstream from Huayuankou in Henan Province, an extensive alluvial basin stretches across both sides of the river. Created by sediment-laden floodwaters, the lower Yellow River alluvial basin, better known as the North China Plain, encompasses 300,000 square kilometers of land (see Map 6-1).

Because of the long flat slope of the riverbed in the lower reaches, as much as a quarter of the river's sediment load (0.4 billion tons) is deposited annually. As a result the riverbed is 4–8 meters higher than the surrounding land surfaces. The lower Yellow River has thus become a divide which bisects the North China Plain into two smaller plains: the Huang-Huai between the Yellow (Huang in Chinese) and Huai rivers and the Huang-Hai between the Yellow and Hai rivers. These two plains are the catchments of the Huai and Hai rivers respectively; the Yellow River has only a small catchment area of about 20,000 square kilometers in the North China Plain.

The combination of an abundant supply of high-riding river water surrounded by low-lying fertile soils makes the North China Plain well suited to gravity irrigation systems. Seepage flow from the river also supplies groundwater aquifers in the plain. From the dawn of agricultural production in China millennia ago, farmers have known how to build canals and dig wells for irrigation. But it was not until the founding of the People's Republic of China that water was diverted from the river for irrigation. In 1952 the first large-scale irrigation system—the People's Victory Canal—was completed. Designed to divert water from the lower reaches of the Yellow River near Huayuankou, the canal irrigated more than 40,000 hectares of farmland to the north of the river in its first years of operation. In recent years more than 1.5 million hectares throughout the North China Plain have been irrigated with water diverted from the Yellow River. If the waters diverted from the Huai and Hai rivers and their tributaries are included, the total irrigated area is estimated at 3 million or more hectares. During the growing season's period of peak water demand, say March to May, the average minimum discharge of the Yellow River at the Huayuankou gauging station is 300 cubic meters per second. In contrast, most of the tributaries to the Huai and Hai rivers have negligible or zero discharge. Surface water irrigation has been slow to develop on the plain because of the problem of silt deposition and the uncertainty of supply (as in times of drought). Groundwater irrigation, however, developed rapidly from 1972 to 1980, spurred on by successive droughts in the early 1970s. Now the area of the North China Plain that is irrigated by groundwater is more than double that by surface water.

Natural Conditions

The North China Plain has a temperate, semihumid, monsoonal climate. The average annual temperature varies from 16° C in the south to 10° C in the north. Rainfall totals vary from year to year and from place to place. The Huang-Huai Plain receives about 700–900 millimeters a year, the Huang-Hai about 500–600 millimeters. Although the North China Plain as a whole averages more than 600 millimeters of rain annually, large fluctuations are common from year to year. There may be as little as 200–300 millimeters in very dry years or as much as 1,300–1,500 millimeters in extraordinarily wet years. In addition, 70–

Map 6-1. *The North China Plain*

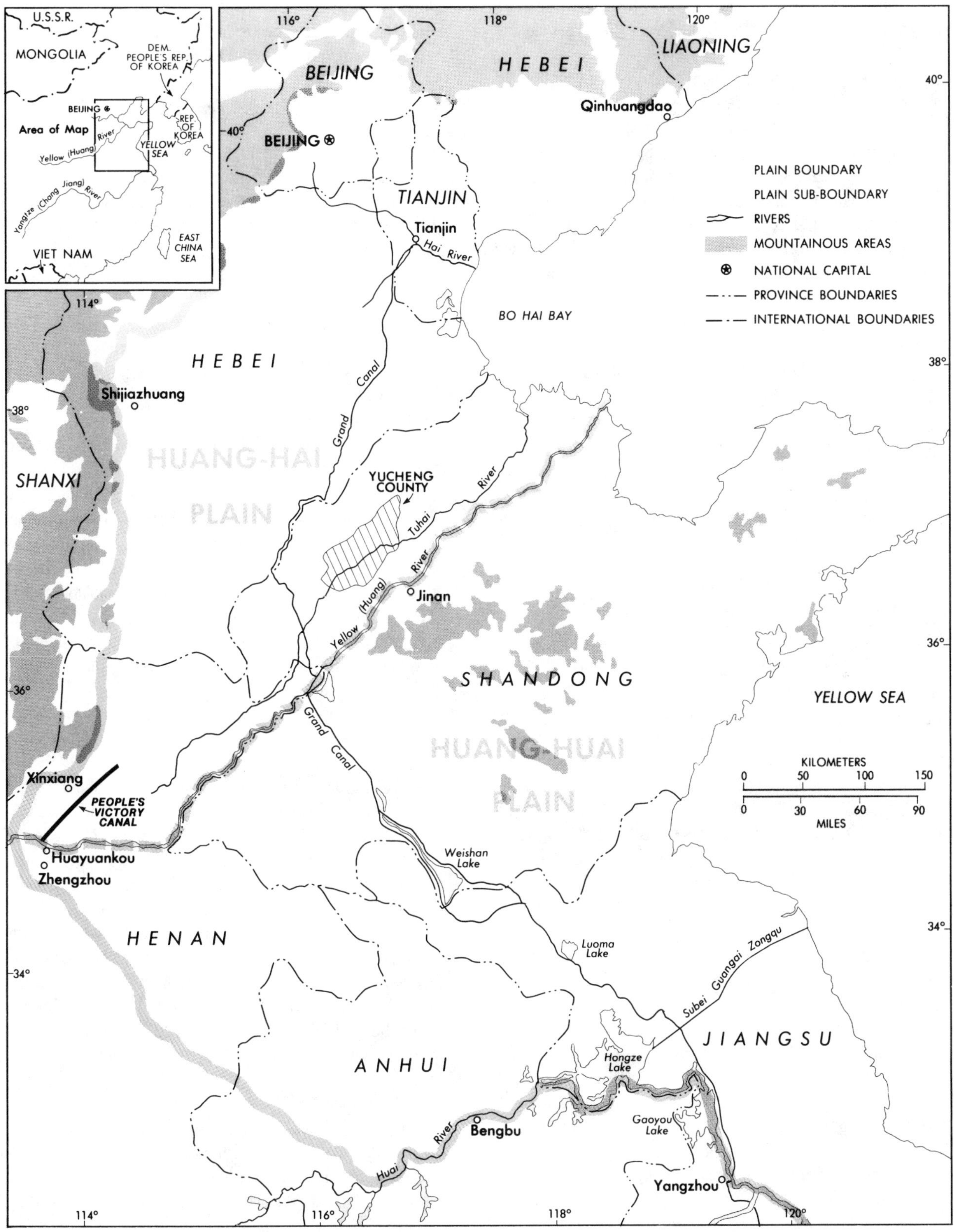

80 percent of the annual precipitation occurs between June and September. As a result the plain is subject to both drought and waterlogging. These two disasters are the main factors behind the low and unstable crop yields of the North China Plain.

Soil salinity is another problem that limits agricultural production in the area. There are about 2 million hectares of saline cultivated land in the plain, most of which are concentrated on the Huang-Hai Plain. Since soil salinity is closely related to the rise of the groundwater table in alluvial plains, measures have been taken to keep the water table at a desired depth. Although it is well-known that drought may promote soil salinity, in the lower Yellow River alluvial basin waterlogging contributes to the salinity problem by raising the groundwater table above the so-called critical depth. The hydrogeological cycle governing the North China Plain is waterlogging in the summer and fall, followed by salinity the next spring. Drought, waterlogging, and salinity can exist simultaneously and interact with one another. Using groundwater for irrigation offers a comprehensive solution to all three problems. By lowering the groundwater table, groundwater pumping creates room in the subsoil for the infiltration of excessive surface water.

Well and Well-Canal Irrigation

In the past decade almost 1.4 million tubewells were constructed in the North China Plain. As a result the groundwater table has been lowered to a depth of 3–4 meters in most places, and the area of saline land has been reduced considerably. In 1980, 370,000 tubewells were in operation in Hebei Province on the Huang-Hai Plain, and their total pumpage of roughly 10 billion cubic meters irrigated about 2.5 million hectares of land. In Shandong Province, which encompasses parts of both the Huang-Hai and Huang-Huai plains, total pumpage in 1980 was about 6 billion cubic meters, and 2 million hectares of land were irrigated. The area of saline land in the two provinces together was reduced by about 25 percent between 1960 and 1980. Although the tubewell pumping of groundwater was not the only cause of this reduction, it was an important one. In 1982 total pumpage for the North China Plain as a whole was 25–35 billion cubic meters and the total irrigated area was roughly 8 million hectares, 1.3 million of which were located in canal irrigation systems known as "well and canal combined" (or well-canal) irrigated areas. The high crop yields in these areas are maintained through conjunctive use of surface water and groundwater, which in effect offers double insurance of adequate water supply. The advantages afforded by well and well-canal irrigation systems have not gone unnoticed. The trend in further development of the North China Plain is toward conjunctive use. In terms of the plain as a whole, conjunctive use schemes that alternate canal irrigated areas with well irrigated areas or use canals at upstream locations or in the highlands and wells in downstream areas or lowlands should also be considered.

Some Achievements and Experiences

Well irrigation in the North China Plain has led to impressive increases in agricultural production, especially in the successive dry years from 1978 to 1980. Grain and cotton did particularly well in almost all the well and well-canal irrigated areas, and the increase in wheat yields was outstanding. For example, the increase in wheat yields between 1972 and 1980 was 127 percent in Henan Province, 25 percent in Hebei Province, and 54 percent in Shandong Province. Cotton yields showed similarly impressive gains. At the same time, improvements were made in the water supply systems for rural domestic and industrial uses.

The return on investments in well construction has been satisfactory. According to 1982 data, the average cost per tubewell with a depth of 10–40 meters was 5,000 yuan and the average cost per hectare was 1,350 yuan. The annual net income per tubewell was 1,500–2,000 yuan. The payout period was three to five years for shallow wells (10–40 meters deep) and five to ten years for deep wells (41–300 meters deep). Average annual income per hectare was estimated at 550–625 yuan. The benefit-cost ratio for a unit of tubewell or a unit of irrigated land is much greater than 1 in most cases and is considered acceptable by most farmers. The beneficial effects of groundwater pumping on salinity and waterlogging are also important.

Politics. In alluvial plains with monsoonal climates, the combined use of wells and canal systems works well to reduce salinity and waterlogging and meet the needs of irrigated agriculture. As a result the water conservancy authorities are pursuing this form of conjunctive use policy throughout the North China Plain. To encourage the farmers to work hard and become self-reliant, the central and local governments have provided financial support to those who drill wells for irrigation. According to data for 1973–81, government subsidy accounted for about 25 percent of the 10 billion yuan invested in the construction of tubewells; the remaining 75 percent came from the farmers.

Technology. To make sure that pumping wells and related on-farm facilities are constructed properly, the provincial bureau of water conservancy should prepare a development plan based on investigation of the groundwater resource and field experiments in the pilot areas. Well-drilling teams composed of technicians and trained laborers are organized by the water conservancy bureau of every county to construct wells in collaboration with the farmers. New drilling machines, such as the "big bowl drill" and the "water jet drill," have been created one after another. One set of water jet drill devices used in Henan Province could dig a well 30–40 meters deep within twenty-four hours, and the construction cost was estimated at 1,500 yuan.

Operation and maintenance. The main objectives of the management of well irrigation are to lower irrigation cost, lengthen well life, and keep wells in good operating condition. In China, whoever constructs a well is the owner and the user and is responsible for its operation and maintenance. Assistance is available from the governmental agencies. Lining head ditches, leveling land surfaces, and adjusting irrigation scheduling for different kinds of crops are important functions of water management.

The increasing shortage of surface water resources and overexploitation of groundwater combine to make efficient use of irrigation water vitally important in the basin. In general there are two approaches: reducing on-farm losses of water and adopting improved irrigation methods, such as subsurface pipelines and drip irrigation.

Exploitation and recharge. Exploitation of shallow groundwater in alluvial plains should be accompanied by recharge to prevent excessive drawdown of the water table and mining of the groundwater resource. Combined development of canal and well irrigation is of course an effective means to achieve this end. Emphasis should be put on recharging the groundwater with the infiltration of surface water. A preliminary evaluation of water resources in the lower Yellow River basin estimates the annual recharge to shallow groundwater at 47–49 billion cubic meters—about one-half of which comes from rainfall, one-quarter from canal seepage and deep percolation of irrigation water, and the rest from seepage flows from rivers and mountainous areas. The proverb "Supplementing the groundwater source with rivers and canals and reaping a good yield with wells" describes one of the experiences of water management in the lower Yellow River alluvial basin.

Problems

Some problems have occurred in connection with the development of well and well-canal irrigation in the North China Plain.

Drawdown of the water table. In some regions overexploitation has resulted in large-scale drawdowns of the groundwater table. For example, in Hebei Province there exist thirty-one cones of depression of the groundwater table with a combined area of 1,200 square kilometers, or 20 percent of the area of alluvial plain in the province; about one-half of this is caused by pumping water for agriculture.

Poor quality and deterioration of wells. According to a rough estimate, 20 percent of the existing wells are not in good condition. The main problems are sand filling, caused by improper design and poor construction materials, and collapse of well casings made of sandless concrete.

Low efficiency and high consumption. In some places the cost of irrigation is high owing to low capacity or yield and improper irrigation methods as well as inefficient operation. The consumption of energy (fuel and electricity) differs greatly from place to place, leading to differences in irrigation costs. In some places the cost of one application of water per hectare may be as low as 3–4.5 yuan, and in others as high as 9–10.5 yuan. Therefore, further development should emphasize cost reduction rather than volume targets. The overall plan is to improve the management of existing well irrigation and to drill more wells in canal irrigated areas to permit the efficient conjunctive use of surface and groundwaters.

Outlook

For the near future the development of groundwater for agriculture in the North China Plain will focus on the following:

- *Renewing existing wells.* Approximately two-thirds of the existing wells need to be renewed. This will be done by the farmers with well irrigation under the direction of water conservancy agencies.
- *Drilling wells in canal irrigated areas.* Drilling wells in the irrigation systems diverting water from Yellow River is a priority.
- *Improving water management practices.* There is an urgent need to restrict the exploitation of deep groundwater for agriculture.
- *Reclaiming saline shallow groundwater, reusing drainage water, and cultivating salt-tolerant crops.* This is quite a task in central Hebei Province, where the area of saline groundwater amounts to more than one-half the total area.
- *Implementing the Interbasin Water Transfer Project from the Yangtze River to the Huang-Hai Plain.* The objective of this huge project of hydraulic engineering is to solve the water shortage problem in the northern part of the basin. In the first stage about 5 billion cubic meters of water are planned to be transferred to the north of the Yellow River each year. A large portion of this will be used

to meet the industrial and domestic needs of large cities, including Tianjin, and the remainder will be used for agriculture and to recharge the groundwater.

• *Enacting legislation on groundwater use and development.* In 1981, for the purpose of strengthening the management of well irrigation, the Ministry of Water Resources and Electric Power approved and issued "Regulations for Tubewell Operation." The regulations have not yet been incorporated into legal procedure, however, and so cannot act as a formal law for groundwater management. For this reason concerned ministries will submit a new document, "Regulations of Groundwater Resources Management," to the state council.

Efficient Conjunctive Use of Surface and Groundwater in the People's Victory Canal

Cai Lingen

The People's Victory Canal irrigation system is a representative example of the conjunctive use of surface and groundwater resources in the lower Yellow River alluvial basin. Its irrigated areas have reached 40,000 hectares, about 27,000 of which are irrigated by both pumping and canal diversions. The system diverts about 490 million cubic meters of water from the Yellow River every year and pumps 120–150 million cubic meters of groundwater (about 50 million cubic meters for industrial and domestic use) from 5,000 wells.

The irrigation system is located in the western part of the lower Yellow River basin, in Xinxiang Prefecture, Henan Province (see Map 6-2). The land is a stepped plain, the south of which is higher than the north. The predominant soil is medium loam, with some salinity in the root zone. Before irrigation, the total area of saline and alkaline soils (located in the northern part of the system) was 7,000 hectares and the depth to the water table was 2–3 meters.

Average annual precipitation in Xinxiang city is 618 millimeters, 70 percent of which occurs between June and September. Average annual evaporation is 1,800 millimeters. The great variation in rainfall, both from season to season and from year to year, is the principal cause of drought, flooding, and waterlogging.

The People's Victory Canal was completed and began diverting water from the Yellow in 1952. Since then agricultural production has increased. Grain yields increased from 1.33 tons a hectare in 1951 to 3.02 in 1958; at the same time, cotton production rose from 0.2 tons per hectare to 0.6. However, because of the overapplication of water, seepage losses from canals, and inadequate drainage, the groundwater table rose year after year, and secondary salinization developed rapidly in the irrigated areas. By 1960 the depth to the water table was about 1.3 meters, and the area of saline land totaled 19,000 hectares. The problem was so serious as to jeopardize the continued operation of the system.

In the early 1960s the system was reformed; new drainage ditches and canals were excavated, wells were dug, and diversion of water from the Yellow was strictly controlled. The reformation effort, which lasted three to five years, gave new life to the People's Victory Canal system. Recently, the average yield of grain crops in the irrigated areas has exceeded 8 tons per hectare; that of cotton exceeded 0.8 ton per hectare. The region now ranks as one of the most stable and productive of agricultural areas using water diverted from the Yellow River.

Water Resources

Water for the People's Victory Canal irrigation system comes from three sources: precipitation, river flow diverted from the Yellow, and groundwater in the system. The water requirements of the major crops and the effective rainfall are analyzed in Table 6-1, based on multiyear irrigation experiments in the area. Effective rainfall is 114 million cubic meters (that is, rainfall is at least 114 million cubic meters annually, 75 percent of the time), which is not enough to meet the crop requirements. Irrigation is therefore needed for crop production.

The Yellow River is the main source of water for irrigation. During the dry season the canal headgate diverts 50–60 cubic meters of water per second from the river. The average annual diversion is 490 million cubic meters (in-

Map 6-2. *The People's Victory Canal System*

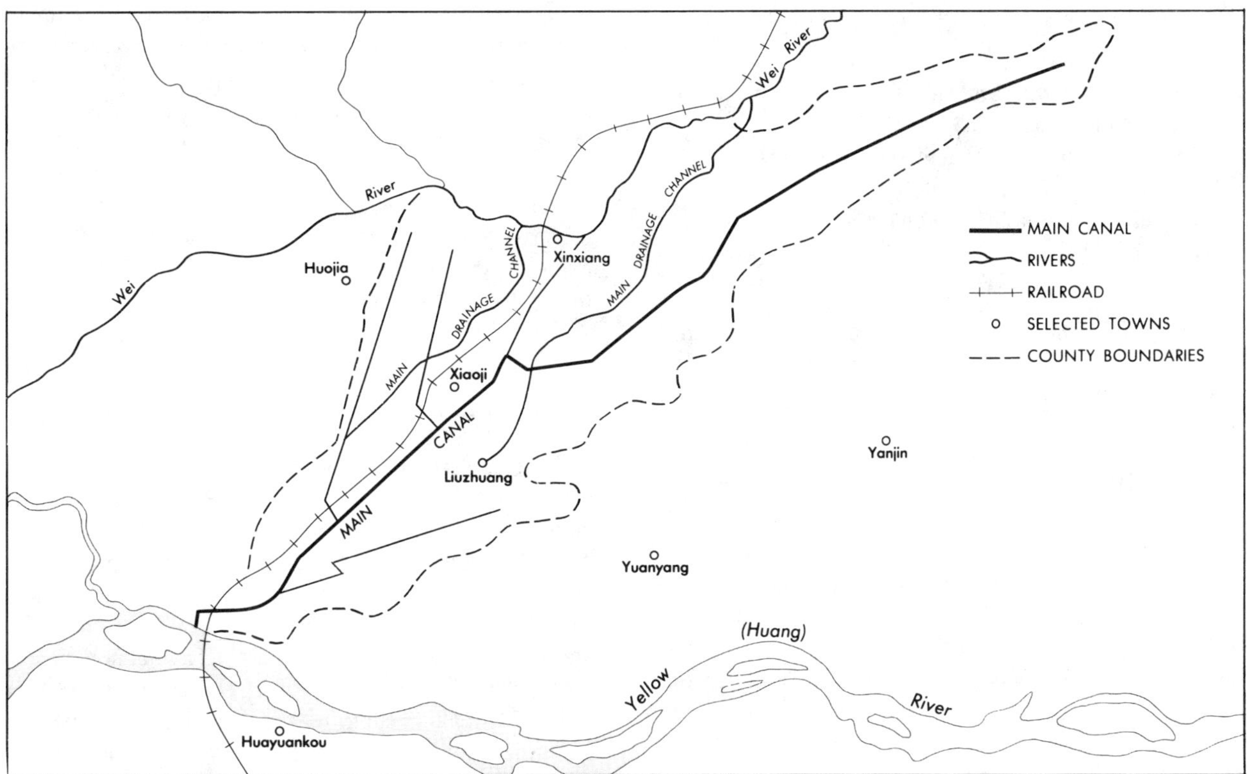

cluding water for industrial use). The quality of the river water is good; salinity is about 0.4 grams per liter. The riverbed is 4–8 meters higher than the land surface of the area, which facilitates gravity irrigation, utilization of the silt that the water carries, and the leaching of salt from the soil. There are, however, several problems with the water that is diverted from the Yellow. First, the high sediment content causes deposition, which decreases the flow capacity of the irrigation canals and drainage ditches. Second, canal seepage losses and deep percolation raise the water table, resulting in secondary salinization and waterlogging. Third, during the period of peak demand for water in the growing season, the supply is limited to the small runoff of the river. Developing the groundwater and using it in combination with the river water is considered the solution to these problems.

Groundwater is plentiful within a depth of 50 meters. Aquifer lithology is mainly silt, sand, and fine gravel. The thickness of the groundwater aquifer varies from 20 to 40 meters, with a transmissibility coefficient of 1,050–1,250 square meters per day, permeability of 25–30 meters per day, specific capacity of 200–1,000 cubic meters per day, and specific yield of silt in the upper aquifer of 0.055–0.065. Groundwater salinity is less than 1 gram per liter in all but a fraction of the irrigated area. The overlay of the phreatic aquifer is sufficiently pervious. The main source of groundwater recharge is seepage of irrigation water; infiltration of rainfall is a minor source. Precipitation in

Table 6-1. *Water Requirements of Major Crops and Effective Rainfall*

Crop	*Area (hectares)*	*Water requirement (cubic meters per hectare)*	*Total water requirement (millions of cubic meters)*	*Rainfall (millimeters)*	*Effective rainfall (cubic meters per hectare)*	*Total effective rainfall (millions of cubic meters)*
Wheat	29,333	4,650	136.4	120	1,080	31.68
Corn	20,667	3,150	65.1	264	1,320	27.28
Cotton	10,667	4,050	43.2	444	3,015	32.16
Rice	10,000	12,750	127.5	324	2,265	22.65
Total	—	—	372.2	—	—	113.77

—Not applicable.

1979 was of about 75 percent frequency. Groundwater recharge in 1979 over a total area of 789.2 square kilometers was as follows:

Infiltration from rainfall
- Yearly rainfall: 407.1 millimeters
- Ratio of infiltration: 0.23
- Amount of supply: 73.9 million cubic meters

Infiltration from irrigation water
- Irrigation water: 486.3 million cubic meters
- Ratio of seepage: 0.50
- Amount of supply: 243.2 million cubic meters

Rainfall infiltration of 73.9 million cubic meters is not enough to satisfy crop needs and should be supplemented with canal irrigation.

Types of Conjunctive Use

Before water was diverted from the Yellow, wells 6–7 meters deep were used for irrigation in some areas. These wells were abandoned after the canal system came into operation. At the beginning of the 1960s, in an effort to boost agricultural production, many tubewells were constructed to be used in combination with canal irrigation. The shallow pumping wells, as they are called, have a depth of 30–40 meters, and their casings, up to 0.8 meter in diameter, are made of sandless concrete.

Most of the shallow pumping wells are located on both sides of the sublateral. The yield of each varies from 40 to 200 cubic meters per hour. Some tubewells with depths of 30–100 meters have also been constructed for industrial and domestic use. These are located along the main canal and near the towns and cities.

There are three categories of conjunctive use of surface and groundwater in the People's Victory Canal system: using both wells and canals for irrigation, recharging the groundwater with water from the canal system, and using groundwater for nonagricultural purposes.

Well-Canal Irrigation

About 27,000 hectares are irrigated by a combination of canal water and well water. The ratio of canal water to well water used for irrigation each year is about 5 to 1. Generally, more of the irrigation water used in the upper part of the system comes from the canal system and more of the water used in the lower part of the system comes from wells. The presowing irrigation of major crops is usually supplied by canal water; during the growing season, however, usually both well and canal water are used.

Groundwater Recharge

Since 1960, about 10,000 hectares of land within the canal system have been irrigated with well water rather than canal water. Because natural recharge of the groundwater system is not enough to meet irrigation demands, it is supplemented with horizontal recharge from the upper canal system and artificial recharge from excess canal water. For example, in the Liuzhuang-Xiazhuang region that is irrigated solely with wells pumpage was 9.1 million cubic meters in 1979; the artificial and horizontal recharge was estimated at 6.5 million cubic meters—approximately 70 percent of the pumpage. In addition, canals are being constructed in some areas where the water table has fallen following several years of dry weather. These will be used to supplement the irrigation with wells.

Groundwater Pumping for Industrial and Domestic Use

In the central part of the system and along the sides of the main canal, where recharge to the groundwater is quite large, water for industrial and domestic use comes not from canal supplies but from tubewells. For example, in Xiaoji People's Commune, 321 wells within an area of 21 square kilometers supply water for industrial and domestic use as well as for agriculture. Of a total pumpage of 15.1 million cubic meters from February to September 1981, 7.3 million cubic meters were for industrial and domestic use and 7.8 million for irrigation. The quantity used for irrigation was more than the entire amount diverted from the Yellow River that year. The groundwater level has been kept at 3–4 meters below the surface for a long time, which helps to control the salinity.

The Role of Conjunctive Use

The conjunctive use of surface and groundwater by well-canal irrigation in the People's Victory Canal is implemented to achieve a number of goals:

- *To rationalize the use of water resources.* Runoff from the Yellow River is subject to significant fluctuation. Minimum flows, even no flow at all, are common during the season of peak water demand. In recent years the runoff during the May-June period has been approximately 2 billion cubic meters. But more than 3 billion cubic meters are needed to meet the demand for water. The People's Victory Canal irrigation system is helping to alleviate the conflict between supply and demand by using the water from the river more efficiently and supplementing it with groundwater. Historical data indicate that more than one-half of the water diverted from the Yellow each year was lost, and a large portion of it infiltrated through the soil into the water table. Rough estimates put the mean annual replenishment of the groundwater, including the natural recharge from precipitation, at 355 million cubic meters. Before the groundwater was developed, the water table was close to the surface and a considerable amount of the replenishment was lost through evaporation; this led to a build-up of salt on the land surface. With well irrigation, the pumping lowers the water table, which reduces evaporation and increases the amount of water available for use. For example, in Xiaoji People's Commune in the very dry year of 1981 (annual rainfall of 360 millimeters), the large volume of water withdrawn from the groundwater aquifer was enough to satisfy the needs of agricultural and industrial production. And in the lower part of the People's Victory Canal system, in the area irrigated solely by wells, excess canal water is diverted into specially constructed ponds and drainage ditches to be stored in the aquifers and put to use later by pumping.

• *To reduce waterlogging.* During the rainy season, excess rainwater needs to be removed speedily to prevent damage to the crops. But this is expensive. With groundwater pumping, the water table is lowered and the recharge by infiltration is increased. For example, during the rainy season of 1972, when the total rainfall over a ten-day period was 426 millimeters, approximately 500 hectares of cropland in canal irrigated areas became waterlogged, but there was no waterlogging in the areas irrigated by wells.

• *To prevent secondary salinization.* Surface drainage alone is not enough to keep the water table at the critical depth. In the People's Victory Canal system, this depth is about 2 meters in loamy soil. By pumping water from wells for irrigation, this depth can be maintained easily. In recent years water levels throughout the system have been maintained at 2–4 meters below the surface, and a large portion of saline and alkaline lands has been reclaimed for agricultural production.

• *To reduce sedimentation.* The Yellow River carries one of the largest sediment loads in the world. Approximately 500 million tons of sediment is diverted with water into the People's Victory Canal system every year, and some of this is deposited in the irrigation canals and drainage ditches. The cost of cleaning the silt deposits is high. Groundwater development would reduce the amount of surface water diverted and thus alleviate the problem of sedimentation.

• *To ensure timely irrigation.* Experiments with wheat show that if the time of three applications of water in the spring is delayed 12–18 days, the yield of wheat per hectare will be reduced by about 1.5 tons. Rice needs even more frequent applications of water than wheat. Thus by ensuring that crops can get the water they need at the time that they need it, supplementing canal irrigation with wells is an effective way to ensure good harvests.

• *To increase the irrigation benefit.* The net benefit to irrigation of the People's Victory Canal system is outstanding. On the basis of thirty years of data, the costs of water supply are as follows:

	Millions of yuan
Present worth of capital investments	19.35
Annual operating and maintenance costs	2.12
Present worth of total cost	60.73
Annual cost	3.12
Mean annual water supply	307.69

Using Miaozhuang production brigade as a representative example of average production levels within the system, the benefit-cost ratio is 1.93, the annual net benefit is 1,469 yuan per hectare, the present worth of the total net benefit during the service life of the project (seventy-five years) is 28,631 yuan per hectare, and the annual irrigation cost as a proportion of agricultural output value is 2.6 percent.

All Is Not Wells in North China: Irrigation in Yucheng County

James E. Nickum

> Since the means of production are under public ownership, the state may . . . regulate all economic operations in the country in a unified way, avoiding the anarchy typical of a capitalist economy and the enormous waste of . . . resources resulting from it.
>
> —Xue Muqiao, *China's Socialist Economy*

We may in theory agree with this statement by Xue Muqiao, one of China's leading economists. In chapter 1, O'Mara points to centralized control over a resource by a single management body as one of three theoretically possible solutions to the problem of physical externalities. Centralizing property rights should permit a planned so-

Table 6-2. ***Comparison of North China Plain and Yucheng County***

Item	*North China Plain*	*Yucheng County*
Precipitation (millimeters)	500–800	616
Cumulative temperature (>0°C)	4,200–5,000	4,951
Frost-free days	180–220	200
Hours of sunlight	2,100–2,800	2,640
Climate	Monsoonal	Monsoonal
Natural disasters	Drought, flooding, salinization	Drought, flooding, salinization
Soil parent material	River alluvium	River alluvium
Percentage of cultivated area that is saline	18.5	35.4
Principal crops	Wheat, maize, cotton, soybeans	Wheat, cotton, maize, peanuts
1980 grain yield (tons per hectare)	3.8	3.3

Note: Data in the first four rows are annual.
Source: Ren and Nickum (1984).

Figure 6-1. ***Precipitation and Evapotranspiration in Yucheng County***

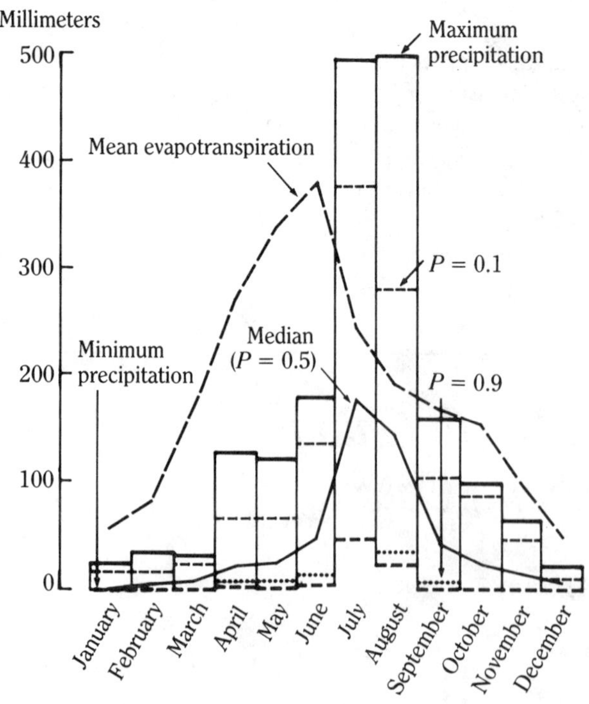

Note: Data are from 1951 to 1980 (thirty years). *P* values are values of Weibull's *P*. Where there is no separate line for $P = 0.9$, the value for $P = 0.9$ coincides with the minimum.
Source: Chinese Academy of Sciences, Institute of Geography, Beijing.

cialist economy to centralize resource use efficiently. Yet in practice, resource abuse seems to be at least as pervasive in China as in nonsocialist systems.[1]

This study surveys one example of resource use in China—the development of irrigation on the North China Plain, particularly in Yucheng County in northwestern Shandong Province. This area has developed both surface and groundwater sources and has dealt with the problems of salinization resulting from excessive reliance on surface water for irrigation.[2] Although Yucheng County has succeeded somewhat in restraining the increase of salinization, it has done so without an effective incentive to encourage the use of wells. Furthermore, water management policy at the state level has developed less as a well-planned, centralized approach through a single agency than as the outcome of a political process involving multiple agencies with different objectives and constituencies.

Background

The North China (Huang-Huai-Hai) Plain consists of the alluvial basins of the middle and lower Huang (Yellow), the Huai, and the Hai rivers. Its 509,000 square kilometers (5.3 percent of the nation's total area) encompass the entire province of Shandong and the centrally administered municipalities of Beijing and Tianjin, virtually all of Hebei Province, eight out of ten prefectures in Henan, and four prefectures plus two counties in the northern (Huaibei) portions of Jiangsu and Anhui provinces.[3] Flat and monsoonal, the North China Plain is prone to drought in the spring and fall, flooding in the summer,[4] and salinization of the soil, mostly in the spring but year-round as well.

Yucheng County, situated in the middle of the North China Plain 30–40 kilometers north of the Yellow River near Ji'nan, represents much of the plain in climate, cropping, hydrology, natural disasters, and economic conditions (see Table 6-2). Its water regime, shown in Figure 6-1, exhibits many features common to North China:

- Concentration of precipitation in the summer. In July and August the rainfall tends to be concentrated in storms and creates a significant drainage problem compounded by a virtual absence of surface storage facilities.
- Extreme year-to-year variation in precipitation. Even the wet months of July and August have variation coefficients for rainfall of 0.62 and 0.74 respectively. Drier months vary more. August precipitation has ranged from 22.7 millimeters in 1968 to 497.3 millimeters in 1951. The Chinese claim that the North China Plain faces drought nine years out of ten. Figure 6-1 shows that extremely wet years (fewer than one year in ten) are significantly wetter than the median (Weibull's $P = 0.50$), especially in August but also throughout the period from April

to September. However, extremely dry years tend to be closer to the median.

• High potential evapotranspiration. Especially in the spring, drought is aggravated by secondary salinization caused by evaporation from the phreatic zone of shallow aquifers. Drought and salinization may also occur in the fall, although to a lesser degree.

Yucheng County, like most of the North China Plain, thus faces a water regime that has a random, highly variable pattern in the short run and is prone in the long run to degradation of soil and groundwater when improperly exploited for human purposes. Defining and measuring improper use is difficult, and the absence of many significant data, especially on salinization, further complicates the task of water management.

Irrigation Comes to the Plain

Although China has been characterized as a "hydraulic society" by some historians (notably Wittfogel 1957), the North China Plain was rarely irrigated below the piedmont until recent decades. With the silt-heavy Yellow River restrained by artificial embankments and the Hai and Huang river systems marked by extremely variable flow,[5] agriculture on the North China Plain traditionally relied on rainfall. Only about 7 percent of the cultivated area in the three core provinces of Hebei, Henan, and Shandong was provided with irrigation facilities in 1949 (Nickum 1987).[6] With a few exceptions, such as the People's Victory Canal, built in 1950–52 to divert water from the Yellow River along the Taihang foothills into the Wei River (see Cai's contribution to this chapter), the vast majority of crops on the plain relied on rainfall for moisture until the 1970s.

Although Chinese planners have long recognized the necessity for an integrated approach to reflect the complex interrelationships of drought, flooding, and salinization, water management policy in practice has tended to have a single focus, often stimulated by the most recent disaster. This myopia has affected both policymakers and water users, the latter especially in their disregard of drainage maintenance. In the early 1950s the state concentrated its resources on flood control along the major rivers, especially the Yellow and the Huai, and did little to support water management on the farms. Then it switched its focus to conquering drought.

Irrigated areas first significantly increased during the Great Leap Forward (1958–60) and the massive water control campaigns accompanying the collectivization of agriculture in the previous two years. On the North China Plain, irrigation was developed by sinking relatively shallow wells, most of them fewer than 6 meters deep, and by diverting surface flows, especially in some areas near the Yellow River. The lack of complementary drainage facilities, the disruption of existing drainage systems by poorly planned canal lines, and the silting up of channels caused severe flooding and salinization. As a result, from 1962 to 1965 all Yellow River diversions except the People's Victory Canal were stanched.

After a major flood in the Hai River in 1963, the central government focused on constructing large surface drains in the North China Plain north of the Yellow River. This program, which included the Tuhai River that bisects Yucheng County,[7] constructed gates along the main drains to allow storage in the dry season.

During the 1960s China's rapidly expanding petroleum output resulted in the spread of pump irrigation, at first for surface water, but later increasingly for tubewells. A major drought in 1968 further propelled tubewell construction. Most of the 2 million tubewells now on the North China Plain were drilled in 1971–74. Although tubewell construction in North China was not included in the state plan until 1973 (Shuili Dianli Bu 1984b, p. 21), state grants of funds and pumps encouraged localities to dig tubewells as early as 1969.

By the late 1970s, groundwater irrigated 11.3 million hectares, or one-third of the total area, in China's seventeen northern provinces and municipalities, which works out to one well for less than six hectares (the area now farmed by eight to ten families). Groundwater was the exclusive source for irrigating 8.7 million hectares. Less than one-quarter of the irrigated area had access to both ground and surface sources (Zhang 1983, p. 2).

By providing dry season (mostly springtime) irrigation, tubewells have not only increased winter wheat output significantly, but have also lowered the water table before the summer rains. With both ditch and well drainage available, water was diverted once again from the Yellow, especially after the early 1970s.[8] In 1982, 10.6 cubic kilometers of Yellow River water were diverted to irrigate 17.3 million hectares in Henan and Shandong (Li 1984, p. 7).

The pattern of irrigation development in Yucheng County reflects the above trends (see Figure 6-2). Until the mid-1960s only a minuscule portion (ranging from 0.8 percent in 1949 to 13.6 percent in 1958) of the county's cultivated land was recorded as irrigated. Irrigation first significantly increased during collectivization in 1956, presumably owing to the digging of hand-operated shallow wells. The Yellow River diversions of the Great Leap Forward do not seem to have added appreciably to Yucheng's irrigated area. In fact, the area under irrigation declined between 1958 and 1961, despite a boomlet in tubewell construction in 1959. Fruitless construction often characterized the Leap: all 452 tubewells added in 1959 were apparently inoperative by 1962.[9]

Upstream diversions seem to have had a deleterious effect on Yucheng. The area of cultivated land affected by salinization tripled, far exceeding the irrigated area.[10] When the Tuhai River system was dredged and straight-

Figure 6-2. *Irrigated and Saline Areas in Yucheng County, 1949–83*

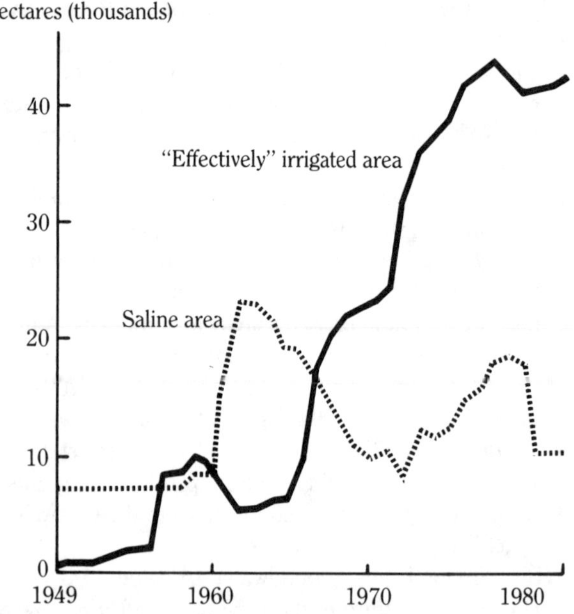

Source: Chinese Academy of Sciences, Institute of Geography, Beijing.

ened and a multilevel drainage system installed in lowland areas during the 1960s, the saline area declined steadily. The present drainage system is said to be capable of safely drawing off a twenty-four-hour rainfall of over 100 millimeters. The first significant increase in the irrigated area in Yucheng occurred more or less simultaneously with the decline in salinization. Both phenomena resulted from the rapid expansion of diesel-powered tubewells after 1967, about two years before most of the rest of the plain.

Beginning in 1972, water from the Yellow River was diverted through Yucheng County by the Panzhuang Canal, the largest of the three major diversions in Dezhou Prefecture. Use of this supplemental water did not much affect either Yucheng's irrigated area or the more restrictive category of "high, stable yield fields" (HSYF), but it did accompany a progressive expansion of the saline area. After 1980, HSYFs increased dramatically, and the saline area declined. The reason was partly that the scheduling of releases through the Panzhuang Canal was changed, but mainly that the area planted in cotton, a summer crop requiring much less irrigation than wheat, increased significantly.

Disjunctive Use of Multiple Sources

As a result of this pattern of development, the conjunctive use of surface and groundwater sources was not a big issue in most parts of the North China Plain until the past decade. Even today most places rely exclusively on one or the other. Irrigators who have a choice strongly prefer surface sources:

> In the course of our investigations of various provinces in recent years, we found a number of districts all proclaiming "[we] link wells and canals" . . . "[we] guarantee the harvest with wells while using the river as a supplemental source" and the like, and they had plans and test sites, but very few actually did anything. Most districts still "do not use groundwater when surface water is available," and numerous wells lie idle. (Zhang 1981, p. 18.)

Tubewells in Yucheng County are used at less than one-third of their capacity for the following reasons in approximate order of increasing importance (Ren and Nickum 1984).

- *Water quality.* Farmers prefer surface water because it is less saline, warmer, and more silt-laden than groundwater.
- *Technical inefficiency.* Well development has been rushed and poorly monitored, and the state supply system has been inadequate. In addition, few farmers in Yucheng County have had training in maintenance, well cleaning, or machinery repair. Technical inefficiency in well use has resulted in such problems as excessive water pressure loss and cavitation.

> The majority of [tubewells in north China] were dug without strict attention to accessory design or model selection. Equipment was of poor quality, not all kinds of supplies were available, and management was poor. Because of all this many pumps were not equipped rationally, leading to low equipment efficiency, high energy consumption, high operating costs and inadequate economic benefits. (Shuili Dianli Bu 1984b, p. 21.)

One official of the county electricity bureau is attempting to persuade users of electrically powered tubewells to change over from a six-inch pump with a 7.5-kilowatt motor to a more frugal but adequate four-inch pump with a 5.5-kilowatt motor. The farmers have been reluctant to switch, in part because of the lack of experimental data, but also quite possibly because the 7.5-kilowatt motor is better for pumping river water.

- *Management organization.* Despite an occasional lapse into Soviet-style gigantomania in other areas, China has not gone in for the large, publicly operated agricultural tubewells used in Pakistan's SCARP program (O'Mara 1984, pp. 37–50). Before the recent decollectivization (under the "production responsibility systems"), wells were most commonly operated by the production brigade (since 1984 superseded by the "natural village") or production team. Despite their rather small scale, however, China's collectives were not always exemplary tubewell managers:

> Before the responsibility system [in Yucheng], the expenses of well irrigation were all borne collectively by

> the production teams. In that case, no one gave much regard for using water sparingly. After the pump was turned on, they let the water flow uncontrolled over the land and even, on occasion, across the roads. (Ren and Nickum 1984).

Since the production responsibility systems became popular in the 1980s, China's collectives have for the most part withdrawn from the day-to-day operation of tubewells. Nearly all of the tubewells in Yucheng County have been contracted to individuals or households. Many pump sets are now privately owned. The more direct link between performance and reward has given well operators an incentive to hold costs down. But the change in use rights is unlikely by itself to induce anyone who has a choice (that is, who has mobile pumping equipment and surface as well as groundwater sources) to opt for the aquifer. The economics overwhelmingly favor surface pumping.

• *Costs.* Zhang (1981) touches on two separate economic issues—the smaller discharge of wells and the difference between investment costs and operation costs.

> The main reason [wells are not used is that they] yield less water than rivers. In particular, a large number of surface projects are invested in by the state, which collects low fees, whereas the costs borne by the local collectives are relatively high for using groundwater. (Zhang 1981, p. 18.)

Because of the difference in discharge alone, the marginal cost to the user is far lower for water that appears in the ditch:

> We have a total of 12 wells, all dug over a decade ago. Now that river water is available, the wells are no longer used for ordinary irrigation. A well can only irrigate 3–5 *mu* [a *mu* is 1/15 hectare] a day; river water can irrigate 15–20 *mu* with the same pump. (Interview with brigade leader, 13 March 1984.)

This is not to say that surface water flows cheaply, however. The maintenance costs of surface diversions are high, especially from the silt-laden Yellow River. The average economic cost of Yellow River water diverted to Yucheng County, including the desilting of the channels, is about 0.03 yuan per cubic meter ($0.01+; Zuo and others 1984, p. 97).[11] This cost has probably increased substantially because the wages for desilting have jumped in recent years. It is not clear whether the cost of pumping from the ditches is included in the above figure. Yucheng has virtually no gravity irrigation. Even with these adjustments, tubewells seem to cost more. Tests at the Yucheng Experimental District, where pumping may be expected to be relatively efficient, indicated an average cost of 0.06 yuan per cubic meter for tubewell water (0.009 yuan per cubic meter for energy, 0.051 for labor; Du 1984, p. 53).

Irrigators in Yucheng, whether collective or private, have clearly not been eager to use tubewells when surface supplies are available. Of the more than 5,000 tubewells dug in Yucheng over the years, one-third are officially recorded as unusable, either because they were improperly dug or because they have not been properly maintained. Many of the remaining wells are incorrectly reported as usable. At least one brigade knowingly dug worthless wells into its saline aquifer in the late 1960s to acquire otherwise unobtainable pump sets from the state supply system. It then used the pumps to draw water from a nearby river.

Policy Options

Although well irrigation costs more, many of China's decisionmakers worry that neglecting wells in areas where they could be used will likely aggravate salinization problems. Yucheng County has received a loan from the World Bank to reduce salinization. Most portions of the county, notably including the experimental area, have been zoned to rely exclusively on groundwater, both to tap Yucheng's rich aquifers and to control the water table (see Map 6-3).

Map 6-3. *Water Use Zones of Yucheng County*

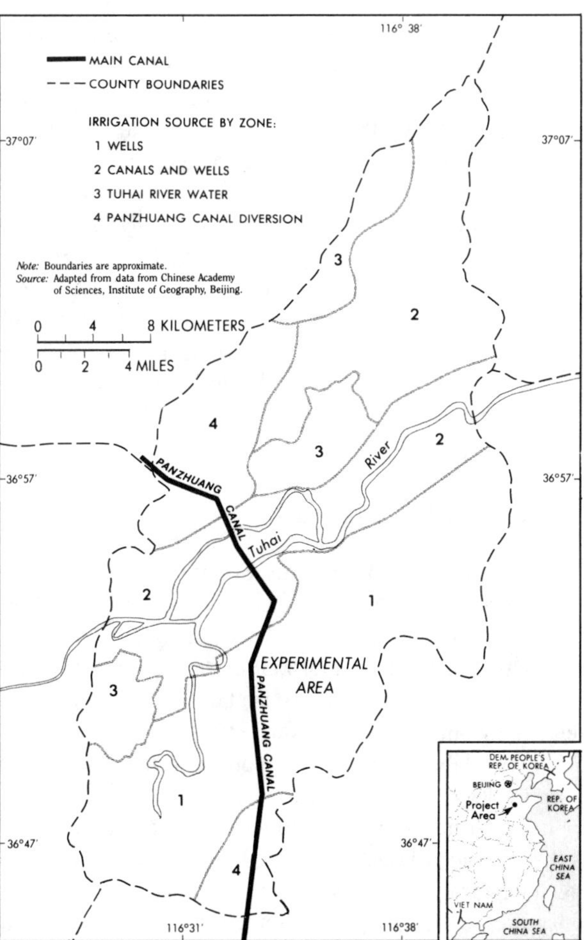

The problem is how to persuade the local irrigators to go along.

The state has adopted three approaches: increasing the cost to the user of surface water, providing subsidies for well use, and denying access to surface water to areas zoned for exclusive groundwater use. Only the last approach seems to have been clearly effective.

Surface Water Charges

Until very recently, surface water in most of China was provided by the state without charge to the user. Although some reservoirs levied water fees, few if any diversion channels did. With the reduction in central government subsidies for irrigation since 1978, many water control bureaus in the prefectures turned to user charges for revenue. Dezhou Prefecture, for example, which includes Yucheng, levies 2 yuan per *mu* (about $11 per hectare) a year for areas watered by the Panzhuang Canal. The prefecture collects this from the county, which includes it in the general tax and fee deductions made when the farmers deliver their grain and cotton to the country procurement stations. The county turns 70 percent over to the prefecture, keeps 20 percent for its water control bureau, and gives 10 percent to the communes (now towns or townships).

Until 1984, the charge did not vary with use and therefore served more to recover costs than to encourage economy. Dezhou authorities indicated that they would begin in 1984 to charge the counties for the volume of water delivered in the main canal. In principle, this could have led to more sparing use of water, since it is the county that requests water from the canal authorities. But there was no evidence of a shift within Yucheng County from an acreage charge to a volumetric one. In any case, inadequate control and measurement of surface flow virtually rules out volumetric charging below the main canal. There are only three measuring devices in Yucheng County, all along main channels. At the farm level, it seems, water charges provide no incentive to use groundwater.

Subsidies for Well Irrigation

Beginning in 1984, Yucheng County was to implement the following measures in areas zoned for well irrigation, according to my notes from a Yucheng county briefing on February 28, 1984: exempt well irrigation zones from the water fee, give them priority supplies of diesel fuel and electricity, and release the farmer using well water from obligatory labor (*yiwu laodong*) on the Yellow River diversion works.

In a later interview, a top county leader mentioned a few other methods being considered, mostly such expensive technological fixes as using more advanced equipment to increase well discharge and lining the canals from well to field. In general, none of these measures promises much success without more effective, and intrusive, supervision and control than are now apparent. Most measures of reducing the cost of well irrigation also reduce the cost of surface pumping. The reference to obligatory labor is a curious provision, since I was informed that no one in Yucheng is subject to such levies.

Withholding Surface Water

In practice, the primary means of ensuring that those who have wells or who are in "well-use zones" do not use surface water is to withhold it, through scheduling and, more drastically, through engineering measures:

> The county controls the supply of surface water to the well irrigation districts. Some places do not have enough wells. When they dig them, the county water conservancy bureau will control their surface supply. Now they still need surface water. The county should not supply surface water to those places with fully equipped wells. (Interview with Yucheng county leader, June 4, 1984.)

> The number 6 branch [of the Panzhuang Canal] was levelled. The experimental station is set up for well irrigation, but the farmers there did not use their wells. This affected the operations of the experimental station. As this is a poor area, the farmers complained [about the levelling] to the staff and workers of the Huang He Diversion Management Office. [Their complaints] were reflected in a difference of opinion among the leaders about well vs. surface irrigation policies. (Interview with Engineer Wu, Yucheng Water Control Bureau, May 15, 1984.)

In the latter instance, the state was able to centralize property rights over diversion water. The affected irrigators had no enforceable rights. The state gave water to them when it built the number 6 branch and took water away when it tore the branch down. This characterization obscures important conflicts of interest among state agencies, however.

The Politics of Diversion

The problem of optimizing the conjunctive use of surface and groundwater has provoked major debates within Yucheng County and between the county and the Panzhuang Irrigation District Management Office, which is under the Dezhou Prefecture Water Conservancy Bureau. Interestingly, the farmers affected by the closing of the number 6 branch complained not to the county but to the Panzhuang office. This office had an interest in retaining the number 6 branch and had also planned a number 8 branch, never built because of county resistance.

Surface water proponents seem to regard the concern over salinization largely as a red herring, probably raised by the county because it is unwilling to pay for a gravity delivery system below the main canal. Currently, the drains are used to store water, but this impedes drainage and requires an extra stage of pumping. Panzhuang's chief manager would just as soon "pump the river dry" if he could. He probably could. The Yellow ceased flowing in its lower reaches in seven of the ten years from 1972 to 1982 (Tong and Zhao 1982, p. 8).

The predominant view in the county, backed by the central research units which operate the experimental station, is that salinization can be avoided only by relying exclusively on wells in about one-third of the county (see Map 6-3). This view has prevailed in the removal of the number 6 branch and in the lining of a small stretch of the Panzhuang Canal which flows along the high border of the experimental area. The entire remainder of the Panzhuang Canal is unlined.

A Final Word

It is difficult to evaluate the efficiency of Yucheng's approach to surface and groundwater use, given a natural, political, and economic environment that is so variable and so imperfectly measured. Still, the bottom line is that crop yields and farm incomes have increased markedly since 1979, while the saline area has declined significantly.[12] Something is working despite, perhaps even because of, agency conflicts of interest and a certain amount of anarchy among irrigators.

Notes

1. See, for example, Smil 1984, especially chapter 3. For a comparison with the Soviet Union, see Goldman 1972. The other two approaches listed by O'Mara are corrective taxes or subsidies and the assignment of clearly defined property rights, including liabilities for damage and rights to compensation.

2. The principal irrigation-related externalities on the North China Plain are salinization and groundwater overdraft. The focus here is on the former, as conjunctive use of surface and aquifer sources is a more likely issue in areas prone to salinization. Cai's contribution to this chapter on the People's Victory Canal also concerns an area with abundant groundwater and a proclivity toward salinization. Aquifer depletion is a serious problem in the northern half of the plain, especially near large cities. In 1983 there were seventy-two recorded cones of depression with a total surface area of 23,921 square kilometers (Li 1984, p. 9).

3. *Zhongguo Nongye* 1980, p. 349. For a more detailed description of the hydrological conditions of the North China Plain, see the Huang's contribution to this chapter. For even more detail, see Biswas and others 1983. There is some disagreement in Chinese sources over the extent of the North China Plain. Zhang (1981, p. 11) and Huang, for example, use a somewhat smaller figure of 300,000 square kilometers, derived at least in part by deducting the Shandong highlands.

4. The Chinese term *lao*, sometimes translated as "waterlogging," is here rendered as "flooding." It refers to the flooding of a field which is directly related to local precipitation and a high water table.

5. The coefficient of variation of annual surface runoff exceeds 0.8 in most of the North China Plain. Up to 70 percent of annual runoff occurs in the summer months (Biswas and others 1983, p. 107).

6. In most other provinces for which we have figures, the irrigated area declined between 1932 and 1949, mainly a period of war, except in Hebei and Shandong, where it increased somewhat (19 percent and 47 percent respectively), although from a very low base.

7. The Tuhai and Majia rivers have their own basins, but are relatively small and lie between the Yellow and the Hai basin proper. They are therefore administratively included in the domain of the Hai River Commission and were widened, deepened, and straightened in the Hai River rehabilitation program of 1964–70.

8. In 1973 an apparent change in the upstream operating procedures of the Sanmen Gorge Reservoir to allow the early spring season storage of 1.2–1.4 cubic kilometers of water also promoted downstream diversions by augmenting the dry season flow in the Yellow River (Shuili Dianli Bu 1984a, p. 68).

9. In net terms at least. The number of tubewells, all powered by diesel engines, was thirty-seven in both 1958 and 1962.

10. Although the official reporting category is "saline-alkaline area" (*yanjian di*), in this area alkaline soil is rare. An indication of the imprecision of salinization figures and, perhaps, their relative unimportance in the early years is that the same figure was reported for all years between 1949 and 1957.

11. Shandong Province's annual cost for desilting when diverting 9 cubic kilometers of Yellow River water is estimated to be 80.50 million yuan (about $28.75 million). Each 100 cubic meters of water contains approximately 2.3 cubic meters of silt (Zuo and others 1984, p. 97).

12. Other factors presumably involved in the decline in the saline area are a string of consecutive dry years (more precisely, given Yucheng's precipitation patterns, the absence of a very wet year) and the 33 percent decline in the sown area of winter wheat, the most frequently irrigated crop, between 1978 and 1982 (from 35,440 to 23,910 hectares).

References

Biswas, Asit K., Zuo Dakang, James E. Nickum, and Liu Changming, eds. 1983. *Long-Distance Water Transfer.* Dublin: Tycooly.

Du Wei. 1984. "Nanshui beidiao dongxian gongcheng dui huanjing yingxiang xitong fenxi" [A systems analysis of the environmental effects of the eastern route of the project to transfer Chang Jiang water northward]. Master's thesis, Institute of Geography, Chinese Academy of Sciences, Beijing, October.

Goldman, Marshall I. 1972. *The Spoils of Progress: Environmental Pollution in the Soviet Union.* Cambridge, Mass.: MIT Press.

Li Boning. 1984. "Wei gonggu he fazhan nongcun dahao xingshi yao dali jiaqiang nongtian shuili jianshe" [It is necessary to make great efforts to strengthen on-farm water works so as to consolidate and develop the excellent situation in the rural areas]. *Zhongguo Shuili* [Water conservancy in China] 1 (January): 7–10.

Nickum, James E. 1987. *Irrigated Area Statistics in the People's Republic of China.* Washington, D.C.: International Food Policy Research Institute.

O'Mara, Gerald T. 1984. *Issues in the Efficient Use of Surface and Groundwater in Irrigation.* World Bank Staff Working Paper 707. Washington, D.C.

Ren Hongzun and James Nickum. 1984. "China's Responsibility System and Farmer Participation in Water Management: The Example of Yucheng County, Shandong." Paper presented at the Expert Consultation on Irrigation Water Management, Yogyakarta, Indonesia, July 16–21.

Shuili Dianli Bu Bangong Ting Xuanchuan Chu [Propaganda Department, General Office, Ministry of Water Conservancy and Electric Power]. 1984a. *Xiandai Zhongguo Shuili Jianshe* [Hydraulic construction in modern China]. Beijing: Shuili Dianli Press, September.

Shuili Dianli Bu Nongtian Shuili Si [Farmland Water Conservancy Department, Ministry of Water Conservancy and Electric Power]. 1984b. "Jijing jieneng jieshui he waqian gaizao" [Saving water and energy through refurbishing tube wells]. *Zhongguo Shuili* 2:21–22.

Smil, Vaclav, 1984. *The Bad Earth: Environmental Degradation in China.* Armonk, N.Y.: M.E. Sharpe.

Tong Linliang and Zhao Minzhong. 1982. "Chongfen liyong Huang He shui ziyuan wei beifang jingji jianshe fuwu" [Thoroughly utilize the water resources of the Huang He to serve economic construction in the north]. *Zhongguo Shuili* 2:8–9.

Wittfogel, Karl A. 1957. *Oriental Despotism.* New Haven, Conn.: Yale University Press.

Xue Muqiao. 1981. *China's Socialist Economy.* Beijing: Foreign Languages Press.

Zhang Tianceng. 1981. "Huabei pingyuan di shuili jianshe yu fazhan nongye di tujing" [Water construction and the path to develop agriculture on the North China Plain]. *Nongye Jingji Wenti* [Agricultural economics] 7:11–18. Zhang is affiliated with the National Resources Comprehensive Survey Committee of the Chinese Academy of Sciences.

Zhang Weizhen. 1983. "The Development of Groundwater for Irrigation in China." Wuhan: Wuhan Institute of Hydraulic and Electric Engineering.

Zhongguo Nongye Dili Zonglun [Survey of China's agricultural geography]. 1980. Economic Geography Research Office of the Institute of Geography, Chinese Academy of Sciences. Beijing: Kexue Press.

Zuo Dakang, Liu Changming, Xu Yuexian, and Du Wei. 1984. "Huang He yi bei diqu dongxian yinjiang wenti di tantao" [An exploration of issues in diverting Chang Jiang water along the eastern route into districts north of the Huang He]. *Dili Yanjiu* [Geographical research] 3 (June): 92–98.

Appendix A. Abstract of the Hydrogeologic Map of the North China Plain

Fei Jin

One way to improve agricultural production in the North China Plain is to store surface water underground. Hydrogeological conditions are a factor in determining whether a given area is suitable for this purpose. In light of the program to divert water from the Yellow and Yangtze rivers into North China along the eastern channel route, the objective of this hydrogeologic map of the Huang-Huai-Hai Plain is to determine roughly the areas in which surface water could be stored and estimate the storage capacity of these areas.

Map coverage extends from the ground surface to the lower boundary of the shallow freshwater aquifer in the plain. Areas considered suitable for storing surface water are those in which the total thickness of all sand layers of the profile exceeds 15 meters, or the value of $\Sigma\mu_i h_i$ (μ_i = the specific yield of the ith sand layer, h_i = thickness of the ith sand layer) exceeds 1.2, and all the sand layers (or the principal sand layer) of the profile are buried near the ground surface, with only a thin impermeable covering. In places containing saline groundwater, an area with isolated saline groundwater distribution and sand layers that meet the aforementioned conditions would be suitable for underground storage of surface water. The amount of water that could be stored underground is that amount which could be stored from the groundwater table with the minimum evapotranspiration loss to the depth to which the groundwater table is expected to be lowered during the annual exploitation, minus the natural replenishment of groundwater.

In all there are seventy-four areas suitable for underground storage of surface water in the North China Plain. The total amount that can be stored throughout the plain is 9.84 billion cubic meters a year. The coastal plain is not suitable for this purpose. Thus, on the whole, the prospects for storing surface water underground in the North China Plain are considerable and should not be overlooked. But the prospects for underground storage in the eastern part of the plain are not bright and do not favor the program to divert water from the Yangtze and Yellow rivers into North China along the eastern channel route.

Appendix B. Conjunctive Use of Surface and Groundwater in the Lower Reaches of the Yellow River

Jiang Ping

The conjunctive use of surface and groundwater is considered an effective solution to the scarcity of water resources for irrigation. Although the efficiency of some of the irrigation projects studied in China ranges from 35 to 63 percent, most of the systems are much less efficient. A large percentage of irrigation water is lost through deep percolation. This not only is a waste of precious water resources but also raises the water table and aggravates saline conditions.

As a remedy, wells are used in conjunction with surface irrigation systems. To maintain the water balance in the system, use of the surface and groundwater supplies is coordinated in any of three ways:

- At the level of the production team, irrigation water delivered by the distribution network is allocated to water as extensive an area as possible, with the remaining land watered with groundwater managed by the production team.
- Irrigation water delivered by the distribution network is allocated to water the land within a certain reach of a canal, with the remaining land watered with groundwater managed by the production team.
- Farmland is watered with a composite supply of surface and groundwater. Groundwater is pumped up and combined with the surface water in the canal.

There is another reason for developing conjunctive use of surface and groundwaters in the North China Plain: to alleviate the problem of sedimentation in the canals and drains. Each cubic meter of irrigation water from the Yellow River and its tributaries usually carries no less than 10–20 kilograms of silt. Inevitably, the silt accumulates in the canals and drains, impeding the flow of water. The greater the volume of water diverted for irrigation, the greater the problem of sedimentation in the canals and drains. Watering some crops with groundwater would reduce the quantity of water diverted and thus minimize the problem.

A study of irrigation management in the lower reaches of the Yellow River focused on the People's Victory Canal system in Henan Province. The system diverts about 490 million cubic meters of water from the Yellow River each year, but only 167 million cubic meters of this is productively diverted. A field study showed that about 81 percent of the water diverted for irrigation was lost—42 percent to seepage and percolation from conveyance and distribution systems, 14 percent to infiltration into farmland, and 25 percent to runoff through the drainage system. In addition, sediment build-up in the canals and drains amounts to 9.6 tons of sediment per hectare per irrigation in this area.

Estimates indicate that the water intake could be reduced to 280 million cubic meters by improving on-farm application and reducing water losses through the system. This would still leave the loss of 113 million cubic meters, or 40 percent of the water diverted, to deep percolation. With full utilization of percolated water as a second source of supply, water intake from the river could be cut to 140 million cubic meters, which would also reduce the amount of silt carried into the system.

Field investigations indicate that 12,741 hectares of land were irrigated during one 226-day irrigation season (February 16 to September 30, 1981). It has been estimated that the groundwater table would have risen 0.48 meter if this area had been irrigated with surface water only. In fact, however, only 5,661.2 hectares, or 44 percent, was irrigated with surface water; the other 56 percent was watered with groundwater. As a result the groundwater table rose only 0.23 meter. With the utilization of groundwater, the diversion of surface water was reduced by 56 percent, as was the diversion of silt.

Nevertheless, the merits of conjunctive use of surface and groundwater in the Yellow River basin are closely tied to the history of irrigation development in the area. Three factors peculiar to local conditions are particularly significant.

- The main canal, which is 53 kilometers long and has a flow discharge of 80 cubic meters per second, was designed as an earth canal. At present, lining or reconstructing the canal to eliminate seepage would require additional capital investment.
- In the early 1960s when diversion of surface water was halted owing to secondary salinization in the irrigated area, many wells were dug for irrigation. There are 4,250 wells with an overall discharge of 200 million cubic meters a year. Farmers can use these wells for irrigation.
- The introduction of paddy rice in this area sharply increased the demand for water. By using wells as a supplementary source of water for irrigation, farmers can reduce the volume of water diverted from the river and thus improve the operation and maintenance of the existing system.

Comment

Bruce Stone

Huang, Cai, and Nickum each discuss conjunctive water use systems that have been developed in China since the 1950s. While Huang discusses the North China Plain in general, Cai and Nickum focus on specific parts of the plain—the People's Victory Canal irrigation system and Yucheng County in Shandong. All three refer to the problem of increasing soil salinity, which in this region is characterized by a high water table, seasonally and sub seasonally concentrated precipitation, and collections of salts dangerously near the soil's surface. Huang and Nickum note that the area is subject to a not infrequent natural cycle of periods of excessive rainfall, which raise the water table, followed by very dry periods during which the salts are drawn further toward the root zone by the strong pull of evaporation. Cai and Nickum also emphasize man-made causes: overirrigation, irrigation development without complementary drainage development, and canal seepage. All three discuss the role that tubewell construction and operation have played in reducing salinization of farmland in the region and in generally increasing farm yields.

It cannot of course be inferred that the rapidly increasing crop yields are caused primarily by desalinization efforts. Clearly, the dissemination of high-yielding crop varieties, expanded and improved irrigation, and rapidly increasing use of manufactured fertilizers are directly responsible (Stone 1983, 1985, 1986a). But moving away from predominant dependence on surface irrigation and focusing increasingly on drainage since the late 1960s have allowed irrigation to expand without repeating the regional salinization disasters of 1959–61.

Only Nickum refers to the roles that adjusted cropping patterns and a succession of flood-free years may have played in reducing salinity: he notes a 33 percent decline in winter wheat area in Yucheng County between 1978 and 1982 and points out that winter wheat is the most frequently irrigated crop. Wheat also requires irrigation (with the risk of overirrigation) during the driest months, when a high water table can most readily result in salinization on the North China Plain.

During the 1960s and the early 1970s, the Chinese government was unable to purchase enough wheat from the countryside for North Chinese cities and had problems transferring grain among provinces and counties. Consequently, the government engaged in a variety of now controversial policies, such as aggressively promoting the cultivation of winter wheat and enforcing local self-sufficiency. High-yielding wheat varieties and surface and groundwater irrigation raised the attractiveness of increasing the wheat-sown area, while the policy of local self-sufficiency, bolstered by rural trade restrictions, crowded out all but the highest-yielding staple crops in this impoverished region and severely affected the production of economic crops, especially cotton.

Further supported by direct controls on sown area, price incentives for wheat cultivation, and fertilizer allocations in exchange for wheat sales, these policies increased wheat area (and average yields) in Shandong and Hebei from a trough of 3.74 million hectares in 1962 to a peak of 6.8 million hectares in 1976–77 (Stone and others 1985). They may also have accelerated soil salinization.

North China's salinization history during the past three decades, sketched broadly by Nickum, may merit clarification. Some salinization occurred naturally because of excessively wet summers—especially those which resulted in a high water table and subsurface accumulations of salt and were followed by particularly dry springs. From 1958 to 1960, however, salinized farmland increased from 1.9 million hectares to more than 3.2 million hectares as a direct result of the overly ambitious and poorly designed and administered Yellow and Huai River irrigation efforts. The area of salinized farmland was reduced to 1.4 million hectares during the succeeding decade by completely retiring some farmlands, ceasing surface irrigation in many areas, and developing drainage, reservoir, and tubewell irrigation systems (Stone 1983, p. 204).

Grain production in North Chinese provinces did not exceed 1957 levels, however, until the late 1960s or early 1970s, despite a much increased share of planted area. Total cultivated area in these provinces appears never to have regained the 1956–57 levels. More ominously, the net decline in salinized farmland that began about 1962 was reversed in the early 1970s, presumably owing to the renewed cultivation of partially desalinized lands, the acceleration in irrigation development during the late 1960s and early 1970s, and perhaps the increase in irrigated winter wheat cultivation. Not only did the area of saline farmland increase to 1.9 million hectares, with an additional 4.7 million hectares in danger of salinization, but the degree of salinity increased on 80 percent of existing saline farmland during the 1970s (Stone 1983, pp. 202–5).

Although available data are somewhat inconclusive, it appears that salinity decreased in much of North China during the late 1970s and early 1980s. Areas with increased soil salinity have been described as primarily canal seepage zones. But they also include lowlands with high water tables adjacent to overirrigated areas (Stone and others 1987). Massive commitments to import wheat were made to provide food security for Chinese cities, and restrictions on rural grain trade were relaxed, which allowed greater local flexibility in cropping.

Prices were raised for all farm products, but especially for cotton, and cotton sales were rewarded with particularly high allocations of chemical fertilizers. A very high-yielding, low-grade cotton cultivar (Shandong number 1) became available; with price incentives favoring quantity over quality and with state-guaranteed purchases, cotton producers prospered. As the rewards for cotton increased, cotton cultivation was also directly promoted through area planning, despite generally decreasing state interference in farmers' production decisions after 1978 (Stone 1985, 1986a, and 1987).

As a result, cotton area increased 43 percent between 1977 and 1984, while yields grew 114 percent (SSB 1983, pp. 159 and 163; SSB 1985, p. 264). This growth was concentrated in North China: cotton, even in 1984, was sown over an area totaling only 63 percent of Chinese wheat area, but its proportion relative to wheat in North China was twice that amount (Zhongguo Nongye 1985, pp. 147 and 150) and was even greater in many saline-prone areas targeted for cotton. The move to cotton was accompanied by a recovery of soybeans, sorghum, and other economic crops and vegetables, which led to a decline in winter wheat area in Hebei and Shandong of around 650,000 hectares, or almost 10 percent of the 1977 wheat-sown area (Stone and others 1985). These statistics do not fully reflect the proportionate reductions of winter wheat in saline-prone areas of these provinces which, to some extent, were undertaken deliberately, as part of agricultural regionalization efforts.

The extent to which farm production growth depended on the control of soil salinity is unclear for North China as a whole and even for the specific locations discussed. Nor does the material provided indicate how much recent progress in decreasing soil salinity can be attributed to policies and institutions that encourage conjunctive use of surface and groundwater sources and how much to the serendipitous succession of consecutive dry years beginning in 1977 or to shifts in cropping patterns. More thorough and up-to-date analysis would be useful.

Clearly, North China's efforts to construct drainage facilities and tubewells have allowed irrigation to expand without a repeat of the massive soil salinization of the early 1960s and have promoted China's recovery from that disaster. Current policy represents an improvement especially in contrast to the disastrous policies of a previous period. Similarly, current agricultural successes in North China have resulted in part from the elimination of earlier counterproductive policies. Those policies had kept prices for farm products punitively low, restricted access to fertilizers, enforced local self-sufficiency while forgoing gains from specialization, weakened the link between work and compensation for individual farmers and farm families, and overpromoted winter wheat development in environmentally inappropriate areas (Stone 1980, 1987; Lardy 1983a).

Does China's current institutional structure have anything to offer other countries? Will it continue to support China's own growth in farm production (and more questionably, grain production) and its efforts to control soil salinity? Can the current structure succeed in expanding irrigation and reducing salinity as well as the previous structure? For each of these questions, Nickum offers some justification for doubt. However, China has faced risks and limitations in developing North China and has given its most serious problems more attention than many other developing countries have given theirs. Chinese leaders have also demonstrated a somewhat greater willingness to initiate institutional change.

It may be useful here to summarize sources of externalities relating to irrigation development in China and to point out some potential conflicts of interest.

- *Upstream and downstream irrigation.* Upstream users may exhaust water supplies during periods of scarcity and dump excess water during periods of plenty. Such external costs are typical of many systems throughout the world; they exist in North China, as in the Indus Basin and the southwestern United States, with the added dimension of salinization risk. This behavior imposes unequal costs on upstream and downstream users. Consequently, well development is concentrated in downstream locations despite the higher average costs for tubewell irrigation than for surface irrigation. The fragmentation of accounting units that is part of current reforms increases the complexity of acceptable resolution processes for this and other sources of externalities.
- *Seepage-induced salinization.* Farms immediately adjacent to major canal works not only have borne the larger costs of losing land to irrigation development, but also have subsequently suffered unforeseen soil deterioration from seepage. They often have, however, greater security of access to irrigation flows.
- *Surface and groundwater irrigation.* Water fees, where they exist, remain a very minor portion of production costs in North China. Subsidized diesel fuel and electricity are allocated to some farmers and are relatively unavailable to others (Lardy 1983b, pp. 62–64; Stone 1987). In some areas that lack adequate surface alternatives, relative prices are such that farmers have little disincentive to overpump. In urban areas, fuel pricing is complex, but again there appears to be little disincentive to overpump in sensitive areas. As elsewhere, collection of water fees poses a problem in China. Yet Ren (1987) provides evidence of almost 50 percent compliance with water charges within five years of their initiation in Dezhou Pre-

fecture, Shandong. This rather impressive record of compliance is probably due to rapidly rising farmer income during the period, especially for cotton cultivation. Conversely, in other areas, individual farmers have little incentive to pump from a large aquifer when the risks of a high or rising water table suggest external benefits from doing so. This is more likely to be the case in regions serviced by both surface and groundwater facilities and is accentuated where relative prices of energy are high and its dependability low. Attempts have been made to address this issue, but Nickum notes problems of applicability and execution. As decisionmaking becomes more decentralized under current rural policy, the choice of instruments for addressing problems with surface and groundwater irrigation may become more restricted. Ren's (1987) study of Yucheng County, Shandong, illustrates the importance of location-specific policy formulation, even within county jurisdictions.

• *Crop selection.* Farmers in historically drought-stricken regions with new irrigation facilities, such as North China, typically overirrigate when water becomes available. External costs for each watershed or aquifer area (especially in the lowest-lying areas) are related to the crop choices of individual farmers. For example, the potential long-term costs of salinization resulting from winter wheat cultivation will be imperfectly reflected in season-to-season production costs and revenues. These external costs will vary by year, by the type of facilities employed, and by the crop selections of other farmers. Similarly, there may be external benefits associated with crop choice. These costs and benefits are partially, but not completely, congruent with the externalities of upstream-downstream use and surface-groundwater use.

• *Water and soil quality control.* During the 1960s and 1970s labor was organized locally to develop drainage for salinization control, to level and terrace land, and to construct irrigation systems at extremely low, daily compensation. Such labor-intensive methods were overused, especially for large, poorly designed, poorly administered, or geographically distant projects. They were heavily attacked during the mid-1970s and officially repudiated (Stone 1980). Currently, grain prices are still artificially depressed by government purchasing organizations, although purchasing is no longer a complete government monopoly. So Chinese society reaps external benefits through lower urban grain prices from all agricultural investment, especially poorly remunerated farmland capital construction works, including the massive and continuous collection and application of organic manure. With decreasing state investment in these kinds of activities, increasing opportunity costs of labor, and increased but still poorly developed interrural trade, it is unclear how this work will be accomplished and the extent to which it needs to be (Stone 1986a, 1986c). Policy initiatives to provide subsidies for these efforts, which help to compensate for artificially depressed prices, are financed from local funds generated by a roughly 25 percent tax on township enterprise profits and some state remittance of county industry profits, and the work is to be performed by specialized labor groups and companies (Stone 1986b and Zu 1986). The success of these initiatives remains to be seen.

• *Water authorities.* Water authorities in China historically developed as flood control organizations that paid little attention to irrigation. During the twentieth century, irrigation developed initially as an auxiliary to flood control. During the latter part of the century China expanded irrigation through the construction of basic facilities. But little attention was paid to water use or drainage within the command areas. The need for drainage facilities was typically addressed only after problems developed, often costly ones. These deficiencies as well as more chronic problems of underutilization and poor maintenance were pointedly criticized during the late 1970s (Stone 1980). The current emphasis is on increasing the use of existing facilities and promoting more efficient water management among farmers, but Huang, Cai, and Nickum give us little indication of how this is happening. According to a study by Ren (1987), there is little dynamism in farmers' water management organizations. This problem may not be overcome until such organizations have genuine authority, such as decisionmaking power over the use of water fees. By contrast, the World Bank has noted both substantial farmer participation and control over water fees in irrigation districts of the Northwest, such as He Tao in Inner Mongolia and the Yellow River area of Ningxia.

• *Regional and provincial conflicts.* Huang brought up the question of Nanshui Beidiao, the controversial Interbasin Water Transfer Project that would move water from the Yangtze River (Chang Jiang) by canal to the North China Plain for urban, industrial, agricultural, and aquacultural use. The introduction of large additional quantities of water onto the plain not only accentuates many of the externalities discussed above, but introduces conflicts among the provinces and within basins. The project would adversely affect the aquacultural product value along the Yangtze shallows and in the shallow Jiangsu and southern Shandong lake system through which one of the transfer routes would pass. In dry years, salt already pollutes the farmland and underground aquifers in the Shanghai municipal area. The transfer of such a large volume of water is expected by some to exacerbate this problem, which will, in any event, become more serious with increasing development in the basin. The lack of control over surface and groundwater development in North China following the period of aggressive state promotion may threaten continued agricultural growth. The risk of salinization with the introduction of large quantities of additional water has already been discussed. But water is chronically scarce in the Tianjin municipal area and nearby localities; riverbeds are dry during much of the irrigation season, even for important rivers such as the Huai and Hai; and groundwater levels have fallen rapidly or aquifers have collapsed in some other areas of the plain (for example, Heilonggang and Shijiazhuang). The data and methods used to analyze the costs and benefits of such a huge undertaking are crude, although they could hardly be described as precise elsewhere, and large costly errors are possible. The structure for resolving these disputes is also limited, particularly because the project does not fully depend on central government support. The provinces of Jiangsu and Henan are developing the transfer within their own boundaries, forcing the state to take a position if access is to be preserved for more seriously imperiled provinces and municipal areas to the north. The current trend of planning and discussion advocates strict limitation of the use of transferred water for agricultural purposes, to the extent that this can be accomplished administratively.

In conclusion, to resolve these conflicts China will need better data and data analysis and more flexible administration than it has had in the past. Signs for the future are favorable: some Chinese policymakers, research institutions, and planners are addressing and evaluating, however imprecisely, many of the externalities discussed. China in the mid-1980s welcomes innovation much more than it did during most of the 1960s and 1970s, and more than many other societies do today. One should not be surprised to discover neglected responsibilities and unforeseen complications during such periods of adjustment. But the next decade will reflect the results of important current tests for water management in North China.

References

Lardy, Nicholas R. 1983a. *Agriculture in China's Modern Economic Development.* Cambridge: Cambridge University Press.

———. 1983b. "Agricultural Prices in China." *World Bank Staff Working Paper* 606, Washington, D.C.

Ren Hongzun. 1987. "North China Plain Irrigation Water Management Issues: An Examination of Yucheng County." Paper prepared for the Institute of Geography, Chinese Academy of Science, and the International Food Policy Research Institute, Washington, D.C. May. Processed.

SSB. State Statistical Bureau of China. 1983. *Statistical Yearbook of China, 1983.* Hong Kong: Economic and Information Agency.

———. 1985. *Statistical Yearbook of China, 1985.* Hong Kong: Economic and Information Agency.

Stone, Bruce. 1980. "China's 1985 Foodgrain Production Target: Issues and Prospects." In A. M. Tang and Bruce Stone. *Food Production in the People's Republic of China.* IFPRI Research Report 15. Washington, D.C.: International Food Policy Research Institute.

———. 1983. "The Chang Jiang Diversion Project: An Overview of Economic and Environmental Issues." In Asit K. Biswas, Zuo Dakang, James E. Nickum, and Liu Changming, eds. *Long Distance Water Transfer: A Chinese Case Study and International Experiences.* Dublin: Tycooly.

———. 1985. "The Basis for Chinese Agricultural Growth in the 1980s and 1990s: A Comment on Document No. 1, 1984." *China Quarterly* 101:114-21.

———. 1986a. "Chinese Fertilizer Application in the 1980s and 1990s: Issues of Growth, Balance, Allocation, Efficiency and Response." In U.S. Congress, Joint Economic Committee. *China's Economy Looks Toward the Year 2000,* vol. 1, *The Four Modernizations.* Washington, D.C.: U.S. Government Printing Office.

———. 1986b. Unpublished field notes from Yucheng and Lin counties, Shandong. Presented at the International Seminar on Rural Employment Promotion Strategies co-sponsored by the International Labour Organisation and the Chinese Ministry of Labor and Personnel and the Rural Development Research Center, Beijing, April 1–13.

———. 1986c. "Food and Agriculture in the Context of Rural Employment Promotion Generation Problems in China and Other Developing Countries." Paper presented at the International Seminar on Rural Employment Promotion Strategies co-sponsored by the International Labour Organisation and the Chinese Ministry of Labor and Personnel and the Rural Development Research Center, Beijing, April 1–13. Processed.

———. 1988. "Relative Foodgrain Prices in the People's Republic of China: Rural Taxation through Public Monopsony." In John Mellor and Raisuddin Ahmed, eds. *Agricultural Price Policy for Developing Countries.* Baltimore, Md.: Johns Hopkins University Press.

Stone, Bruce, Charles Greer, Tong Zhong, Clark Friedman, Mary McFadden, and Melanie Snyder. 1985. "Agro-Ecological Zones for Wheat Production in China: A Compendium of Basic Research Materials." Washington, D.C.: International Food Policy Research Institute. Processed.

Stone, Bruce, Ren Hongzun, Jiang Dehua, and Chuck Gitomer. 1987. "Water Management and Salinity Control in North China." Paper prepared for a seminar at the International Food Policy Research Institute, Washington, D.C., April 20. Processed.

Zhongguo Nongye Nianjian Bianji Weiyuanhui [Agricultural Yearbook of China Editorial Committee]. 1985. *Zhongguo Nongye Nianjian 1985* [Agricultural Yearbook of China 1985]. Beijing: Nongye Chubanshe [Agricultural Publishing House].

Zu Guobu. 1986. "The Economic and Social Situation and the Orientation of Development in the Countryside of China." Paper presented at the International Seminar on Rural Employment Promotion Strategies co-sponsored by the International Labour Organisation and the Chinese Ministry of Labor and Personnel and the Rural Development Research Center, Beijing, April 1–13. Processed.

Part III

Analytical Methods and Applications

7

A Review of Groundwater Management Models

Steven M. Gorelick

In the past two decades the field of groundwater hydrology has turned toward numerical simulation models to help evaluate groundwater resources. The application of finite difference and finite element methods to groundwater flow equations has permitted complex, real world systems to be modeled. Numerical simulation models have enabled hydrogeologists to develop a better understanding of the functioning of regional aquifers and to test hypotheses regarding the behavior of particular facets of groundwater systems. The simulation method has provided a framework for conceptualizing and evaluating aquifer systems. Models have become tools to evaluate the long-term impacts of sustained water withdrawals, groundwater–surface water interaction, and the migration of chemical contaminants.

Clearly, simulation as both a method to explore hydrogeologic problems and a tool to predict effects upon groundwater systems will continue to be essential to hydrologists and to water managers. Simulation models are often used, however, to explore groundwater management alternatives. In such cases a model is executed repeatedly under various design scenarios which attempt to achieve a particular objective, such as isolating a plume of contaminated groundwater, preventing saltwater intrusion, dewatering an excavation area, or obtaining a sustainable water supply. Such an approach often sidesteps rigorous formulation of groundwater management goals and fails to consider important physical and operational restrictions. Determining the proper objective function in a groundwater management model is often difficult but is an essential aspect of management modeling and should not be avoided. Optimal management alternatives are unlikely to be discovered using only simulation techniques. What is required is not a simulation model alone, but a combined simulation and management model. Joint simulation and management models have recently been developed. A combined model considers the particular behavior of a given groundwater system and determines the best operating policy under the objectives and restrictions dictated by the water manager.

This review concerns the joint use of groundwater simulation models and optimization methods. Only those management models which simulate groundwater hydraulics or groundwater solute behavior by solving the governing partial differential equations—that is, distributed parameter models—will be discussed. Of particular interest are the uses of linear and quadratic programming along with numerical simulation. Such techniques of mathematical programming are powerful tools for those simulation studies in which management alternatives are being explored. Models using other approaches to groundwater management (Chaudry and others 1974; Cummings and McFarland 1974; Taylor and Luckey 1974; Bockstock, Simpson, and Roefs 1977; Khepar and Chaturvedi 1982) will not be discussed here. A guide to a broad range of groundwater models, including management models, was presented by Bachmat and others (1980).

Methods are first described for aquifer and water supply management. A classification of groundwater management models is presented that divides the approaches into two categories: hydraulic management models and policy evaluation and allocation models. Next, methods for the management of groundwater systems used for both waste disposal and water supply are discussed. These are categorized independently of the above groundwater management models as steady-state and transient groundwater quality management models. This description of methods and applications presented in the literature will perhaps aid hydrogeologists in determining whether they should employ a combined simulation and management model.

Figure 7-1. *Classification of Groundwater Management Models*

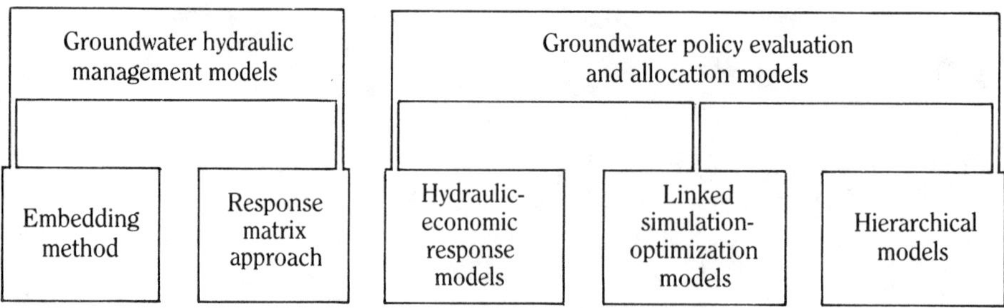

Discussion of areas missing from the literature, in addition to problems with existing methods, point the way to new research directions.

Classification of Groundwater Management Models

Studies combining aquifer simulation with management models may be grouped into two categories (see Figure 7-1). The categories distinguish between models primarily concerned with management decisions about groundwater hydraulics and those concerned with policy evaluation as well as the economics of water allocation. In the first category, models are aimed at managing groundwater stresses such as pumping and recharge. These models treat the stresses and hydraulic heads directly as management model decision variables. The physical decision variables, such as pumping rates, may be interpreted as surrogate economic variables. These are implicit economic considerations. Some models may even contain explicit economic factors, such as pumping and well costs. In any case, models that are principally concerned with managing aquifer hydraulics will be referred to as groundwater hydraulic management models.

The second category involves models that can be used to inspect complex economic interactions such as the influence of institutions upon the behavior of an agricultural economy or complex groundwater–surface water allocation problems. Although these models do not explicitly determine regional groundwater policy, they can be used in policy evaluation. These models are generally characterized by multiple optimizations, one for each subarea in a region, and have a strong economic management component. Furthermore, the series of optimizations may be followed by external checks, simulations, or additional optimizations. These models will be referred to as groundwater policy evaluation and allocation models. Both categories of models employ the optimization techniques of linear or quadratic programming. Such techniques attempt to optimize an objective, such as by minimizing costs or maximizing well production, and are subject to a set of linear algebraic constraints limiting or specifying the values of variables such as local drawdown, hydraulic gradients, or pumping rates.

In both categories the simulation model component of the management models is based upon the equation of groundwater flow in saturated media (Cooper 1966; Pinder and Bredehoeft 1968; Remson, Hornberger, and Molz 1971). For the nonsteady two-dimensional heterogeneous anisotropic case the equation is

$$\frac{\partial}{\partial x_i}\left(T_{ij}\frac{\partial H}{\partial x_j}\right) = S\frac{\partial H}{\partial t} + W \qquad i, j = 1, 2 \tag{7-1}$$

where

T_{ij} = transmissivity tensor, L^2/T
H = hydraulic head, L
W = volume flux per unit area, L/T
S = storage coefficient, L^0
x_i, x_j = cartesian coordinates, L
t = time, T.

While the vast majority of groundwater simulation and management modeling studies have been concerned with two-dimensional flow, some physical systems must be treated in three dimensions. Clearly, management modeling approaches are not limited to one- and two-dimensional equations. As experience is gained with three-dimensional models, application of management methods should be extended.

Groundwater Hydraulic Management Models

Groundwater hydraulic management models incorporate a simulation model of a particular groundwater system as constraints in the management model. Management decisions as well as simulation of groundwater behavior are accomplished simultaneously with two techniques. The embedding method treats finite difference or finite element approximations of the governing groundwater flow equations as part of the constraint set of a linear programming model. Decision variables are hydraulic heads at each node as well as such local stresses as pumping rates and boundary conditions. The response matrix approach uses an external groundwater simulation model to develop unit responses. Each unit response describes the influence of a pulse stimulus (such as pumping for a brief period) upon hydraulic heads at points of interest throughout a system. An assemblage of the unit responses, a response matrix, is included in the management model. The decision variables in a linear, mixed integer, or quadratic program include the local stresses such as pumping or injection rates and may include hydraulic heads at the discretion of the modeler.

Embedding Method

The embedding method for the hydraulic management of aquifers uses linear programming formulations that incorporate numerical approximations of the groundwater equations as constraints. Aguado and Remson (1974) initially presented this technique for groundwater hydraulic management. Their work demonstrated with one- and two-dimensional examples that the physical behavior of the groundwater system could be included as an integral part of an optimization model. Finite difference approximations were used in both steady-state and transient problems. In all examples, physical objective functions maximized hydraulic heads at specified locations. Constraints were placed upon heads, gradients, and pumping rates.

The examples treated confined and unconfined aquifer hydraulics. For the confined case the governing equation is linear, and the resulting finite difference approximations were treated as linear constraints. For the unconfined aquifer case the steady-state equation was treated as linear with respect to the square of the hydraulic heads (see Remson, Hornberger, and Molz 1971). This linear form was then amenable to treatment using the embedding approach.

For the transient confined case the governing equation was discretized over space and time. A set of equations for each time step was included in the linear program as constraints. This approach can result in an extremely large constraint matrix. If a groundwater model consists of 1,000 nodes and thirty time steps are taken, there will be 30,000 decision variables corresponding to hydraulic heads over space and time. An equal number of constraints will represent the finite difference equations.

The transient unconfined case is truly nonlinear and was treated by Aguado and Remson (1974) using the predictor-corrector method of Douglas and Jones (1963). A succession of linear programming problems was solved for the corrector step while the predictor step was solved by direct numerical simulation. The latter approach is limited in that management is possible only from one time step to the next. Objectives spanning long management time horizons are not possible. Aguado (1979) has described the embedding method in detail.

A real world example was presented for determining the optimal steady-state pumping scheme to maintain groundwater levels below specified elevations for a dry-dock excavation site (Aguado and others 1974). The embedding approach was used, as was the linear finite difference form of the Boussinesq equation. The solution to the dewatering problem indicated that the minimum total pumping could be accomplished by developing the maximum number of wells as close to the excavation as the finite difference grid would permit. This dependence of the management solution upon the grid used in the numerical approximation technique was due to the failure of the model to account for setup costs incurred for each well introduced into the solution.

In a later study setup costs were included by reformulating the problem as a fixed charge problem where costs of pumping and the fixed costs of well installation were considered (Aguado and Remson 1980; also see the discussions by Elango 1981 and Padmanabhan 1981, as well as the reply of Evans and Remson 1982). The fixed charge problem must be solved as a mixed integer linear programming problem. In such a problem, certain variables (those corresponding to the setup charge for well installation) were integers. Each integer took a value of one when setup charges were to be considered and a value of zero when no well was to be constructed. When penalties were assessed for well installation and development, the solution lost its strong dependence upon the finite difference grid. Wells were no longer located at every node bordering the excavation area.

The dry-dock dewatering model without fixed charges was subjected to sensitivity analysis to determine which local hydraulic conductivities were most critical to the optimal management of the system (Aguado, Sitar, and Remson 1977). The approach taken was to vary the hydraulic conductivities systematically throughout the domain and observe changes in the minimum total pumping rate as indicated by a series of separate linear programming solutions. The approach was demonstrated using both finite difference and finite element approximations of the groundwater equations.

Alley, Aguado, and Remson (1976) considered aquifer management under transient conditions. Again, the embedding technique was used. The finite difference form of the confined aquifer equation was treated as constraints in a linear program. Their illustrative problem sought to maximize hydraulic heads while obtaining a specified flow rate from wells within the system. The transient behavior was treated by creating successive management models, one for each time step. The optimal solution from one time step was treated as an initial condition for the next management model. In comparison, Aguado and Remson

(1974) lumped all time step models as constraints in one linear program. Their approach enables short-run aquifer management goals to be included in the context of long-run management, but could result in a very large matrix.

Alley, Aguado, and Remson (1976) required optimization of several smaller single time step models. The deficiency of this approach is that management decisions in the short run may contradict long-run management requirements. In fact, stepwise optimization may cause certain long-run management goals to become infeasible. For example, suppose an objective for the first ten-day time step was to maximize well production. To accomplish this, water levels were drawn down to their specified lower limit. The result is that well production immediately following the tenth day would necessarily be less to prevent further water-level declines. The decision made during the first time step would severely limit further decisions.

Willis and Newman (1977) solved a transient embedding problem involving minimization of pumping costs subject to exogenous water demands. The management model was formulated as a problem in optimal control and solved using mathematical programming. The management problem consisted of a nonlinear objective function subject to linear constraints. A finite element model was used to discretize equation (7-1). An algorithm (Tui 1964) which solves a succession of linear programming problems was used to solve the optimization problem.

Schwarz (1976) suggested incorporating as linear programming constraints a series of mass balances among cells which compose an aquifer. In this case the simulation model was identical to that of Tyson and Weber (1964). According to Schwarz (1976, p. 385), "the transformation model represents the properties of the aquifer by water-level transformation equations for discrete cells. These are continuity equations which have to be expressed for all the cells and for all the time intervals included in the planning horizon."

This approach is essentially the same as the embedding approach in that a model which was discretized over space and time was included as linear programming constraints. The difference is that the transformation model depended upon a multicell aquifer model, while the embedding models employed finite differences or finite elements.

The embedding technique has been demonstrated for cases involving hydraulic gradient control using injection and pumping wells. Molz and Bell (1977) applied the technique to a hypothetical case involving the steady-state control of hydraulic gradients to ensure the stationarity of a fluid stored in an aquifer. Their linear program contained constraints which consisted of the confined groundwater flow equation in finite difference form. They constrained gradients to fixed values in the region of their five-point well scheme. The objective function was to minimize the total of pumping plus injection. Solutions were achieved for their eighty-one–node model with minimal computational requirements.

Remson and Gorelick (1980) demonstrated the embedding approach to contain a plume of contaminated groundwater in the context of other regional management goals, including the dewatering of two excavation areas and the obtaining of water for export from the system. The objective was to minimize pumping. The solution selected those nodal locations where either pumping or injection wells should be located. Furthermore, it determined the optimal pumping rates and gave the resulting steady-state hydraulic head distribution over the ninty-nine active nodes.

Although management models have been solved successfully, no large-scale application of the embedding approach has ever been reported. From my experience with the method, numerical difficulties are likely to arise for large-scale problems if commercial linear programming solutions are used. The commercial codes try to maintain the sparsity of the matrix upon which linear programming operations are performed (the basis). Lower-upper triangular decomposition of the basis is used. Unfortunately, there are numerical difficulties with lower-upper basis factorization when banded matrices are involved. The discretization of groundwater flow equations using finite differences or finite elements yields a banded matrix. The numerical difficulties are not severe for small problems. Although those who have applied the embedding approach have noticed some numerical problems in using the commercial linear programming packages (see Evans and Remson 1982; Gorelick 1980), solutions have been achieved. For larger systems the solutions may blow up. Elango and Rouve (1980) have experimented with the finite element and linear programming method and hypothesized that problems may arise when a large number of equality constraints are considered.

In addition to the possible numerical difficulties, commercial codes are not specifically designed to take advantage of the matrix structure generated when the flow equations are discretized. The effectiveness of alternative linear programming algorithms on problems relating to an underlying discretized partial differential equation has been discussed by O'Leary (1978). This work concluded that both stable matrix updating methods and iterative linear systems solvers may be used in solving such management problems.

Response Matrix Approach

The second technique in groundwater hydraulic management is the response matrix approach. Incorporation of a response matrix into a linear program was initially proposed in the petroleum engineering literature by Lee and Aronofsky (1958). They developed a linear programming management model which sought to maximize profits from oil production. A response matrix was used to convert pumping stresses linearly into pressure changes in a petroleum reservoir. The pressure response coefficients were developed using an analytic solution to the flow equation (Van Everdingen and Hurst 1949). Constraints ensured that reservoir pressures were maintained above one atmosphere, limited total production to the reservoir capacity, and guaranteed that oil purchased from an outside source did not exceed the pipeline capacity. Their solution gave optimal production rates for each of five potential sources during four time periods of two years each. Aronofsky and Williams (1962) extended this work to the scheduling of drilling operations.

Wattenbarger (1970, p. 995) presented a method to maximize total withdrawals from a gas reservoir. He noted that "for any number of wells in a reservoir, it is necessary to know only each well's own drawdown curve and interference curves for each pair of wells in a reservoir. The pressure behavior at each well can be calculated through superposition for any rate schedule."

It was suggested that the equation for real gas flow could be expressed in linear form as an approximation and solved using the finite difference method. The drawdown and interference curves could then be calculated and collected into a single transient influence matrix (that is, a response matrix). To develop the matrix, one numerical simulation run was required for each well in the system. In that study a gas withdrawal scheduling problem was formulated as a linear programming management model. The objective was to maximize well production. Constraints based upon the response matrix served to maintain pressures above critical values at each well. Additional constraints established the maximum reservoir withdrawal rate by forcing total production not to exceed projected gas demand. Wattenbarger made one final important comment. A constraint matrix including the response relations will generally be extremely dense for linear programming problems. If off-diagonal entries (those representing well interference) of the response matrix are relatively small (little interference), they may be neglected and set to zero, thereby decreasing the matrix density. This technique can make large-scale problems much more manageable.

The response matrix approach soon became established in the groundwater literature. Deninger (1970) considered maximization of water production from a well field and proposed that drawdown responses be calculated using the nonequilibrium formula of Theis (1935). The linear programming formulation sought to maximize total well discharge. Constraints were written which limited drawdowns and accounted for pump and well facility limitations. Of greater significance, Deninger also presented a management model formulation to minimize the cost of water production. The objective function was nonlinear because water production costs were assumed to be directly proportional to the products of the rates of discharge and the lifts. Both discharge rates and lifts were unknown prior to solution. The use of quadratic programming was suggested to solve this problem, but no solutions were presented.

Maddock (1972a) solved the nonlinear problem of minimizing pumping costs. A constraint set was developed as a response matrix, which the author called an algebraic technological function. The response coefficients showed drawdown changes induced by pumping at each well. The finite difference program which calculated the response coefficients appears in an open file report (Maddock 1969). The quadratic objective function minimized the present value of pumping costs. An example considered a complex heterogeneous groundwater system in which discharge from three wells was scheduled for ten seasons. Constraints guaranteed meeting semiannual water targets and set upper limits on the pumping capacity of each well. A standard quadratic programming package (Karash 1962) was used to solve the problem.

Maddock (1974b) developed a nonlinear algebraic technological function for the unconfined aquifer case in which saturated thickness varies with drawdown. The solution of the nonlinear partial differential equation was given by an infinite series of solutions to linear partial differential equations. Drawdown was treated as a finite sum of a power series. The ratio of drawdown to saturated thickness was shown to be the key to determining the appropriate number of terms in the series. With this technique a quadratic program was formulated in which all nonlinearities appeared in the objective function. The constraints remained linear.

Substantial caution must be used in calculating pumping costs based upon local drawdowns given by numerical simulation models. Finite difference and finite element models are able to account for the influence of complex geometry, heterogeneous aquifer parameters, as well as complex patterns of discharge and recharge. Regional models generally do not give accurate values of in-well drawdowns. Rather, they give average drawdown values over discrete cells or elements. These average values may be drastic underestimates of actual in-well values, thereby giving rise to low well production cost estimates.

Two techniques have been developed to obtain estimates of in-well drawdowns using simulation models in conjunction with analytic solutions. In one technique (Akbar, Arnold, and Harvey 1974; Trescott, Pinder, and Larson 1976; Williamson and Chappelear 1981) hydraulic heads are computed using simulation followed by a correction to account for local drawdown. Corrections are based upon analytic solutions for radial flow conditions. In the second technique (Cavendish, Price, and Varga 1969; Hayes, Kendall, and Wheeler 1977; Charbeneau and Street 1979) the analytic drawdown correction is added to the original finite element formulation, and a single system is solved. Although no groundwater management models have incorporated these techniques to improve local drawdown estimates, they should be considered for studies involving limitations upon pumping costs and in-well drawdown.

Mixed integer programming has been used in conjunction with a response matrix to determine the optimum location of wells. Rosenwald (1972) and Rosenwald and Green (1974, p. 44) identified "the best location for a specified number of wells so that the production-demand curve is met as closely as possible." An external finite difference model was again used to generate a transient response matrix which described pressure changes caused by pumping. This response matrix was used to develop constraints identical in form to those of Wattenbarger (1970). Integer constraints were added, however, which specified the number of wells permitted to enter the solution. The management model maximized production within the demand constraints by selecting the best wells among the potential well sites. The branch and bound method was used to obtain solutions.

Optimal well selection was also examined by Maddock (1972b), who developed a mixed integer quadratic programming model that minimized pumping costs plus fixed costs for well and pipeline construction. The quadratic portion of the objective (pumping times lift) was made separable by a transformation that enables solution by a combination of mixed integer and separable programming. This study illustrated sensitivity and error analysis to evaluate the effects on planning activities of uncertainty in economic and hydrologic factors.

Rosenwald and Green (1974) described a gas reservoir management case. The governing equation was nonlinear, and linear superposition (required in the response matrix approach) did not yield an adequate approximation. Attempts were made to develop external iteration procedures using the information from the linear programming management model to improve the solution. In the first method adjustments were made to the allowable drawdown. A revised linear program was executed, but validation with the nonlinear simulation model indicated that the revised permissible drawdowns were overcorrected. Increased pressure drop limits were used and a new management model executed. Solution verification indicated better agreement, but some significant errors remained. Rosenwald and Green (1974, p. 51) mentioned that "one should be able to repeat the routine until closer agreement is obtained." The second iterative method used the results of the initial linear program to adjust the response matrix. The unit pumping rates used to generate the response matrix were revised and a new response matrix developed. Again, the updated linear programming management model results did not satisfy the nonlinear flow equation but did improve the solution.

Schwarz (1976) described groundwater hydraulic management models using influence equations. An influence matrix, compiled from solutions of numerical or analytic models or from the results of pumping tests, served as linear programming constraints. This is identical to the response matrix approach described previously.

Bear (1979) in his chapter, "Application of Linear Programming to Aquifer Management," reviewed the works of Maddock (1972a) as well as a report by Schwarz (1971). Management formulations and results demonstrated steady-state pumping allocation for a two-cell aquifer and for a twenty-five-cell aquifer using a transformation equation and an influence equation, respectively. Both Schwarz (1976) and Bear (1979) presented a graphical solution to the steady-state two-cell aquifer management model. Another example showed a transient two-cell aquifer management model which maximized the present worth of net benefits derived from operating a water supply system. The water may be pumped from the two cells or imported from outside the region. The project was to be developed in four stages of five years each.

Variations of the response matrix approach have been applied to the management of aquifer-stream systems. Taylor (1970) demonstrated with a very simple small problem how the conjunctive use of groundwater and surface water along the Arkansas River in Colorado could be managed using linear programming. A set of approximate response curves was developed showing the average effect of aquifer stresses upon streamflow. The response curves were averaged over large cells which were roughly parallel to the river and extended along the entire stream reach. The objective was to minimize stream depletion during two summer months. Constraints limited pumping in each cell and specified the demand during each month from each cell.

Morel-Seytoux (1975a,b) presented various linear programming formulations and solutions involving conjunctive surface water–groundwater management. The discrete kernel (response coefficient) generator of Morel-Seytoux and Daly (1975) was used to develop response matrix constraints that predicted the reduction in return flow along a river reach owing to groundwater pumping. Models were used to explore four management strategies. The first was to maximize total pumping during a growing season subject to the restriction that flow to the stream could not fall below a specified rate. A second model added a constraint that pumping during all weeks must be equal. A third model sought to minimize the need for water storage to meet demand with available supplies. Finally, a stochastic storage strategy used sequential solutions which minimized the need for storage. Weekly updates of actual rather than expected river flow were used to calculate conditional expected flow for the remaining weeks. Illangasekare and Morel-Seytoux (1982) have linked discrete kernel (influence coefficient or response coefficient) approaches for an aquifer and for a stream using a linear stream-aquifer interaction relationship. The resulting stream-aquifer simulation model could be incorporated into a mathematical optimization problem.

Applications of the response matrix approach to groundwater hydraulic management involving large numbers of decision variables (pumping rates at well sites) for regional aquifers have been demonstrated. Larson, Maddock, and Papadopulos (1977) constructed a management model to determine the safe yield of a groundwater system near Carmel, Indiana (referred to as the Carmel aquifer). Their management model maximized the steady-state pumping rate by selecting from 199 potential well sites. Restrictions placed a lower limit on the pumping rate at each active well site and limited the number of permissible active wells. These restrictions led to a series of integer variables specifying whether a well exists at a particular site (integer variable equals one) or no well exists (integer variable equals zero). Other constraints forced pumping rates below design capacity and limited drawdown to one-half the initial saturated thickness.

The Carmel aquifer is unconfined and therefore was described by the nonlinear Boussinesq equation. To linearize the system, steady-state drawdowns for a confined aquifer were simulated and adjusted for unconfined conditions (Jacob 1944). The linear response matrix was created using the U.S. Geological Survey aquifer simulation model of Trescott, Pinder, and Larson (1976). Aquifer parameters calibrated by Gillies (1976) for the Carmel aquifer were used. The mixed integer linear programming management problem was solved using the MPSX-MIP package of IBM (1971). In the solution, twenty-six well sites were selected, and the distribution of pumping rates was determined. Drawdowns predicted at the well sites using a nonlinear simulation model compared favorably with the results using the linearized and adjusted drawdowns. It was concluded that the linearization did not seriously affect the results.

Willis (1983) applied the response matrix approach to the agricultural Yun Lin basin, Taiwan. The linear programming problem was to determine the optimal pumping scheme for three consecutive periods in order to meet agricultural water demands. The objectives were to maximize the sum of hydraulic heads and minimize the total water deficit. Constraints were placed on local groundwater demands, hydraulic heads, and well capacity. Of key interest is the use of quasilinearization to solve the equation describing transient unconfined flow. The term involving hydraulic heads squared was expanded with a generalized Taylor series, which yielded a linear approximation requiring iteration to achieve a solution to the original Boussinesq equation. The iterative procedure used the head distribution from one linear programming solution to update the linearized response constraints for the next solution. If a suitable starting solution was chosen, convergence was attained in about four iterations (Robert Willis, oral communication, 1982).

This linearization procedure yields an approximate solution to the equation describing unconfined flow, since it relies upon a single average transmissivity distribution during each pumping period. Because transmissivity is a continuous function of time, results from the quasilinearized management model should be checked using direct simulation with a predictor-corrector and small time steps. Alternatively, many short pumping periods could be included in the management model, but this would increase the dimension of the constraint matrix. The simulation model used a finite element network containing 101 nodes. Each iteration of the linear programming problem, which had 225 constraints and 438 decision variables, required about thirteen minutes of central processing unit (CPU) time and 400,000 (400K) bytes of core on a Cyber 170.

Heidari (1982) recently applied the response matrix approach to groundwater hydraulic management in the Pawnee Valley of south-central Kansas. A groundwater model for that area was developed in previous studies by Sophocleous (1980), and the aquifer parameters were used in conjunction with the computer program of Maddock (1969) to generate a transient response matrix. The unconfined system was treated as a confined aquifer as an approximation, and drawdowns were corrected using the relationship of Jacob (1944). The response matrix was used in a linear program maximizing pumping rates over time. Total pumping during each time period was forced to meet demands. Each pumping rate was limited by water rights. Drawdown at any time was constrained to a fraction of the total saturated thickness. Different solutions were obtained under maximum drawdown fractions ranging from 0.1 to 0.25. To reduce the number of decision variables, wells were clustered into sixty-one well fields. Models were run for a five-year planning horizon and a ten-year planning horizon. In each case the time horizon was broken down into five equal management time units of one and two years, respectively. The models consisted of 610 decision variables (pumping rates and drawdowns) and were subject to 615 constraints. The linear programming models were solved using MINOS (Saunders 1977) on a Honeywell 60/65 computer. Each solution took approximately ten minutes of CPU time and occupied 137K words of core storage (Manoutchehr Heidari, written communication, 1982). By considering various sets of constraints in several model runs, Heidari (1982) has demonstrated that the response matrix approach is applicable to a real world system and is a valuable tool for evaluating groundwater management strategies.

Comparison of Hydraulic Management Methods

This review has pointed out some of the advantages and disadvantages of each of the two groundwater hydraulic management methods. In the embedding approach, the discretized flow equations are included in the linear program as constraints, and a complete simulation model is solved as part of the management model. Hydraulic heads throughout the entire spatial domain and at each time step are treated as decision variables. Hydraulic stresses over space and time are additional decision variables. The embedded simulation model yields a great deal of information regarding aquifer behavior. Management rarely involves all the hydraulic heads over space and time, however. Many of the decision variables and constraints will thus be unnecessarily contained in the linear programming model.

For computational economy and avoidance of numerical difficulties, the embedding approach should currently be restricted to small steady-state problems. For cases in which hydraulic heads are constrained at many locations and many well sites are considered, the embedding approach will be more efficient than the response matrix approach. The embedding approach may become a more valuable tool if future research results in a variation of the revised simplex method geared to solving linear programming problems which contain embedded discretized equations. If decomposition methods can be developed in conjunction with the embedding approach, solution of larger transient problems may become feasible.

In the response matrix approach, solutions to the flow equations serve as linear programming constraints. This approach yields incomplete information regarding system functioning but is generally a more economical method. The developing of a response matrix requires the solution of one external simulation model for each potentially managed well or other excitation, such as local boundary condition, and may require a large initial expenditure of computational effort. The resulting response matrix, however, is a highly efficient, condensed simulation tool. Constraints are included only for specified locations and times. Unnecessary constraints or decision variables are not incorporated into the linear programming management model. The response matrix approach can therefore handle large transient systems efficiently.

Both methods have been valuable in selecting potential well sites and in determining optimal pumping rates to control drawdown and hydraulic gradients. The field applications have attempted to quantify the safe yield using a management model. This was done by maximizing well production while limiting drawdown. For all practical purposes, simulation alone cannot accomplish this. The determination of safe yield under different system-specific restrictions regarding pumping and drawdown is an important application of groundwater hydraulic management models.

Groundwater Policy Evaluation and Allocation Models

Groundwater policy evaluation and allocation models are valuable for complex problems where hydraulic management is not the sole concern of the water planner. These models inspect water allocation problems involving economic management objectives. Such models have been applied to large-scale transient problems which study how an agricultural economy responds to institutional policies and to optimization of conjunctive water use.

Three types of models have been developed for groundwater policy evaluation and allocation problems. Hydraulic-economic response models are a direct extension of the response matrix approach to problems in which agricultural and surface water allocation economics play a key role. Linked simulation-optimization models use the results of an external aquifer simulation model as input to a series of subarea economic optimization models. Information and results from each planning period are used for management during the next period. Because the simulation and economic management models are separate, complex social, political, and economic influences can be considered. Hierarchical models use subarea decomposition and a response matrix approach. Large and complex systems can be treated as a series of independent subsystems, and multiple objectives can be considered. They are particularly valuable for large-scale optimization where detailed hydraulic management is required in the context of a complex water allocation problem.

Hydraulic-Economic Response Models

The hydraulic response matrix approach has been extended to include agricultural-economic or surface water allocation management components. Each of these models is formulated as a single optimization problem in which hydraulic and other controllable activities are considered. Maddock (1974a) presented the first such combined groundwater and economic management model. A conjunctive stream-aquifer simulation model using finite differences was used to develop an algebraic technological function (response matrix). The planning model was developed for a hypothetical water management agency with the objective of minimizing discounted expected value of water costs supplied by the stream-aquifer system. Constraints regarded expected water demand, downstream water rights, water spreading for aquifer recharge, and stream-aquifer water transfers. Within the context of this quadratic programming problem, water demand was considered a stochastic process. Demands could be represented by a Markov process or could be serially independent. A single planning model solution provided operating rules for groundwater withdrawals, stream diversions, water spreading, and return flow to the stream. Sensitivity of the model to variation in parameters, such as demand variance, was demonstrated. Including other stochastic components, such as groundwater hydraulic parameter uncertainty, would clearly be valuable.

Maddock and Haimes (1975) used the response matrix approach along with agricultural-economic considerations to study a tax-quota system for groundwater management applied to a hypothetical farming region. The scheme employed a quota in which each user was assigned an annual draft (resulting from the optimization model) and a tax which would be assessed if this draft limit was exceeded. Farmers using less than the quota could be entitled to a rebate. A quadratic programming problem was formulated which maximized combined net revenues on the farms and thereby determined pumping quotas. Linear constraints restricted drawdowns and pumping rates using a hydraulic response matrix. Additional linear constraints limited the acreage planted and guaranteed that irrigation water requirements were met. Response coefficients were developed, and the optimization model was solved for a five-year planning horizon. The optimal (primal) solution gave the pumping quotas for each well in the system. The Lagrange multipliers were shadow prices corresponding to the cost of savings per unit of overpumping or underpumping in relation to the quotas. These costs were then used to assess taxes and rebates depending upon actual deviations in pumping relative to the quotas.

Economic management of an aquifer-stream system was the topic of Morel-Seytoux and others (1980). In an approach similar to that of Maddock (1974a), they demonstrated a link between an economic model and a hydrologic model in one quadratic programming formulation. In a hypothetical system, water for irrigation was supplied from groundwater and a river diversion. The objective of the model was to minimize the total cost of water. Constraints were placed upon water availability, water rights, and land availability for each crop. Optimal crop yield as well as crop prices were known, and crop yield was assumed to vary with the degree of irrigation. Pumping costs were related to groundwater drawdown, which can be computed using the response matrix approach. The model consisted of a quadratic objective function relating to water costs and a series of linear constraints. Although no solution was presented, the authors noted that one could be achieved by using existing techniques.

Linked Simulation-Optimization Models

Linked simulation-optimization models were developed to study the impact of institutional changes upon groundwater use (for early works see Fiering 1965 and Martin, Burdak, and Young 1969). Bredehoeft and Young (1970) explored the effects of two policy instruments (a tax and a quota) upon groundwater basin management. They considered a complex hypothetical basin in which a management scheme was desired for the temporal allocation of groundwater to agricultural users. The management objective was to maximize the net (agricultural) economic yield of the basin. An economic management model was formulated as a series of linear programming problems where crop acreage and the annual quantity of water pumped were decision variables. One such model was formulated for each agricultural subarea, and separate models were solved for each five-year management interval. An external numerical simulation model provided input to the linear programs as estimated annual cost of pumped water for each subarea. The finite difference model simulated transient unconfined flow for the regional aquifer.

A sequential computational link was developed between the economic models and the groundwater simulation model. Preliminary subarea water demands were computed using the linear program. Then calculations and checks were made regarding the profitability of reinvestment in wells, well life limits, and economic failure for each subarea. Pumping requirements were adjusted and the aquifer simulation rerun. The resulting hydraulic heads (and depth to water values) were used to compute estimated future pumping costs for the next five-year interval in the economic linear programming management models. The procedure was repeated within a fifty-year time horizon.

Although the simulation model was not included in the linear program as a series of constraints, the behavior of the groundwater system was accounted for through the sequence of aquifer simulations. There are limitations to this approach. Maddock (1972a, p. 129) noted of the linked model approach:

> The annual optimization of cropping patterns does not guarantee that changes in cropping patterns over the design horizon occur in an optimal fashion. The assumption that lifts are constant annually and are updated only at the end of the season may lead to an underestimation of pumping cost. The assumption that pumping is uniformly distributed over the wells does not use information available from the distributed parameter groundwater model. Local conditions may favor more pumping from some wells than others, particularly if the aquifer must be modeled with nonhomogeneous parameters, such as transmissivity and storage coefficient.

Although these deficiencies are significant, the linked model approach is concerned primarily with economics and evaluation of institutional policy instruments rather than hydraulics. In comparison with the response matrix approach, the linked model allows greater economic complexity. Social and legal factors can be integrated into the management model. Moreover, the hydraulic nonlinearities (unconfined conditions) do not enter into the management model because the hydraulic simulation model is a separate component. Nonlinear programming need not be used.

The linked model approach was extended to management of groundwater and surface water systems by Young and Bredehoeft (1972). They studied a fifty-mile (eighty-kilometer) reach of the South Platte River in Colorado in which all groundwater ultimately came from the river.

Their study considered an institutional framework which would tend to minimize the effect upon river flow of groundwater pumping for irrigation. River flow would be affected if groundwater were used late in the irrigation season or if wells distant from the river were used. They examined the policy implications of a basin authority for partial centralization of control, allowed for the development of new water supplies, and required readjudication of water rights so that surface water rights could be transformed into groundwater rights.

A model was developed which linked economic management with numerical simulation. As in the former study, numerous linear programming models and periodic groundwater simulation models were executed. The goal of the basin authority was assumed to be maximization of the average annual economic yield. A linear program was developed which maximized net revenues at the beginning of the irrigation season (the planning stage model). Constraints considered water acquisition, crop production alternatives, and irrigable land limitations. A sequence of computations then determined irrigation return flow, flow diversions, and crop acreages. Monthly subarea operating models were formulated as linear programs in which surface water use and groundwater use were decision variables. The constraints related to irrigation requirements and local net revenues were maximized. After each subarea operating model was solved, an aquifer simulation was executed for the nongrowing season. The planning stage model was then run for the next irrigation season and the process repeated for a ten-year time horizon.

Bredehoeft and Young (1983) have recently updated their 1972 model by considering the influence of uncertain surface water supplies. Their study was again patterned after a reach of the South Platte River in Colorado and used the linked model approach. They argued that in addition to maximizing net revenue the farmers tended to limit the variance in income caused by short water supplies. Farming economics in the study area were such that relying on uncertain surface water supplies was unwise. Increasing well capacity so that all acreage could be irrigated by wells significantly reduced the variance in income. As in this study, the management modeling approach illustrates quantitatively that the development of groundwater systems may be motivated by a desire to ensure against periodic short water supplies. The study provided a more comprehensive framework to evaluate management institutions.

Daubert and Young (1982) used the linked model approach to evaluate the influence of two groundwater management policies along the lower South Platte River in Colorado. Their model was similar to that of Young and Bredehoeft (1972), but used the aquifer-stream response model of Morel-Seytoux (1975b) and contained a legal submodel which distributed surface water according to water rights and groundwater according to assumed management policies. The management model was first concerned with a groundwater policy that effectively prohibited groundwater use during drought periods because surface water rights were satisfied before groundwater could be used. The second policy regarded groundwater as common property and limited pumping only by the well capacity. Results showed that while unrestricted groundwater use resulted in enhanced net income from irrigated farming, it imposed large "external costs on surface-water users who must cope with declining flows" (Daubert and Young 1982, p. 84).

Hierarchical Models

The hierarchical approach was developed to model and optimize large-scale water resource systems. The general approach is described by Haimes (1977). Yu and Haimes (1974, p. 625) first employed this approach to optimize a complex regional groundwater allocation problem. They reasoned that a

> decomposition and multilevel approach breaks down a complicated regional resource management problem into smaller local level problems, each of which is optimized before attempting to optimize the overall regional problem. The process is iterative and requires coordination and feedback from the solution of the higher-level (overall) problem to the local level optimizations.

Yu and Haimes were concerned with planning to improve the allocative efficiency of regional water resources, particularly groundwater. The regional problem was broken down into one problem for each subregion. Each subregional problem was formulated as a nonlinear mathematical programming problem which minimized net water supply costs, which were considered to be functions of water importation, pumping, and water levels. The aquifer model was a two-dimensional asymmetric polygonal finite difference system (MacNeal 1953). Subregional model solutions were fed to a second level representing coordination by a regional authority. The authority ensured consistent boundary water levels across the individual polygonal regions, determined tax rates, and determined the optimal artificial recharge operation rate to minimize regional water supply costs. The interaction between the regional authority and subregions was accomplished using an iterative decisionmaking process. This three-stage process is fairly involved and was described in a detailed example by Yu and Haimes (1974). By structuring the model in this manner, decentralized decisions were tied together to solve a complex regional groundwater allocation problem by imposing a regional pumping tax to provide revenue for artificial recharge.

A multilevel management model for a complex groundwater–surface water system was presented by Haimes and Dreizen (1977). Perhaps the most comprehensive of the groundwater management models, it uses both the response matrix approach and a multilevel model structure to optimize a complex, conjunctive use system. The strategy was to solve a series of optimization models employing response matrices. In the first stage, optimization models were executed which maximized each water user's net benefit. The problems were formulated as quadratic models in which decision variables were local pumping quantities, artificial recharge activity, and surface water use. Constraints specified water requirements, lift and pumping limitations for each well, recharge facility capacity, and surface water allocations. Response matrices were developed with the aquifer simulation program of Maddock (1969).

In the study, individual subarea models were solved using the multicell simulation technique described in another article (Dreizen and Haimes 1977). This technique is particularly valuable for management models involving large regions where many wells must be considered. The second stage of the multilevel optimization scheme determined drawdowns throughout the system and flows induced from the river into the aquifer. This process was accomplished by summing the influences resulting from the solutions obtained for each area from the first stage in the optimization. Next, operation of a surface water reservoir was considered. Net streamflow to the reservoir was calculated, reservoir overflow was checked, and reservoir operation costs were determined. These factors were then entered into an optimal surface water allocation (quadratic) program in which the reservoir capacity and periodic allocation limits served as constraints. A sample solution was presented over a six-year planning horizon. The optimal solution indicated a well-pumping plan, a recharge plan, and a surface water use plan.

A hierarchical model has been developed by researchers at the World Bank to study groundwater and surface water policies for the Indus Basin in Pakistan. Although the Indus Basin study is incomplete and results cannot be placed in the public domain, the structure of the hierarchical, multilevel programming model was described by Bisschop and others (1982, p. 32). The simulation model employed the transformation matrix approach with fifty-three irrigated polygonal regions.

> In addition to the above-mentioned water constraints, each polygonal model has embedded in it a single farm level model to characterize the agricultural production system of the area. Such a farm level model simulates the resource allocation choices of a single representative farmer which determine the production and disposition of 11 crops and 4 livestock commodities. Exogenous resource limitations are imposed on land, labor and canal water.

The original model consisted of 20,000 linear constraints, but various simplifications reduced the problem to fewer than 8,000 constraints.

A method was developed to solve the Indus Basin model as a two-level programming problem. One level concerned the farmers and their individual objectives. The individual farm level objectives were summed, and the aggregated farm income was maximized. The second level concerned the government which must account for the long-term consequences of water allocation decisions. The solution procedure involved creation of an augmented linear program that served to fix certain policy variables pertaining to government taxes and subsidies. These were then used to solve a second problem to determine decisions on the part of the farmers.

Applying this procedure may result in politically unacceptable water management policies. Water allocations might ignore certain regions within the country. This could be overcome by executing revised models with additional political constraints and by then evaluating their impact upon the objective function values. Alternatively, explicit constraints relating to historic water rights have been incorporated into the model to eliminate the need for repeated solutions (Gerald T. O'Mara, oral communication, 1982).

The few studies that have developed groundwater policy evaluation and allocation models demonstrate the capability of aiding in the management of real world groundwater–surface water systems. Such models may serve as important tools to analyze the economically motivated behavior of water users. The efficacy of basinwide regulation and water management policies may be evaluated before being implemented. Examples indicate that various levels of approximating groundwater behavior can be used. In cases where goals are primarily economic, a simulation model which is fully external to the optimization model may suffice. If pumping schedules and the like are needed, a response matrix component must be incorporated into the optimization model.

Complex large-scale problems for which detailed hydraulic information is needed must use a multilevel optimization approach. The choice of the appropriate procedure depends upon the objectives of each particular investigation, the scale and complexity of the problem, the uncertainty regarding the system's functioning, and the level of detail required of the solution (how the results will be used). Methods in this category of groundwater management models also enable complex scheduling and allocation problems to be inspected. Furthermore, the influence of institutional policies upon the operation of conjunctive use systems can be studied and evaluated.

Groundwater Quality Management

The joint use of numerical simulation and linear programming has been applied to groundwater pollutant source management. The underlying management problem here is use of an aquifer for both waste disposal and for water supply. Such problems have involved managing waste disposal activities while maintaining water quality at specified locations. Combined groundwater simulation and management models have been developed to contend with these problems. Simulation models in the case of groundwater pollutant source management involve solution of the advective-dispersive equation (Reddell and Sunada 1970; Bear 1972; Bredehoeft and Pinder 1973; Konikow and Grove 1977). For the general linear case of two-dimensional transport with linear decay and sorption of a single dissolved chemical constituent in saturated porous media, the governing equation is

$$(7\text{-}2)\quad R_d \frac{\partial C}{\partial t} = \frac{\partial}{\partial x_i}\left(D_{ij}\frac{\partial C}{\partial x_j}\right) - \frac{\partial}{\partial x_i}(CV_i) - \frac{C'W}{\phi b} - \lambda R_d C \qquad i, j = 1, 2$$

where

C = concentration of the dissolved chemical species, M/L^3
D_{ij} = dispersion tensor, which is a function of V_i, L^2/T
V_i = average pore water velocity in the direction i, L/T
b = saturated aquifer thickness, L
ϕ = effective aquifer porosity, L^0
C' = solute concentration in a fluid source or sink, M/L^3
W = volume flux per unit area, L/T
R_d = retardation factor, L^0
λ = first-order kinetic decay rate, $1/T$
x_i, x_j = cartesian coordinates, L
t = time, T.

Different types of management models have been developed for steady-state pollutant distributions, which often represent a worst-case pollution scenario, and for transient cases involving solute redistribution. The remainder of this review cosiders steady-state management models and then transient models.

Steady-State Management Models

The first model aimed at managing the disposal of wastes in aquifers was that of Willis (1976a; also see Willis 1973). In this study the aquifer was considered as a component of a regional waste treatment system. The objective of the model was to minimize the costs of surface waste treatment. Both dilution water costs and treatment plant costs were considered. The treated effluent would be injected into an aquifer at predetermined sites. The assimilative capacity of the aquifer was therefore part of the treatment system. Restrictions were placed upon water quality at supply (pumping) wells and recharge wells.

A solute transport model served as a series of linear constraints in the planning model. In the simulation model no temporal variations could occur in either the hydraulic head or solute distributions. Solute transport simulation proceeded in two steps. First, the steady-state form of equation (7-1) was solved using the finite difference method. In that simulation model the locations of the wells and their respective pumping and recharge rates were assumed to be known prior to solving the management model. This significant assumption was necessary because the linear solute transport simulation model required a known, constant velocity field. In all studies involving pollutant source management, the velocity field(s) must be known (or estimated) prior to use of the pollutant source management model. The ramifications of this restriction are discussed later in this review.

Second, the steady-state solute distribution was determined for each constituent as part of the management solution. Willis (1976) chose to neglect dispersion, and therefore only advective transport and linear chemistry were modeled. The finite difference method was used to discretize the resulting equation for each constituent, and water quality constraints were then based upon the discretized equations. The solute transport simulation model was first formed as a finite difference coefficient matrix. The inverse of this matrix was then computed, and relevant portions were included in the management model as constraints.

The former management model was demonstrated for a hypothetical case involving two injection wells and two pumping wells. The model considered several unit processes for the waste treatment plant that involved primary, secondary, and various forms of advanced waste treatment. Wastes contained organic matter, coliform, suspended solids, and conservative constituents. First-order kinetic reactions were assumed. The management model was formulated and solved as a mixed integer programming problem where integer variables corresponded to unit processes. The solution determined the optimal unit treatment process and the most cost-effective volume of imported dilution water.

Futagami, Tamai, and Yatsuzuka (1976; also see Futagami 1976) incorporated the discretized solute transport equation as constraints in a linear programming management model. These studies considered the general problem of large surface water body pollution and not, specifically, groundwater pollution. The two systems are similar, however, and both are described by solutions to equation (7-2). Different models were constructed using finite differences and finite elements to solve for the steady-state solute distribution. They considered advection, dispersion, and linear decay as well as pollutant sources (both controllable and uncontrollable loads). Their linear programming model sought to maximize total waste disposal under local waste load restrictions. The physical behavior of the system was accounted for by embedding the finite difference (FD) or finite element (FE) equations into the linear program (LP) as constraints. Futagami and others called their approaches the FELP and FDLP methods after the numerical method used in conjunction with linear programming. The greatest significance of this work was in the linear programming solution algorithm.

Futagami (1976) was able to take advantage of the special structure of the constraints represented by the discretized transport equation to develop a more efficient computational algorithm using the simplex method. An initial basic feasible solution (the starting point for the solution of a linear programming problem) was developed without the introduction of artificial variables. Because the embedding approach leads to an enormous constraint matrix for large (usually transient) systems, the efficiency gained by this approach may become important in some applications. However, the primary problem of treating each concentration in space and in time as a decision variable still remains, even with their special simplex method implementation.

Gorelick and Remson (1982a) and Gorelick (1980) employed the embedding approach using the steady-state finite difference form of equation (7-2) to maximize waste disposal at two locations while protecting water quality at supply wells and maintaining an existing waste disposal facility. Both concentrations and waste disposal fluxes were treated as decision variables. After achieving a solution, the model was subjected to sensitivity analysis using parametric programming. In this way, waste disposal tradeoffs at the various facilities could be inspected. As noted previously, the groundwater hydraulic system must be modeled before the waste disposal management problem can be solved. This study described an iterative procedure which handled changes in the groundwater velocity field induced by waste water injection. A second problem was to identify all sites suitable for waste disposal. The linear programming management model was manipulated so that the optimal value of the dual variables represented unit source impact indicators. With interpretation the solution of two linear programming problems led directly to identifying all feasible disposal sites.

Transient Management Models

Pollutant source management problems may frequently involve the migration of plumes of contaminated water. In such cases water quality must be protected for a long time. A waste disposal decision today may not manifest itself as polluted groundwater at down-gradient wells or streams for many decades. Methods to contend with the dynamic management of groundwater pollutant sources have been developed.

Both Willis (1976b) and Futagami (1976) have suggested that the embedding method use the finite element discretization of equation (7-2) to solve transient management problems. Willis (1976c) employed the embedding method to determine effluent disposal standards for food processing wastes using spray irrigation. The finite element model considered one-dimensional advective and dispersive transport, adsorption, and first-order kinetic decay in the unsaturated zone. When steady flow was assumed, the transient solute transport model was linear. Five disposal cycles were considered for a twenty-four-element configuration. For such small systems the embedding method is applicable. For the general transient case, however, embedding will result in an extremely large constraint matrix. This problem of dimensionality was discussed in conjunction with hydraulic management models in the section on embedding method in this review. In essence, there will be at least one constraint and one decision variable for each concentration value over space and time. Furthermore, the solution of the discretized system is contingent upon concentrations at every node in the system; no values can be eliminated from the management model. Clearly, for large systems or problems over long time frames, the embedding approach as it is currently implemented will not be suitable.

Gorelick, Remson, and Cottle (1979) considered the management of a transient pollutant source. Their model determined the maximum permissible concentration in a river which lost water to an aquifer. Water quality at agricultural supply wells would be protected by ensuring that crop chloride tolerances were not violated. They recognized the problem of storing and manipulating a large time- and space-discretized matrix when using the embedding approach and developed an alternate method that did not require linear programming.

By taking advantage of the mathematical structure of the transient constraint matrix, Gorelick, Remson, and Cottle developed a recursive pollutant source management formula. Concentrations throughout the system were expressed as a function of the river water quality. The management solution was quite efficient and provided valuable information regarding the effects of successive water quality constraints, as well as the travel times of the solute plume peaks from the source to the location of the supply wells. The method is limited to systems in which the concentrations throughout the system can be expressed as a function of a single parameter. Controllable pollutant sources must therefore be linearly related if more than one disposal period or more than one pollutant source is involved. Transient management problems will frequently have special mathematical structure, and future research to exploit this would be quite valuable.

Willis has discussed a model for aquifer management involving waste injection control. He considered the "groundwater problems associated with the well injection of waste waters in a groundwater aquifer system conjunctively managed for supply and quality" (Willis 1979, p. 1305). The planning problem was broken into a groundwater hydraulic management component (Willis 1977), and a pollutant source management component. By breaking down the problem in this manner, the hydraulic heads as determined from the hydraulic management solution were used to generate the groundwater velocity field required for the water quality management model. This technique uncoupled the management model of equation (7-2) from that of equation (7-1). The method is useful when the hydraulic problem can be solved first and can be treated independently of the water quality component.

The two component problems were formulated and solved as separate linear programming problems. In an illustrative example, hydraulic heads were managed at extraction and injection wells while a water target within each planning period was met and pumping and injection rates were limited. The water quality component maximized the lowest of the waste injection concentrations while meeting a waste disposal load and preserving water quality at all wells during the operational cycle. The example considered four 120-day management periods. Decisions regarding pumping and injection were made at the beginning of each period and remained constant during that period. Transient hydraulic simulations were avoided by using an average groundwater velocity field over the 480-day operational cycle for the water quality simulation-management model.

Each of the two management models regarded the system's behavior using numerical simulation. To handle the problem of dimensionality associated with transient problems, Willis (1979) began by constructing finite element models for equations (7-1) and (7-2). Rather than discretizing the time derivatives, he discretized the system over space alone. Analytic solutions to these systems of space-discretized ordinary differential equations were obtained (Bellman 1960), and the solutions were entered into the linear programming models as constraints. For the hydraulic component, this approach represents an important contribution because it eliminates most of the hydraulic head decision variables for each time step encountered when using the embedding technique. The hydraulic heads at the end of one planning period (a period much longer than a numerical simulation time step) could be expressed as a function of heads and pumping rates at the beginning of that planning period.

For the water quality problem there are some difficulties with this approach. First, evaluating concentrations at the beginning and end of planning periods does not suffice. In transient problems decisions are made regarding levels of waste disposal at potential sites. The optimal selection of sites and waste disposal strengths are unknown, and therefore the arrival time of contaminant plume peaks at water supply wells remains unknown. Contaminant plume peaks will not arrive at each supply well at the beginning or end of a planning (waste injection) period. Solute concentrations at supply wells must therefore be checked at many discrete intervals until the end of the operational (disposal) cycle. Second, numerical problems may arise when solving the solute transport equation as a system of ordinary differential equations. The space-discretized equations form a nonsymmetric matrix and are therefore difficult to solve using the matrix exponential.

Gorelick and Remson (1982b) approached the dynamic management of groundwater pollutant sources by using a concentration response matrix. They considered an illustrative one-dimensional case involving multiple sources of groundwater pollution and the maintenance of water quality at water supply wells over all time. They were concerned with the problem of managing waste disposal activities for several years in such a way that solute concentrations at supply wells would never exceed water quality standards, even after waste disposal activities had ceased.

An external finite difference model was used to construct a unit source concentration response matrix. This matrix provided linear programming constraints and converted solute injection fluxes at potential disposal locations into concentration histories at water supply wells. Any linear solute transport simulation model could be used to generate the concentration response matrix. The concentration response during one injection period could be used to generate the specially structured concentration response matrix for any number of periods as long as the velocity field was approximately the same. The linear programming solution was subjected to parametric programming to analyze the influence upon the optimal solution of various waste injection strategies. Furthermore, a mixed integer formulation and solution enabled restrictions to be placed upon the number of pollutant sources permitted to operate during certain management periods.

Gorelick (1982) used the U.S. Geological Survey solute transport model (Konikow and Bredehoeft 1978) as a component in a groundwater quality management model applied to a complex hypothetical regional aquifer. The concentration response matrix method was used in conjunction with linear programming to maximize waste disposal at several facilities during several one-year planning periods. A limit on solute concentrations at observation wells was enforced from the first arrival of groundwater pollution until the pollutant plumes cleared the observation wells. By formulating the problem as a dual linear programming problem (see Hillier and Lieberman 1974; Wagner 1975; Dantzig 1963) and by using a numerically stable implementation of the revised simplex method (Saunders 1977) a large field-scale problem could be solved. The results of the pollutant source maximization problem indicated that waste disposal was enhanced by pulsing rather than maintaining constant disposal rates at various sites. Solutions to problems with successively greater numbers of waste disposal periods were compared. The marginal impact of the water quality standard imposed at the supply wells was greater for short management horizons than for extended management horizons.

Gorelick and Remson (1982b) and Gorelick (1982) were concerned with managing disposal decisions until the pollutants essentially cleared the system. The water quality problem considered by Willis (1979) differs because the disposal cycle and the time during which groundwater pollution was of concern were synonymous. Furthermore, the advantage of solving the space-discretized form of equation (7-2) by using an analytic solution (provided that numerical difficulties can be overcome) lies in the ability to obtain a solution at any arbitrary time interval rather than at successive discrete time intervals. If analytic solutions are obtained over discrete time intervals in order to ensure water quality preservation at supply wells over the entire problem time frame, the primary benefit of the analytic solution technique is lost. In addition, incorporating transient flow field variations would require numerous evaluations of matrix exponentials at discrete intervals.

In the case considered by Gorelick and Remson (1982b) and Gorelick (1982), pollutant plume peaks arrived over a span of years and at different unknown times at the various supply or observation wells. If a problem dictates the need for checks on concentrations at discrete intervals, a simulation model which gives solutions over discretized time seems most appropriate. Temporal flow field variations could be incorporated by the sequential solution of equations (7-1) and (7-2) when developing the concentration response matrix. Finally, use of the concentration response method is not restricted to a particular simulation model. Any simulation model may be used to generate the concentration response matrix, as Gorelick (1982) demonstrated.

Nonlinearities in Groundwater Quality Management

Although groundwater quality management models have been developed for important steady-state and transient cases, research is needed to solve nonlinear groundwater quality control problems. Nonlinearities arise from management decisions that create unknown groundwater velocity fields as well as from problems involving chemical interactions.

Consider a groundwater reclamation project that attempts to capture a contaminant plume originating from a waste disposal site. A likely management scenario is to intercept the plume using extraction wells. One objective is to minimize pumping costs subject to the restriction that extracted water meets quality standards. The decision variables for this problem are the pumping rates at the plume interception wells. Constraints must incorporate the groundwater flow equation (7-1) and the solute transport equation (7-2), plus limitations on water quality. Because pumping rates are decision variables and are therefore unknown, the velocity field and dispersion coefficients of the transport equation (7-2) remain undetermined. Nonlinear constraints appear as a result of products of unknown concentrations and unknown velocity components which occur in advective and dispersive transport terms. The result is a nonconvex programming problem containing a quadratic objective subject to nonlinear constraints. This problem remains to be solved.

Nonlinearities also arise in saltwater intrusion control problems. The density difference between fresh and saltwater serves as a significant driving force for the migration of solutes. In such cases the groundwater velocity field is a function of solute concentrations. Nonlinearities thus appear in advective and dispersive transport terms. Simulation of this nonlinear system has been successful (Segol and Pinder 1976). Research is needed to develop distributed parameter management models of saltwater intrusion that involve simulation of this nonlinear system.

Commonly, groundwater solute transport is accompanied by chemical interactions in the moving water (homogeneous reactions), as well as complex interactions of solutes with the porous media (heterogeneous reactions). Multicomponent chemical interactions in solute transport are usually nonlinear (Rubin and James 1973; Valocchi, Street, and Roberts 1981; Jennings, Kirkner, and Theis 1982). Numerical simulation often involves solution of the resulting nonlinear equations. Optimal management of such highly nonlinear systems will require major innovations involving the union of numerical simulation with nonlinear programming. The result might be an enhanced ability to evaluate groundwater pollution control using natural chemical and biological degradation.

Conclusions

Distributed parameter models of groundwater hydraulic or solute behavior have been combined with optimization techniques in mathematical programming to inspect a variety of aquifer management problems. Groundwater management models concerned with hydraulics are capable of determining optimal pumping and recharge rates and optimal well locations subject to restrictions upon drawdowns, gradients, and water demands. The union of numerical simulation with linear and quadratic programming, as well as parametric, mixed integer, and separable programming, has been clearly established. The hydraulic nonlinearity associated with unconfined aquifers can be suitably linearized for many problems. Considerations such as energy costs for pumping or fixed well installation costs can be taken into account. A class of groundwater quality management problems has been solved. Sources of conservative or linearly sorbing or decaying pollutants can be managed in such a way that local water supplies remain protected over time. The general limitation of such problems is that the groundwater velocity field must be determined prior to pollutant source management.

Groundwater policy evaluation and allocation models

employing such techniques as a linked model structure, hierarchical optimizations, and system decomposition provide means to solve fairly complicated large-scale groundwater management problems. The influences of policy instruments such as taxes and quotas can be studied, and complex conjunctive water use allocation problems can be solved.

Groundwater studies are frequently initiated because of local or regional groundwater management problems. Although groundwater simulation models have become basic tools of the hydrogeologist, combined distributed parameter simulation and management models have not. Given the variety of demonstrated methods, why have these potentially useful combined models not been adopted? Additional field applications would benefit water managers by providing an efficient means to discover optimal groundwater plans and to evaluate proposed institutional policies. These applications would also provide researchers with feedback regarding practical requirements and needed extensions of existing methods.

Among the many types of problems that remain to be solved, three stand out. First, improved hydraulic management models are needed to account for aquifer parameter uncertainty. Similarly, conjunctive use allocation models could be extended to deal with streamflow variability. Second, the pattern of groundwater management modeling research has been to tackle those cases involving linear constraints and then those with constraints that could be linearized, such as those stemming from the unconfined flow equation. Groundwater and groundwater quality management models that include nonlinear constraints are a key area for research. Such models could be used for management problems of the unsaturated zone, groundwater reclamation, complicated pollutant chemistry, and coastal aquifer protection. Third, a broad spectrum of institutional factors and real system features should be accounted for in policy evaluation and allocation models. In these models, water quality considerations have generally been neglected.

Research that joins groundwater hydraulic and quality simulations with political and economic management factors would be valuable. An objective for future research might be to develop policy instruments that simultaneously avoid excessive water use and water quality degradation. In addition, only economic factors have been seriously considered in studying the response of groundwater users to management strategies. Groundwater policy evaluation models have yet to be constructed that include a comprehensive variety of economic, political, and legal issues. Such models might result in groundwater management approaches that have general application.

Solutions to the above problems may be improved by combining simulation with optimization techniques other than those discussed in this review. Such techniques as dynamic programming, multiobjective programming, stochastic programming, multitime period decomposition, semi-infinite programming, and many techniques in nonlinear programming could be joined with simulation models. The results might enhance the capacity to deal with larger and longer transient problems, analysis of complex management policies, and studies of the influence of uncertainty in system parameters as well as numerical discretization upon management model results.

In addition to mathematical programming techniques, parameter estimation models that use regression methods might be linked with management models. The result could be models that use the best set of hydraulic parameters, that determine optimal management plans, and that give information about the reliability and sensitivity of management solutions. Such innovations would tremendously aid the management of groundwater resources and stream-aquifer systems.

Note

This chapter originally appeared in a somewhat different form in *Water Resources Research*, vol. 19, no. 2 (April 1983), pp. 305–19 (published by and reprinted here with the permission of the American Geophysical Union). The author gratefully acknowledges the fine review comments provided by Thomas Maddock III, Irwin Remson, John D. Bredehoeft, Jacob Rubin, Jared Cohon, and four anonymous reviewers. The use of computer and software brand names in this report is for identification purposes only and does not imply endorsement by the U.S. Geological Survey or the World Bank.

References

Aguado, Eduardo. 1979. "Optimization Techniques and Numerical Methods for Aquifer Management." Ph.D. dissertation, Stanford University, Stanford, Calif.

Aguado, Eduardo, and Irwin Remson. 1974. "Groundwater Hydraulics in Aquifer Management." *Journal, Hydraulics Division, American Society of Civil Engineers* 100 (HY1): 103–18.

———. 1980. "Ground-Water Management with Fixed Charges." *Journal, Water Resources Planning and Management Division, American Society of Civil Engineers* 106 (WR2): 375–82.

Aguado, Eduardo, Irwin Remson, M. F. Pikul, and W. A. Thomas. 1974. "Optimal Pumping for Aquifer Dewatering." *Journal, Hydraulics Division, American Society of Civil Engineers* 100 (HY7): 860–77.

Aguado, Eduardo, N. Sitar, and Irwin Remson. 1977. "Sensitivity Analysis in Aquifer Studies. *Water Resources Research* 13(4): 733–37.

Akbar, A. M., M. D. Arnold, and O. H. Harvey. 1974. "Numerical Simulation of Individual Wells in a Field Simulation Model." *Society of Petroleum Engineers Journal* 14:315–20.

Alley, W. M., Eduardo Aguado, and Irwin Remson. 1976. "Aquifer Management under Transient and Steady-state Conditions." *Water Resources Bulletin* 12(5): 963–72.

Aronofsky, J. S., and A. C. Williams. 1962. "The Use of Linear

Programming and Mathematical Models in Underground Oil Production." *Management Science* 8(3): 374–407.

Bachmat, Y., J. D. Bredehoeft, B. Andrews, D. Holtz, and S. Sebastian. 1980. *Groundwater Management: The Use of Numerical Models.* Water Resources Monograph Service, vol. 5. Washington, D.C.: American Geophysical Union.

Bear, Jacob. 1972. *Dynamics of Fluids in Porous Media.* New York: Elsevier.

———. 1979. *Hydraulics of Groundwater.* New York: McGraw-Hill.

Bellman, R. E. 1960. *Introduction to Matrix Analysis.* New York: McGraw-Hill.

Bisschop, Johannes, Wilfred V. Candler, J. H. Duloy, and G. T. O'Mara. 1982. "The Indus Basin Model: A Special Application of Two-Level Linear Programming." *Mathematical Programming* 20: 30–38.

Bockstock, C. A., E. S. Simpson, and T. G. Roefs. 1977. "Minimizing Costs in Well Field Design in Relation to Aquifer Models." *Water Resources Research* 13(2): 402–26.

Bredehoeft J. D., and G. F. Pinder. 1973. "Mass Transport in Flowing Groundwater." *Water Resources Research* 9(1): 194–210.

Bredehoeft. J. D., and R. A. Young. 1970. "The Temporal Allocation of Groundwater: A Simulation Approach." *Water Resources Research* 6(1): 3–21

———. 1983. "Conjunctive Use of Ground Water and Surface Water for Irrigated Agriculture: Risk Aversion. *Water Resources Research* 19(5): 1111–21.

Cavendish, J. C., H. S. Price, and R. S. Varga. 1969. "Galerkin Methods for the Numerical Solution of Boundary Value Problems." *Society of Petroleum Engineers Journal* 246:204–20.

Charbeneau, R. J., and R. L. Street. 1979. "Modeling Groundwater Flow Fields Containing Point Singularities: A Technique for Singularity Removal. *Water Resources Research* 15(3): 583–94.

Chaudhry, M. T., J. W. Labadie, W. A. Hall, and M. L. Albertson. 1974. "Optimal Conjunctive Use Model for Indus Basin." *Journal, Hydraulics Division, American Society of Civil Engineers* 100(HY5): 667–87.

Cooper, Hilton, Jr. 1966. "The Equation of Groundwater Flow in Fixed and Deforming Coordinates." *Journal of Geophysical Research* 71(20): 4785–90.

Cummings, R. G., and J. W. McFarland. 1974. "Groundwater Management and Salinity Control." *Water Resources Research* 10(5): 909–15.

Dantzig, G. B. 1963. *Linear Programing and Extensions.* Princeton, N.J.: Princeton University Press.

Daubert, J. T., and R. A. Young. 1982. "Ground-Water Development in Western River Basins: Large Economic Gains with Unseen Costs." *Ground Water* 20(1): 80–85.

Deninger, R. A. 1970. "Systems Analysis of Water Supply Systems." *Water Resources Bulletin* 6(4): 573–79.

Douglas, James, Jr., and B. F. Jones. 1963. "On Predictor-Corrector Methods for Nonlinear Parabolic Differential Equations." *Journal of Applied Mathematics* 195–204.

Dreizen, Y. C., and Y. Y. Haimes. 1977. "A Hierarchy of Response Functions for Groundwater Management." *Water Resources Research* 13(1): 78–86.

Elango, K. 1981. "Discussion of Ground-Water Management with Fixed Charges by E. Aguado and I. Remson." *Journal, Water Resources Planning and Management Division, American Society of Civil Engineers* 107(WR2): 583.

Elango, K., and G. Rouve. 1980. "Aquifers: Finite-Element Linear Programming Model." *Journal, Hydraulics Division, American Society of Civil Engineers* 106(HY10): 1641–58.

Evans, Barbara, and Irwin Remson. 1982. "Closure: Ground-Water Management with Fixed Charges by E. Aguado and I. Remson." *Journal, Water Resources Planning and Management Division, American Society of Civil Engineers* 108(WR2): 237.

Fiering, M. B. 1965. "Revitalizing a Fertile Plain: A Case Study in Simulation and Systems Analysis of Saline and Waterlogged Areas." *Water Resources Research* 1(1): 41–61.

Futagami, T. 1976. "The Finite Element and Linear Programming Method." Ph.D. dissertation, Hiroshima Institute of Technology, Hiroshima, Japan.

Futagami, T., N. Tamai, and M. Yatsuzuka. 1976. "FEM Coupled with LP for Water Pollution Control." *Journal, Hydraulics Division, American Society of Civil Engineers* 102(HY7): 881–97.

Gillies, D. C. 1976. "A Model Analysis of Groundwater Availability near Carmel, Indiana." *U.S. Geological Survey Water Resources Investigation* 76–46.

Gorelick S. M. 1980. "Numerical Management Models of Groundwater Pollution." Ph.D. dissertation, Stanford University, Stanford, Calif.

———. 1982. "A Model for Managing Sources of Groundwater Pollution." *Water Resources Research* 18(4): 773–81.

Gorelick, S. M., and Irwin Remson. 1982a. "Optimal Location and Management of Waste Disposal Facilities Affecting Groundwater Quality." *Water Resources Bulletin* 18(1): 43–51.

———. 1982b. "Optimal Dynamic Management of Groundwater Pollutant Sources." *Water Resources Research* 18(1): 71–76.

Gorelick, S. M., Irwin Remson, and R. W. Cottle. 1979. "Management Model of a Groundwater System with a Transient Pollutant Source." *Water Resources Research* 15(5): 1243–49.

Haimes, Y. Y. 1977. *Hierarchical Analysis of Water Resources Systems.* New York: McGraw-Hill.

Haimes, Y. Y., and Y. C. Dreizen. 1977. "Management of Groundwater and Surface Water via Decomposition." *Water Resources Research* 13(1): 69–77.

Hayes, L. J., R. P. Kendall, and M. F. Wheeler. 1977. "The Treatment of Sources and Sinks in Steady-State Reservoir Engineering Simulations." In R. Vichnevetsky, ed. *Advances in Computer Methods for Partial Differential Equations II.* New Brunswick, N.J.: International Association for Mathematics and Computers in Simulation (IMACS).

Heidari, Manoutchehr. 1982. "Application of Linear System's Theory and Linear Programming to Groundwater Management in Kansas." *Water Resources Bulletin* 18(6): 1003–12.

Hillier, F. S., and G. J. Lieberman. 1974. *Introduction to Operations Research.* San Francisco: Holden-Day.

IBM. 1971. "Mathematical Programming System Extended (MPSX) Mixed Integer Programming (MIP) Program Description." *P.N. 5734-XM4,* IBM Program Development Center, Paris.

Illangasekare, Tissa, and H. J. Morel-Seytoux. 1982. "Stream-Aquifer Influence Coefficients as Tools for Simulation and Management." *Water Resources Research* 18(1): 168–76.

Jacob, C. E. 1944. "Notes on Determining Permeability by Pumping Tests under Water-Table Conditions." *U.S. Geological Survey Open-File Report.*

Jennings, A. A., D. J. Kirkner, and T. L. Theis. 1982. "Multicomponent Equilibrium Chemistry in Groundwater Quality Models." *Water Resources Research* 18(4): 1089–96.

Karash, R. 1962. "Quadratic Programming System (QPS) Applications Program Manual, APM-23." Cambridge, Mass.: MIT Information Processing Center.

Khepar, S. D., and M. C. Chaturvedi. 1982. "Optimum Cropping and Ground Water Management." *Water Resources Bulletin* 18(4): 655–60.

Konikow, L. F., and J. D. Bredehoeft. 1978. "Computer Model of Two-Dimensional Solute Transport and Dispersion in Ground Water." *U.S. Geological Survey Technical Water Resources Investigation* 7-C2.

Konikow, L. F., and D. B. Grove. 1977. "Derivation of Equations Describing Solute Transport in Ground Water." *U.S. Geological Survey Water Resources Investigation* 77–19.

Larson, S. P., Thomas Maddock III, and S. Papadopulos. 1977. "Optimization Techniques Applied to Groundwater Development." *Memoirs International Association Hydrogeology* 13:E57–E67.

Lee, A. S., and J. S. Aronofsky. 1958. "A Linear Programming Model for Scheduling Crude Oil Production." *Journal of Petroleum Technology* 213:51–54.

MacNeal, R. H. 1953. "An Asymmetrical Finite Difference Network." *Journal of Applied Mathematics* 11(3): 295–310.

Maddock, Thomas, III. 1969. "A Program to Simulate an Aquifer Using Alternating Direction Implicit-Iterative Procedure." *U.S. Geological Survey Open-File Report* 70211.

———. 1972a. "Algebraic Technological Function from a Simulation Model." *Water Resources Research* 8(1): 129–34.

———. 1972b. "A Ground-Water Planning Model: A Basis for a Data Collection Network." Paper presented at the International Symposium on Uncertainties in Hydrologic and Water Resource Systems. International Association of Hydrological Sciences, University of Arizona, Tucson.

———. 1974a. "The Operation of Stream-Aquifer System under Stochastic Demands." *Water Resources Research* 10(1): 1–10.

———. 1974b. "Nonlinear Technological Functions for Aquifers Whose Transmissivities Vary with Drawdown." *Water Resources Research* 10(4): 877–81.

Maddock, Thomas, III, and Y. Y. Haimes. 1975. "A Tax System for Groundwater Management." *Water Resources Research* 11(1): 7–14.

Martin, W. E., T. Burdak, and R. A. Young. 1969. "Projecting Hydrologic and Economic Relationships in Groundwater Basin Management." *American Journal of Agricultural Economics* 15(5): 1593–97.

Molz, F. J., and L. C. Bell. 1977. "Head Gradient Control in Aquifers Used for Fluid Storage." *Water Resources Research* 13(4): 795–98.

Morel-Seytoux, H. J. 1975a. "A Simple Case of Conjunctive Surface-Ground Water Management." *Ground Water* 13(6): 505–15.

———. 1975b. "Optimal Legal Conjunctive Operation of Surface and Ground Waters." Paper presented at the Second World Congress, International Water Resources Association, New Delhi, India.

Morel-Seytoux, H. J., and C. J. Daly. 1975. "A Discrete Kernel Generator for Stream-Aquifer Studies." *Water Resources Research* 11(2): 253–60.

Morel-Seytoux, H. J., G. Peters, Robert Young, and Tissa Illangasekare. 1980. "Groundwater Modeling for Management." Paper presented at the International Symposium on Water Resource Systems, Water Resources Development Training Center, University of Roorkee, Roorkee, India.

O'Leary, D. P. 1978. "Linear Programming Problems Arising from Partial Differential Equations." In I. S. Duff and G. W. Stewart, eds. *Sparse Matrix Proceedings.* Philadelphia: Society for Industrial and Applied Mathematics.

Padmanabhan, G. 1981. "Discussion of Ground-Water Management with Fixed Charges by E. Aguado and I. Remson." *Journal, Water Resources Planning and Management, American Society of Civil Engineers* 107(WR2): 584.

Pinder, G. F., and J. D. Bredehoeft. 1968. "Application of a Digital Computer for Aquifer Evaluation." *Water Resources Research* 4(5): 1069–93.

Reddell, D. L., and D. K. Sunada. 1970. "Numerical Simulation of Dispersion in Groundwater Aquifers." Hydrology Paper 41, Colorado State University, Fort Collins.

Remson Irwin, and S. M. Gorelick, 1980. "Management Models Incorporating Groundwater Variables." In Daniel Yaron and C. S. Tapiero, eds. *Operations Research in Agriculture and Water Resources.* Amsterdam: North-Holland.

Remson, Irwin, G. M. Hornberger, and F. J. Molz. 1971. *Numerical Methods in Subsurface Hydrology.* New York: Wiley-Interscience.

Rosenwald, G. W. 1972. "A Method for Determining the Optimum Location of Wells in an Underground Reservoir." Ph.D. dissertation, University, of Kansas, Lawrence.

Rosenwald, G. W., and D. W. Green. 1974. "A Method for Determining the Optimum Location of Wells in a Reservoir Using Mixed-Integer Programming." *Society of Petroleum Engineers Journal* 14: 44–54.

Rubin, Jacob, and R. V. James. 1973. "Dispersion-Affected Transport of Reacting Solutes in Saturated Porous Media: Galerkin Method Applied to Equilibrium-Controlled Exchange in Unidirectional Steady Water Flow." *Water Resources Research* 9(5): 1332–56.

Saunders, Michael. 1977. "MINOS Systems Manual." *Systems Optimization Laboratory Technical Reports* 77–31.

Schwarz, Joshua. 1971. "Linear Models for Groundwater Management." *Rep. P.NET/71/062,* Water Planning for Israel Ltd., Tel Aviv, Israel.

———. 1976. "Linear Models for Groundwater Management." *Journal of Hydrology* 28:377–92.

Segol, Genivive, and G. F. Pinder. 1976. "Transient Simulation of Saltwater Intrusion in Southeastern Florida." *Water Resources Research* 12(1): 65–70.

Sophocleous, Marios. 1980. "Hydrogeologic Investigations in the Pawnee Valley." *Kansas Geological Survey Open-File Report* 61.

Taylor, O. J. 1970. "Optimization of Conjunctive Use of Water in a Stream-Aquifer System Using Linear Programming." *U.S. Geological Survey Professional Paper* 700-C: C218–C221.

Taylor, O. J., and R. R. Luckey, 1974. "Water-Management Studies of a Stream-Aquifer System: Arkansas River Valley, Colorado." *Ground Water* 12(1): 22–38.

Theis, C. V. 1935. "The Relation between Lowering of the Piezometric Surface and the Rate and Duration of Discharge of a Well Using Ground Water Storage." *Eos Transactions of the American Geophysical Union* (April): 519–29.

Trescott, P. C., G. F. Pinder, and S. P. Larson. 1976. "Finite-Difference Model for Aquifer Simulation in Two-Dimensions with Results of Numerical Experiments." *U.S. Geological Survey Technical Water Resources Investigation* 7-C1.

Tui, H. 1964. "Concave Programming under Linear Constraints." *Dokl. Akad Nauk SSSR* 159: 32–35.

Tyson, N. H., and E. M. Weber. 1964. "Groundwater Management for the Nation's Future: Computer Simulation of Groundwater Basins." *Proceedings of the American Society of Civil Engineering* 90 (HY4): 59–77.

Valocchi, A. J., R. L. Street, and P. V. Roberts. 1981. "Transport of Ion-Exchanging Solutes in Groundwater: Chromatographic Theory and Field Simulation." *Water Resources Research* 17(5): 1517–27.

Van Everdingen, A. F., and W. Hurst. 1949. "The Application of the Laplace Transformation to Flow Problems in Reservoirs." *Transactions, American Institute of Mining, Metallurgy, and Petroleum Engineering* 186: 305–24.

Wagner, H. M. 1975. *Principles of Operations Research.* Englewood Cliffs, N.J.: Prentice-Hall.

Wattenbarger, R. A. 1970. "Maximizing Seasonal Withdrawals from Gas Storage Reservoirs." *Journal of Petroleum Technology* (Aug.): 994–98.

Williamson, A. S., and J. E. Chappelear. 1981. "Representing Wells in Numerical Reservoir Simulation, 1, Theory." *Society of Petroleum Engineers Journal* 21: 323–38.

Willis, Robert. 1973. "Optimization of the Assimilative Waste Capacity of the Unsaturated and Saturated Zones of an Unconfined Aquifer System." Ph.D. dissertation, University of California, Los Angeles.

———. 1976a. "Optimal Groundwater Quality Management: Well Injection of Waste Water." *Water Resources Research* 12(1): 47–53.

———. 1976b. "Optimal Management of the Subsurface Environment." *Hydrological Sciences Bulletin* 21(2): 333–43.

———. 1976c. "A Management Model for Determining Effluent Standards for the Artificial Recharge of Municipal and Industrial Wastewaters." In Zubir Saleem, ed. *Advances in Groundwater Hydrology.* Minneapolis: American Water Resources Association.

———. 1977. "Optimal Groundwater Resource Management Using the Response Equation Method." In W. G. Gray and G. F. Pinder, eds. *Finite Elements in Water Resources.* London: Pentech.

———. 1979. "A Planning Model for the Management of Groundwater Quality." *Water Resources Research* 15(6): 1305–12.

———. 1983. "A Unified Approach to Regional Groundwater Management." In J. S. Rosenshein and G. D. Bennett, eds. *Groundwater Hydraulics.* Washington, D.C.: American Geophysical Union.

Willis, Robert, and B. A. Newman. 1977. "Management Model for Groundwater Development." *Journal, Water Resources Planning and Management Division, American Society of Civil Engineers* 103(WR1): 159–71.

Young, R. A., and J. D. Bredehoeft. 1972. "Digital Computer Simulation for Solving Management Problems of Conjunctive Groundwater and Surface Water Systems." *Water Resources Research* 8(3): 533–56.

Yu, W., and Y. Y. Haimes. 1974. "Multilevel Optimization for Conjunctive Use of Ground and Surface Water." *Water Resources Research* 10(4): 625–36.

Comment

Robert G. Thomas

Gorelick surveys an impressive range of methods. The models on policy evaluation and allocation are, however, much closer to the concerns of this symposium volume in their analysis of management options for efficiently allocating groundwater.

Are the methods discussed actually being put into practice? Their use is limited so far, which is to be expected since the methods have only recently been developed. Physical measurements on aquifer processes are difficult and expensive. Since these modeling methods efficiently use the limited information from physical measurements to improve groundwater management, the author expects them to be widely used in the future. These computer-based methods have been incorporated extensively into the engineering curricula of the universities in the past decade or so. As more engineering graduates trained in these methods enter into professional practice, the adoption of such methods will accelerate.

8

New Approaches to Using Mathematical Programming for Resource Allocation

Peter P. Rogers, Joseph J. Harrington, and Myron B. Fiering

Many conventional uses of mathematical programming to solve water resource problems—especially irrigation problems—tend to pursue optimality as though it were inherently noble. An objective function is neatly superimposed on such constraints as physical quantities, resulting in an attractive algorithm which reduces the indeterminacy of the problem and which leads inexorably to a global optimum. Unfortunately, such conventional uses do not fully reflect the true concerns of the decisionmakers. Although often used in some vague way by erudite planners, mathematical programming is rarely used to make the critical decisions involved in project planning. Our purpose is not to identify all the abuses surrounding such models and their algorithmic solutions; the social and behavioral sciences have much to contribute in addressing these problems. Rather, our chapter will examine why some of the negotiations involved in planning are inevitably so difficult, and what might be done about it.

Limitations of Models

Using models to optimize resource allocation has been enthusiastically taken up by agricultural experts. The models of linear programming and its derivatives are so commonly used that the invention of a new algorithm was reported on the front page of the *New York Times.* There are problems, however, which receive little attention from practitioners—problems dealing with whether the model is adequate and with how the results are evaluated once the model has been specified and run. These problems are of course related, since how the model is specified largely determines the quality of the results.

Alternative solutions, or multiple optima, arise in the actual practice of linear programming because resource planning problems typically have many available choices, often with little to distinguish one from another. Resource planners who use mathematical programming usually ignore these alternative solutions simply because most machine codes do not generate them, although the codes might indicate which nonbasic variables could be introduced without changing the optimal value of the objective function. Uzawa (1958) suggested an algorithm for generating all alternative optimal solutions, but we have no evidence of its implementation. Gidley (1981) extended it to a wider class of problems, and Harrington (1978) proposed that these alternatives be more closely explored. Gidley (1981) outlined methods for exploring these alternative solutions, making the point that nonunique dual solutions present serious difficulties with many economic interpretations of the shadow prices.

Many solutions have values near the optimal solution. Harrington (1978) and Gidley (1981) explored this phenomenon for a variety of water resource problems, but their results apply as well to more general problems in resource planning and allocation. They found many solutions whose values were within a few percentage points (say one, two, or five) of the global optimum, and they modified standard machine codes to identify these nearly optimal solutions. Since linear combinations of solutions are of course also solutions, the number of nearly optimal solutions is infinite. Wide variation in the decision variables therefore does not necessarily imply large shifts in the value of the objective function.

If the coefficients of the linear program are random variables with known densities, the distribution of the objective function can sometimes be written to cast the constraints in a probabilistic setting, or expected values or some specified quantile can be used to form a deterministic problem with a unique solution. But if the solution or activity vector has stochastic components so that we have a distribution of values for each decision variable, the distribution of the sum of products must be assessed to identify a solution or set of solutions to the original programming problem. That is, the decisionmakers must be prepared to choose between a more stable solution with a smaller return and a more variable solution with a larger return, where return is perhaps the expected value of the objective function, and stability is a characteristic inversely related to its standard deviation.

Another area in which we propose a new approach to mathematical programming involves resilient solutions characterized by clusters of core variables. Suppose a given element of the solution vector attains approximately the same value for nearly all alternative solutions within a few percentage points of the optimum. That variable, which survives under nearly all interesting objective functions, is called a *tight* variable. Conversely, a *diffuse* variable attains widely different values, including zero; its proper specification depends on which model or set of parameters is chosen.

A reasonable negotiating strategy would be to fight to retain the tight variables in the solution, thereby eliminating the dependence of the objective function on them and reducing the dimensions of the remaining decision space which is subject to negotiation. If most of the participants in the decisionmaking process, including those who demonstrate subtle but unarticulated differences in the objective function, can agree on a few of the tight variables to put in (or take out of) the decision vector, the resulting negotiation might influence only a very small fraction of the value of the objective function. The truly important or governing issues might not even be expressed in the formal objective criterion, which might dissipate some of the controversy over the decision and might bring into focus the remaining differences on political and institutional issues which are difficult if not impossible to incorporate in the surrogate function.

Most mathematical models of planning problems require for their solution a vector of decisions X which typically encompasses temporal and spatial variability. Some of its elements reflect distinct decision variables while others reflect changes in the magnitude of a given operating decision of the system. For many problems, the decision vector X can reasonably be identified and implemented directly; that is, an optimal decision could be identified, and some components could then be scheduled for immediate execution while details are planned for the financing and scheduling of other components to be added later.

Suppose a large agricultural irrigation system, for example, is to be constructed, but is to be installed in stages over several years. A complete construction schedule could be identified whereby increments of the canal system would be added at specified times—on the assumption that enough is known about the system, that its model and coefficients are suitably deterministic, and that the effect of the partially completed facility is well enough defined to allow the entire optimal program to be identified at the start. Under these conditions, a complete system could be clearly planned through all its stages.

It is more likely that new technology, additional data, changing climates (political, economic, and meteorological), and effects of the partially completed system will make it more attractive to maintain flexibility and to modify some of the design decisions initially selected as part of the entire construction schedule. We would like to identify those decisions which will remain intact (or nearly so) under a variety of changing goals and conditions. Their property, known as system *resilience,* has been the object of much inquiry over the past few years (Holling 1978; Fiering and Holling 1974; Matalas and Fiering 1977; Fiering 1978, 1982a, 1982b, 1982c, 1982d; Hashimoto and others 1982a, 1982b). Many of these studies emphasize how system resilience applies to ecology and water resource planning, but their concepts can be extended to general systems planning as well.

In principle, each decision variable x_i in the vector X' of decisions at, or near, the optimum has a density function $f_i(x_i)$, where the subscript on the function f indicates that the density function might differ for each decision variable. Some of the densities might be compact while others might be diffuse. Variables with compact densities do not vary significantly over the range of admissible decisions and could comfortably be recommended without further concern. These decisions survive as elements of most optimal or nearly optimal solutions, and the tightness of their densities suggests that modifying them carries a significant penalty for the system criterion function. Variables with diffuse densities constitute the wide range of alternative combinations which might form the solution, and the values of these variables might be correlated so that they appear (or disappear) in clusters or sets. Clearly, the correlation structure among the diffuse decision variables is more amenable to analysis if its dimensions are kept small by first removing from the system all those variables for which the density is tight (and the decision therefore clear).

We have discussed the difficulties involved in specifying a model, defining its parameters, and selecting a solution (preferably selecting one of the many optimal solutions or a more attractive nearly optimal solution instead of the global optimum to which the solution first is directed). Our recent work explores whether it is worthwhile to refine the solution, that is, whether the solution is adequate. We offer some examples of the difficulties.

Examples of Difficulties

The objective function of a nearly optimal solution is often rather flat. This common observation is discussed by Kühner and Harrington (1975a, 1975b), Muhich (1966),

Ortolano (1969), Gupta and Rosenhead (1968), and others. The many feasible solutions might be nearly optimal but might have very different physical features, such as the size, location, and timing of the facilities, or other, more subtle aspects that affect financial subsidies and tax incentives, regulations on environmental quality, shifts in population or consumer preference, measures of equity, and so on. When all of these issues are included, a suboptimal or inferior solution might not only be good enough, but might when closely examined be actually more desirable than the theoretical optimum.

Harrington (1978), on whose work this section largely is based, writes that among the technical circumstances leading to the persistent search for the global optimum has been the ubiquitous use of linear programming since the earliest applications of systems analysis and operations research. Using several river systems as examples, he shows how mathematical programming can lead to the difficulties suggested above.

Harrington's first example is from the well-known book by O'Laogharie and Himmelblau (1974), which analyzes a semihypothetical region based on the Maule River basin in Chile. O'Laogharie and Himmelblau's model includes consumptive use for irrigation and nonconsumptive use for hydroelectric power production. Four dams at specified nodes in the model regulate the flow, and ten unregulated streams provide the physical inputs to the system. The specified schedule of increasing energy and irrigation needs is to be met by building a suitable combination of dams at three possible dam sites, and any of three sizes of dams may be built at each site. The problem is when and where to locate each dam, given the planning horizon of fifty years that has been chosen.

O'Laogharie and Himmelblau used Fulkerson's out-of-kilter algorithm and Little's branch-and-bound algorithm for their calculations. Although the details are unimportant here, O'Laogharie and Himmelblau give the vector of decisions for twelve feasible solutions. They remark on the efficiency of their heuristic algorithm for finding an initial feasible solution (p. 122): "The return from the first feasible solution was within 3.3% of the final optimum." Instead of enjoying this result, perhaps they should have noted the insensitivity of the objective function to a wide range of possible combinations! Reviewing the characteristics of the twelve feasible solutions shows that many permutations of the construction schedule result in very close values of the objective function, making it difficult to understand how a practical decisionmaker can distinguish among such close values. It is likely to be interesting to investigate more fully, along a negotiation frontier, some of the fiscal, institutional, reliability, and equitability characteristics of those feasible solutions that are nearly optimal.

In another example cited by Harrington, Russell and others (1970) studied plans to expand capacity in municipal water supply systems designed to protect against severe and protracted drought; their empirical work was based on data from Massachusetts. Population growth and per capita demand are given by exponential functions, and the present value of the capital cost of an expansion is given by the conventional exponential relationship. The present value of expected drought losses was formulated in terms of demand, safe yield, capacity expansion as a function of time, and a variety of system parameters. The objective function was to minimize the sum of present values of capital costs and expected drought losses, with constraints limiting the solutions to nonnegative values, an upper bound on the interval between capacity increments, and a terminal condition on the total incremental capacity.

Russell's results contain the following statement: "Most striking, perhaps, is the closeness of agreement between the two sets of costs. The difference between the two figures is only once larger than 5% and is often less than 1%." Unfortunately, the related decision vectors are not available (Russell, personal communication, 1976), but the remark suggests that further investigation into the characteristics of such nearly optimal solutions would be useful. In this instance the two solutions referred to by Russell include a random search technique with systematic sweetening or improvement and the more formal nonlinear optimization procedure known as Zoutendijk's method of feasible directions. In any event, the important point is how closely the solutions can be made to agree by simply

Table 8-1. ***Coefficient Matrix for Linear Program Problem***

	Structural vectors					Slack vectors					Right-hand side
z	x_1	x_2	x_3	x_4	x_5	x_6	x_7	x_8	x_9	x_{10}	
0	0.02274	0.02770	0.05862	0.09249	0.09081	1	0	0	0	0	148
0	0.31772	0.27870	0.70812	0.96956	1.00356	0	1	0	0	0	1,800
0	0.05253	0	0.11681	0	0	0	0	1	0	0	182
0	0.02555	0.07523	0.05485	0.21186	0.42324	0	0	0	1	0	234
0	0	0.08370	0	0.30910	0.08650	0	0	0	0	1	234
−1	1.56	0.84	2.79	3.81	2.14	0	0	0	0	0	0

Note: The optimal solution is: $z^* = 8{,}021.112$, $x_1 = 3{,}464.69$, $x_4 = 686.67$.
Source: Babbar (1955).

adapting solutions over time to take advantage of increased and improved information as the decisionmaker accumulates it.

These examples suggest that analysts should be more concerned with using optimization methods to generate nearly optimal alternatives than they currently are. We agree that there are inadequacies—often serious—in representing actual objectives by the stated objective function, and the value of any objective function in a problem of realistic size is likely to be subject to substantial errors. Moreover, the function might be quite insensitive to substantially different designs which accommodate wide shifts, correlated through the constraint matrix, in the decision variables.

One final example is based on a sample problem provided by Babbar (1955) and extended by Dumas (1983) as part of a continuing research project at Harvard University. The matrix of constraints and the objective function are given in Table 8-1. Babbar also provides some variances for the elements of A, b and c. Admittedly, this density of information rarely is available for linear programming models and might have to be estimated rather broadly. A useful research topic for future study is the extent to which the sensitivity of the conclusions presented here is affected by the range of variability expressed by the stability of these parameters. The moments of two of the decision variables are 3,863.57 for x_1, with a coefficient of variation of 44.9 percent, and 712.65 for x_4, with a coefficient of variation of 52.3 percent (see Table 8-2). The probability that x_1 would come out of the basis is 0.0053; x_1 is thus remarkably stable. The probability that x_4 would come out of the basis is 0.022; although less stable, x_4 is still perhaps stable enough to be a variable which the decisionmakers would agree to put into the solution. These results are shown graphically in Figures 8-1 and 8-2.

Conclusion

Our purpose in this chapter is not to attack mathematical programming. Even if we were so inclined, such an effort would be futile. Our point is that systems analysis, or optimization (whichever term one prefers), is frequently not so much misused as underused in its capacity to identify a wide range of alternative solutions whose characteristics might be quite distinct and which might form the basis for negotiation between potential beneficiaries. We propose that systems analysis be used to identify the negotiation frontier, as part of a larger activity which looks carefully at models, particularly the very large ones which have recently come to be accepted without significant consideration of the reliability of their parameters and the relative flatness of their objective functions. Such efforts to use models more humanely and to reduce their scale so that they become less of a computational burden may provide more insight into the decisionmaking process.

Table 8-2. *Stochastic Programming Results*

	x_1	x_4
μ	3,863.57	712.65
σ^2	3.01440e+06	1.38984e+05
CV	4.49379e−01	5.23126e−01
μ_3	1.76018e+10	9.22385e+07

Note: CV = coefficient of variation; μ_3 = skewness.
Source: Dumas (1983).

Figure 8–1. *Probability Density Function of x_1, Optimal Basis*

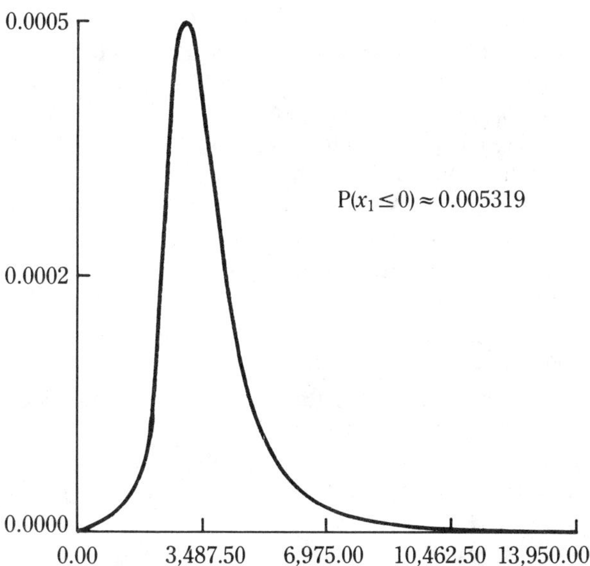

Figure 8–2. *Probability Density Function of x_4, Optimal Basis*

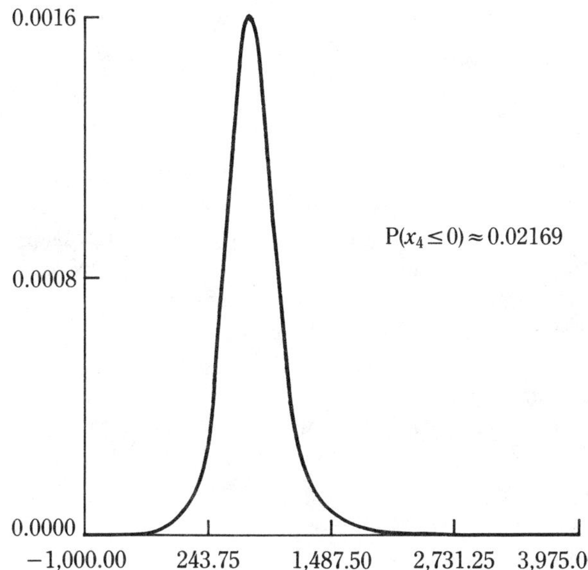

References

Babbar, M. M. 1955. "Distribution of Solutions of a Set of Linear Equations with an Application for Linear Programming." *Journal of the American Statistical Association* 50:854–69.

Dumas, W. R. 1983. Memoranda for Ph.D. dissertation. Harvard University, Cambridge, Mass.

Fiering, Myron B. 1978. "Compressed Policy Analysis." Memorandum, International Institute of Applied Systems Analysis, Laxenburg, Austria. Adapted for inclusion in Holling 1978.

———. 1982a. "A Screening Model to Quantify Resilience." *Water Resources Research* 18(1): 27–32.

———. 1982b. "Alternative Indices of Resilience." *Water Resources Research* 18(1): 33–40.

———. 1982c. "Estimates of Resilience Indices by Simulation." *Water Resources Research* 18(1): 41–50.

———. 1982d. "Estimating Resilience by Canonical Analysis." *Water Resources Research* 18(1): 51–57.

Fiering, Myron B., and C. S. Holling. 1974. "Management and Standards for Perturbed Ecosystems." *Agro-Ecosystems* 1:301–21.

Gidley, James. 1981. "Nearly Optimal Decisions in Water-Resources Planning." Ph.D. dissertation, Harvard University, Cambridge, Mass.

Gupta, S. K., and J. Rosenhead. 1968. "Robustness in Sequential Investment Decisions." *Management Science* 15(2).

Harrington, J. J. 1978. "Nearly Optimal Alternative Decisions in Planning Water-Resource Systems." In Myron B. Fiering, J. J. Harrington, and Peter P. Rogers. *Standards, Optimality and Resilience in Water-Resource Management.* Report 2, Office of Water Research and Technology, Harvard University, Cambridge, Mass.

Hashimoto, Tsuyoshi, and others. 1982a. "Reliability, Resiliency, and Vulnerability Criteria for Water Resource System Evaluation." *Water Resources Research* 18(1): 14–20.

———. 1982b. "Robustness of Water Resources Systems." *Water Resources Research* 18(1): 21–26.

Holling, C. S., ed. 1978. *Adaptive Environmental Assessment and Management.* Laxenburg, Austria: International Institute of Applied Systems Analysis.

Kühner, Jochen, and J. J. Harrington. 1975a. "Mathematical Models for Developing Regional Solid Waste Management Policies." *Journal of Engineering Optimization* 1:237–56.

———. 1975b. "Discussion of 'Capital Cost Minimization of Drainage Networks' by Dajani and Hasit." *Journal, Environmental Engineering Division, American Society of Civil Engineers* 101(2): 270–71.

Matalas, N. C., and Myron B. Fiering. 1977. "Water-Resource Systems Planning." In James Wallis, ed. *Climate, Climatic Change and Water Supply.* Washington, D.C.: National Academy of Sciences.

Muhich, Anton. 1966. "Capacity Expansion of Public Works." Ph.D. dissertation, Harvard University, Cambridge, Mass.

O'Laogharie, D. T., and D. M. Himmelblau. 1974. *Optimal Expansion of a Water-Resource System.* New York: Academic Press.

Ortolano, Leonard. 1969. "Artificial Aeration and the Capacity Expansion of Waste-Water Treatment Facilities." Ph.D. dissertation, Harvard University, Cambridge, Mass.

Russell, C. S., and others. 1970. *Trout and Water Supply: Implications of the Massachusetts Experience for Municipal Planning.* Baltimore, Md.: Johns Hopkins University Press.

Uzawa, Hirofumi. 1958. "An Elementary Method for Linear Programming." In Kenneth Arrow and others, eds. *Studies in Linear and Non-Linear Programming.* Stanford, Calif.: Stanford University Press.

Comment

Nathan Buras

Rogers, Harrington, and Fiering focus on the inadequate use of mathematical programming models in solving water resource problems in general, and irrigation problems in particular. The inadequate use manifests itself in two opposite directions: on the one hand, analysts and planners pursue a unique optimal solution without regard to what may be the true concerns of decisionmakers and managers; on the other hand, models were insufficiently used to identify wider ranges of solutions which might form the basis of negotiation toward an accepted solution. Most of the models so inadequately used are those of linear programming and its derivatives.

An interesting and seldom exhibited attribute of linear programming applied to problems in water resources development is the existence of multiple solutions in the neighborhood of the optimum. Since the value of the objective function for these solutions varies by a few percentage points—often within the accuracy range of the data—these solutions may also be considered as optimal. We may thus be faced with multiple optima which, since we are in a linear programming mode, are associated with nonunique dual solutions. This nonuniqueness is difficult to interpret, particularly for shadow prices.

The authors identify three sources of difficulty in the application of linear programming to problems of resource allocation. One source is misspecification of the model, or the formulation of functional relationships in a selective manner, ignoring relationships which are difficult to quantify, such as those involving social goals and objectives. Variables reflecting these goals may be *diffuse* (to use the authors' terminology) in that they attain widely different values depending on the chosen model.

A second source of difficulty is the uncertainty inherent in the parameters of a linear programming formulation. The cost coefficients entering the objective function, the available resources, and the elements of the matrix of coefficients of the constraining set have some probability densities that are fairly "tight" and impart greater confidence to an optimal solution; others are "diffuse" and thus diminish our confidence in the results. In connection with this aspect of mathematical programming, the authors suggest that one should seek decisions that are *resilient*—that is, those that will remain invariant (or nearly so) under a broad range of changing objectives and constraints.

The third source of difficulty or ambiguity stems from the selection of one of the general optimal (or nearly optimal) solutions that would be most attractive politically, rather than insisting on the global optimum produced by the linear programming model, or vice versa.

By confining their discussion to linear programming models and their derivatives, the authors miss an important source of difficulty and ambiguity—the time element. Because the time element is missing from their chapter, as it is from most linear programming formulations, the dynamics of the system under consideration—resource allocation—are missing as well. As a result, the planning process is incomplete, since the changes that transform the current state of the region (or river basin) into a desired state are completely ignored. Attempts to link a sequence of linear programming formulations, each representing a different stage of development in an integrated system, resulted in enormously large matrices of the constraining set and in serious computational difficulties.

Dynamic programming should be revisited and refined into an effective instrument for planning resource allocation in a dynamic setting. The functional solutions of dynamic programming (as contrasted with the point solutions of linear programming) will help to identify nearly optimal solutions and may provide more insight into the important negotiations of the decisionmaking process.

9

Modeling Efficient Conjunctive Use of Water in the Indus Basin

Gerald T. O'Mara and John H. Duloy

Areas that use surface and groundwater conjunctively may require special policies to coordinate water use by farmers with the response of the stream-aquifer system. This chapter examines alternative policies for achieving more efficient conjunctive use in the Indus Basin of Pakistan. Results are presented from our simulation experiments which analyze conjunctive use in the region's irrigation system.

The Indus Basin has been the subject of a number of studies in the past several decades, as the long-run effects of large-scale canal irrigation in the flat, slowly draining Indus plains—for example, waterlogging and salinization—became increasingly troublesome (Chaudry and others 1974; Fiering 1965; Greenman, Swarzenski, and Bennett 1967; IACA 1966; Lieftinck, 1968; White House 1964; Tipton and Kalmbach Inc. 1967; WAPDA–Harza Engineering Co. 1963; WAPDA 1979). The studies of WAPDA–Harza, White House, IACA, Tipton and Kalmbach, and Lieftinck, Sadove, and Creyke were unanimous in recommending large-scale public tubewell development for vertical drainage for efficient conjunctive use, although the long-term need for horizontal drainage to remove salt accumulations was recognized. These recommendations were incorporated in the government's investment program of the 1960s and 1970s. In retrospect, these studies underestimated both the strength of the incentives for private tubewell investment and the difficulties in implementing and managing a massive public tubewell operation. These difficulties and the Pakistani response have been carefully documented by Johnson (chapter 5, this volume). At present, about three-quarters of tubewell withdrawals are by private agents, and the problem of achieving efficient conjunctive use has been completely transformed from that envisioned by the scenarios of the 1960s (WAPDA 1979).

The problem of efficient conjunctive use is inherently dynamic, and much of the early work was explicitly dynamic (Buras 1963; Burt 1964, 1966, 1967; Bredehoeft and Young 1970; Brown and Deacon 1972; Noel, Gardner, and Moore 1980). Dynamic optimization suffers from the curse of dimensionality, however, and dynamic models must necessarily simplify to the point that significant aspects of real-world applications must be suppressed. This dilemma has led to modeling methods that are not explicitly dynamic, for example, static steady-state models, such as that of Rogers and Smith (1970). Excellent reviews of the state of the art in modeling groundwater and stream-aquifer systems are found in the work of Bachmat and others (1980) and Gorelick (chapter 7, this volume). Modeling that incorporates more real-world detail has uncovered potential conflicts between public and private interests stemming from the physical link created when individual well operators rely on a common aquifer—that is, a physical external diseconomy or externality. This problem has been characterized as a hierarchical or multilevel one (Yu and Haimes 1974). One device for closing the gap between public and private interests caused by the externality is some form of tax or quota on pumping. This solution has been explored in a number of studies (Bredehoeft and Young 1970; Maddock and Haimes 1975; Feinerman and Knapp 1983). This chapter uses a static deterministic formulation, multilevel structure, and a tax-subsidy instrument in analyzing efficient conjunctive use.

This study originated in the World Bank's involvement as the executing agent of a "master planning" effort by the Water and Power Development Authority (WAPDA) of Pakistan. The United Nations Development Programme (UNDP) funded WAPDA's preparation of a Revised Action Programme (RAP) for irrigation investments in the basin, which updated the Action Programme set out by a similar planning effort in the 1960s.

The resulting Indus Basin family of models has demonstrated a capacity for providing answers to a variety of policy-relevant questions—for example, issues of mechanization, technical change, and agricultural price policy, as well as irrigation system management and evaluation of investment projects and programs. To date these models have been used in three project appraisals in the World Bank, and there are several more prospective applications of this type. However, these models may well have more potential use to the government of Pakistan policymakers than to the World Bank. This possibility was noted by the Planning Division of WAPDA, and a team of WAPDA programmers and systems analysts was trained at the Bank to effect the transfer of this modeling technology to Pakistan.

The next section offers a description of the Indus model family structure, followed by a review of model validation. Results are then presented from our simulation experiments which analyze conjunctive use in the Indus irrigation system and assess alternative policies.

Model Structure

In the past, many economic models concerned with policy and planning have been straightforward optimizing models. While admirably direct, this approach neglects an important aspect of the economic policy environment. Models designed for policy analysis normally involve two kinds of agents: policymakers and policy receivers. If the policy receivers are optimizing agents, one is faced with a hierarchical decisionmaking problem. In the case of the Indus Basin models the government plays the role of policymaker, while the farmers play the role of policy receivers. The government decides on water-related investments and surface water allocations and sets (some) agricultural prices, taxes, and subsidies. The farmers, in turn, react to the setting of these policy instruments by using water (both surface and groundwater) and other inputs, making private investments in tractors, tubewells, and so on, and choosing cropping patterns to maximize their own welfare.

The general strategy of the Indus family of models is to separate analytically the two types of decisionmakers by simulating the response of the policy receivers to the actions of the policymakers in the model and to represent the actions of the policymakers by changes in model structure or parameters. There are exceptions to this rule, however, particularly in the policy constraints of physical externalities (for example, surface water–groundwater interactions) that are not recognized by policy receivers.

The basic structure of the Indus Basin model can be visualized as follows. The entire basin is partitioned into fifty-three irrigated regions, referred to as polygons. Each polygon is essentially homogeneous with respect to groundwater and preserves boundaries that are significant to the groundwater-aquifer system. Links in water supply that arise from seepage of surface water to the aquifer and withdrawal of groundwater by tubewells or capillary action, as well as from underflows between polygons, are explicitly modeled for each polygon, thereby interlocking the polygons. Each polygon also receives surface water on a monthly basis from one or more control points of the surface water delivery system. Figure 9-1 presents the schematic diagram of the Indus Basin Irrigation System and identifies the control points where diversions to individual canal commands are made.

To embed the differences in soils, climate, and so on that create regional comparative advantage in different crops, model cropping technologies were specialized for nine agroclimatic zones (ACZs). The mapping of polygons into ACZs is given in Figure 9-2. The data for the differentiation of the basin into ACZ-specific cropping technologies were largely derived from the 2,000-farm sample of the Master Planning Agroeconomic Survey. Table 9-1 gives average cropping patterns for the ACZs as some evidence of the appropriateness of the partition.

The surface water distribution system is superimposed on the complex mapping of groundwater areas (polygons), canal commands, and ACZs by a network formulation. All of the flows of the schematic for the Indus Basin Irrigation System are represented as directed flows along segments, which are the arcs between one control point or node and another.

In addition to the above-mentioned water constraints, each polygonal model has embedded in it a single-farm–level model to characterize the agricultural production system of the area. Such a farm-level model simulates the resource allocation choices of a single representative farmer who determines the production and disposition of eleven crops and four livestock commodities. Exogenous resource limitations are imposed on land and labor. The water supply and demand constraint of each farm-level model includes estimates of water available from rainfall, evapotranspiration from the aquifer, and canals and tubewells. When used to evaluate water allocation policies, canal water allocations are endogenous, as is the volume of private tubewell pumping. The model maximizes the objective function, which is the sum of polygonal farm incomes less polygonal risk premium terms. The risk term essentially linearizes a nonlinear mean standard deviation of income trade-off surface. Moreover, farm income enters into linear constraints for family consumption and this formulation can be shown to be equivalent to maximizing a nonlinear utility or weighting function that places great emphasis on meeting family consumption needs.

All polygonal models have a groundwater balance constraint, which may be deleted in certain solutions. In brief, this constraint forces equality between additions to and withdrawals from the aquifer. The presence of this constraint is crucial to the solutions of the basinwide model with endogenous canal water allocation. Without it the

Figure 9-1. *Structure of the Indus Basin Irrigation System*

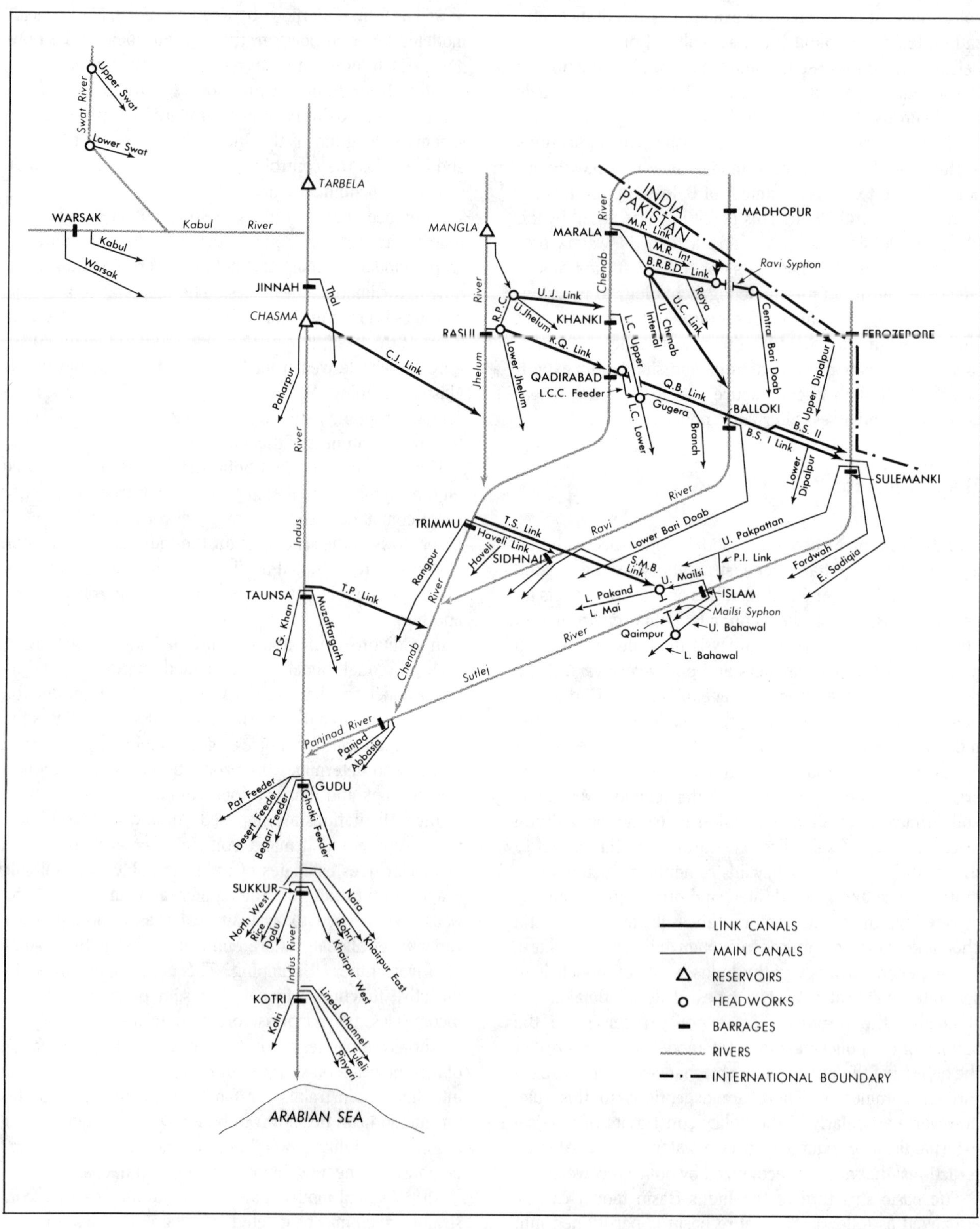

solution is not an equilibrium in the sense that it would be indefinitely sustainable. Because individual farmers do not recognize their individual effects on groundwater equilibrium, which must be maintained over the long run, the government must take into account the long-term consequences of any water allocation scheme and the effect of water-related investments on equilibrium. This expresses precisely the two-level aspect of the Indus Basin model,

Figure 9-2. *Indus Basin Agroclimatic Zones*

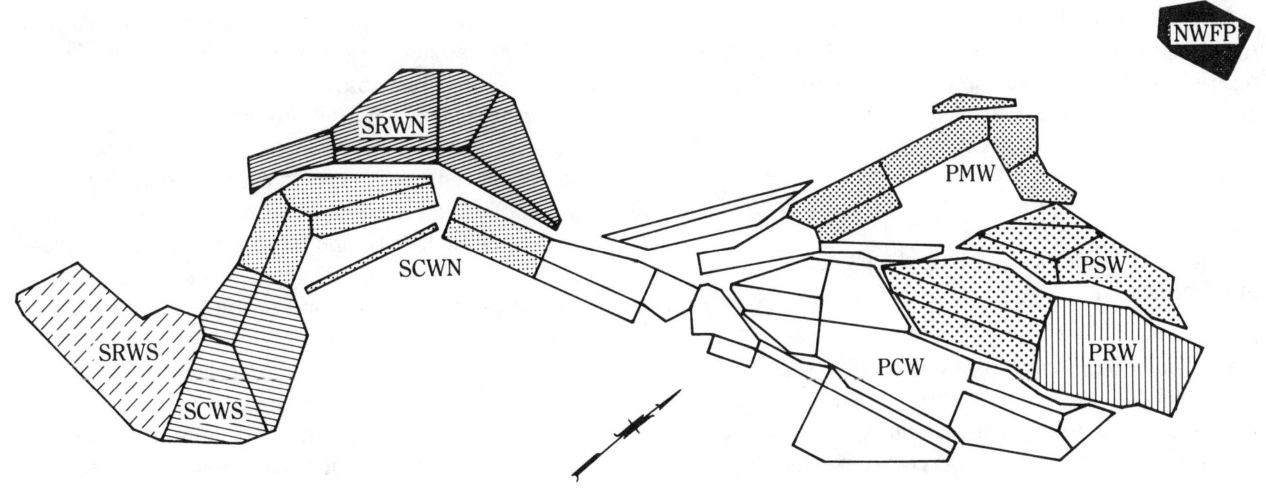

SRWS = Sind Rice-Wheat South
SCWS = Sind Cotton-Wheat South
SCWN = Sind Cotton-Wheat North
SRWN = Sind Rice-Wheat North

PCW = Punjab Cotton-Wheat
PMW = Punjab Mixed Crops–Wheat
PSW = Punjab Sugarcane-Wheat
PRW = Punjab Rice-Wheat
NWFP = Northwest Frontier Province

where some constraints are not formally recognized by the farmers (the policy receivers) even though the government (the policymaker) requires that they be satisfied. How the government might accomplish this task is explained in detail by Bisschop and others (1982). If groundwater balance is imposed, the dual variable (shadow price) corresponding to this constraint is the tax or subsidy which would induce farmers to pump tubewells at the level required for groundwater balance.

The original Indus Basin model was a linear programming problem with more than 20,000 constraints and with an objective function for the basinwide model that is simply the sum of the objective functions of the polygonal submodels. A model of this size exceeds the capability of existing software for linear programs, and it was apparent early on that a special simplification would be needed. By converting the height of the water table in each polygon to a policy instrument, structural simplifications could be

Table 9-1. *Average Cropping Patterns by Agroclimatic Zone, 1972–73 and 1975–76 Cropping Years*
(percentage of cropped area)

Item	*Northwest Frontier Province*	*Punjab cotton-wheat*	*Punjab mixed crops–wheat*	*Punjab rice-wheat*	*Punjab sugarcane-wheat*	*Sind cotton-wheat north*	*Sind cotton-wheat south*	*Sind rice-wheat north*	*Sind rice-wheat south*	*Total*
Rice	0.4	5.2	2.5	30.7	5.0	12.5	8.8	54.8	51.0	14.0
Wheat	34.4	38.5	52.0	41.8	42.1	36.3	41.7	21.2	20.5	38.4
Cotton	0.5	26.2	8.9	2.7	13.3	25.8	30.0	3.3	14.8	17.3
Corn for grain	35.0	1.3	1.0	1.3	4.1	0.3	1.5	0.1	0.5	2.5
Gram	0.1	1.5	7.3	0.8	1.3	5.1	0.1	10.8	0.6	2.5
Sugar	18.7	3.6	5.0	1.9	8.4	3.6	3.7	0.4	5.4	4.6
Rape and mustard	0.6	3.2	5.1	1.4	1.9	5.7	2.6	3.4	2.2	2.9
Kharif fodder	3.6	11.0	10.0	7.0	10.6	3.8	5.4	0.4	1.6	8.3
Rabi fodder	6.7	9.4	8.3	11.1	13.4	7.1	6.2	5.6	3.4	9.4
Area cropped (thousands of hectares)	336	4,745	757	1,604	2,043	837	636	931	345	12,231
Cropping intensity	158	113	107	134	121	105	91	106	83	114

Source: Data for cropping years 1972–73 from Agricultural Census Organization (1972); data for 1975–76 from provincial reports on cropped acreages.

made so that the entire model contains fewer than 8,000 constraints, which is solvable using a large machine and commercial software.

Although this introduction has been brief and has ignored many details, the reader will have gotten some impression of the site, structure, and complexities that are captured by the system. Readers desiring a complete description of the model structure may obtain it from the authors.

Calibration, Water Loss Characteristics, and Validation

The Indus Basin model is a comparative statical model that simulates producer response to policy intervention. It can be used to compare producer response to different environments where the environmental change is wrought by policy intervention in the sense of complete producer adjustment (that is, long-run response) to environmental change. The important function of model validation is therefore not appropriately accomplished when actual historical conditions on a year-by-year basis are used to simulate a dynamic path of producer adjustment. Rather, average conditions in some base period should be used to generate model solutions that can be compared with an average of actual producer responses in the base period, on the assumption that producer responses on the whole are close to long-run equilibrium.

Two time periods were considered for validation runs—1967–75 and 1975–80. The earlier period includes responses after the introduction of the new green revolution varieties of wheat and rice and after the initial use of Mangla Reservoir and prior to the initial use of Tarbela Reservoir. The later period contains the history of post-Tarbela producer responses.

The experiments reported are actually experiments to select a set of water loss parameters that permit the model to acceptably reproduce important aspects of both production response—for example, the cropping patterns and intensities of Table 9-1—and the state of the groundwater aquifer. This procedure presupposes prior calibration of the specification of agricultural technology and producer behavior, which was done previously on a polygon-by-polygon basis. The procedure for calibrating farm-level models is well known and need not be discussed here.

The existence of significant uncertainty about the loss characteristics of the surface water system was unanticipated. It had been assumed, perhaps somewhat naively, that these characteristics, which are subject to measurement, would be known with some precision by the operators of a system with as long a history as the surface water irrigation system of the Indus Basin. This assumption turned out to be incorrect.

We responded to this uncertainty by testing the model with several specifications of system loss characteristics representing a spectrum of plausible scenarios for system performance. To keep the number of validation experiments within reasonable bounds, these scenarios were restricted to specification of a limited number of cases covering the range of likely loss characteristics. The specification of these cases is detailed in Table 9-2. Note especially that the high efficiency assumption for watercourse and field losses is at the level specified in the Indus Special Study (Lieftinck, Sadove, and Creyke 1968), which presents in detail the planning exercise behind the appraisal of the Tarbela Dam project in the mid-1960s. In contrast, the low efficiency assumption for watercourse and field losses is at the level specified in the Revised Action Programme (WAPDA 1979).

Solutions to the model configured for historical simulation with surface water diversions and reservoir capacity appropriate to the base periods 1967–75 and 1975–80 were obtained for the six cases specified in Table 9-2. The validation experiments specify exogenously the level of the water table and the surface water allocation for each polygon. Thus one important validation test is whether or not a given water loss specification of the model acceptably reproduces the observed production response of farmers to the historically given gross canal water supplies. Another is whether or not the calculated net recharge is consistent with available evidence on the state of the aquifer. Important aspects of farmer response include cropping patterns, cropping intensities, and livestock holdings.

Considering both production response and the state of

Table 9-2. *Cases for Sensitivity Analysis of Estimates of Water Loss Parameters*

Specification of losses

A_1 = low canal losses, set at approximately 50 percent of high losses

A_2 = high canal losses, set at 21 percent of pre-Tarbela diversions

B_1 = high efficiency at watercourse and field level, set at level approximating that of Lieftinck report, or 0.65

B_2 = medium efficiency at watercourse and field level, set at approximately 0.50

B_3 = low efficiency at watercourse and field level, set at level approximating that of RAP, or 0.395

Case	*Canal system efficiency (percent)*
A_1B_1	56.8
A_2B_1	50.6
A_1B_2	43.9
A_2B_2	39.2
A_1B_3	34.4
A_2B_3	30.7

Note: A_i = levels of canal losses, $i = 1, 2$; B_j = levels of combined watercourse and field efficiencies, $j = 1, 2, 3$. Diversions of 109.9 billion cubic meters and high canal losses (exclusive of link canals) of 23.2 billion cubic meters are assumed.

the aquifer, our tests clearly pointed to case A_2B_2—that is, high canal losses, medium efficiency for watercourse delivery and field application—as the scenario with the best estimate of water loss parameters among those considered. The water loss parameters of case A_2B_2 were thus accepted as specifying this aspect of the system.

The method employed was to impose a rigorous consistency test of model solutions, with consistency defined as conformity with the observed aspects of the Indus Basin Irrigation System in a base period. Given the logical consistency specified by model structure, the additional requirement of empirical consistency with observations from many independent sources is a stringent test. In fact, the severity of this test permitted estimation of unobserved loss parameters when no other method of estimation was available.

An Application to System Management

One of the virtues of modeling any economic system is the capacity that it creates to simulate counterfactual scenarios of system performance. In particular, this capacity permits the investigator to ask penetrating questions about system efficiency. Of course, any effort to assess the efficiency of resource allocation must specify a criterion by which efficiency is to be measured, and this criterion must be acceptable to the people of the country concerned if the assessment is to be meaningful. It is proposed here that the criterion of efficiency be maximization of the sustainable level of agricultural production from available water, given a vector of prices. The qualification for prices is necessary because agricultural production is quite sensitive to relative prices and the model does not provide solutions that optimize prices.

Analytical Framework

Clearly, a complicated, large-scale simulation model presents problems of interpretation if model solutions incorporate multiple changes. For this reason the sequence of system management experiments has been designed to incorporate only single changes (from some reference case) in each experiment. A number of the specifications of the experimental sequence are essentially imposed by the choice of efficiency criterion—that is, maximization of the sustainable level of agricultural production from available water, given a vector of prices. For example, the water endowment of the system from rim station inflows must be specified with the best available estimates of long-run supply (that is, some appropriate measure of central tendency). For this reason the water endowment at rim stations for the sequence of experiments is specified as the monthly flows of the median season over the period from 1967–68 through 1979–80 at each of the rim stations, with Ravi and Sutlej flows deleted since title to these was given to India in the Indus Waters Treaty of 1960. The resulting estimates (for annual flows) are presented in Table 9-3.

Table 9-3. ***Annual Rim Station Inflows, Indus Basin, 1967–68 to 1979–80***

Rim Station	*Monthly mean, 1975–76 to 1979–80*	*Seasonal median, 1967–68 to 1979–80*
Indus at Tarbela	72.175	72.925
Swat at Chakdara	5.651	5.785
Kabul at Warsak	18.737	19.460
Haro at Gariala	0.973	0.771
Soan at Dhok Pathan	1.759	1.322
Jhelum at Mangla	28.350	27.978
Chenab at Marala	35.774	29.763
Ravi at Balloki	15.471	7.617
Sutlej at Ferozepore	7.804	8.558
Total	186.692	174.179
Total less Ravi and Sutlej	163.417	158.004

Note: Measurements in billions of cubic meters.

For convenience the vector of prices prevailing in 1976–77, the period for which much of the data base was collected, was chosen as the exogenous price vector. Comparison established that relative prices in other years were similar to those prevailing in 1976–77 except for petroleum products and one agricultural commodity whose world price has a very large variance. Several experiments employed the price vector prevailing in 1980–81 to test model sensitivity to the price vector used as well as to indicate the long-run equilibrium effects of the large increase in the relative price of petroleum that occurred in 1979–80.

The sequence of experiments that provides the framework for the analysis of system management is laid out categorically in Table 9-4. Experiment J is the base case in the analysis of system management. The experiments listed to the right of experiment J constitute the main sequence of single-step variations in water allocation policy. Thus, experiments J, Q, S, and T all specify a minimum water allocation based on historic water rights, where these are defined operationally as the mean diversion over the pre-Tarbela (but post-Mangla) period, 1967–75. Experiments J and Q specify historic water rights for each polygon, but S and T specify historic water rights only for each province. Similarly, J and S specify monthly and Q and T specify seasonal water rights. Experiment M substitutes the level of farm income derived from a pre-Tarbela model solution as a lower bound to farm income in place of the historic water rights constraint. Experiments O and P drop all explicit distributional constraints, with O including and P excluding the groundwater balance constraint. Strictly speaking, experiment P does not pro-

Table 9-4. *System Management Experiments*

Water endowment and loss parameters	Pre-Tarbela allocation as lower bound to canal water allocation: Polygonal, Month	Polygonal, Season	Provincial, Month	Provincial, Season	Pre-Tarbela income as lower bound to farm income	No distributional constraints: With GWB[a]	No distributional constraints: Without GWB[a]
Monthly flows based on 1967–80 seasonal median without Ravi and Sutlej rivers							
1976–77 prices	J	Q	S	T	M	O	P
1980–81 prices	K						
Plus 50 percent energy income, 1980–81 prices	L						
With adjustment for watercourse loss, 1976–77 prices	R						

a. GWB = groundwater balance.

vide meaningful long-run water allocation, but it is included to show the effect of dropping the groundwater balance constraint. The experiments listed below experiment J in the second column of Table 9-4 retain J's specification of long-run rim station inflows and the historic water rights constraint in the form of a monthly polygonal lower bound on canal diversions but vary prices or water loss characteristics. Experiments K and L substitute 1980–81 prices for 1976–77 prices, and in addition L increases energy prices by 50 percent and drops the subsidy on fertilizer use. Experiment R differs from experiment J in that losses along watercourses have been adjusted to reflect watercourse improvement or rehabilitation, as specified by the On Farm Water Management Project, a credit to Pakistan from the International Development Association (IDA) of the World Bank.

System Management Experiments

The experiments specified in Table 9-4 were completed with the model configuration in a water-optimizing mode, that is, solved for an endogenous water allocation. Since the experimental solutions maximize farm income subject to farmer preferences for family subsistence requirements and risk aversion, these solutions represent the maximum agricultural production that can be obtained given model specification and farmer preferences and hence correspond to the efficiency criterion adopted. In all experiments, existing stocks of private tubewells and tractors in 1975–76 are given as initial conditions, and these stocks can be augmented by endogenous private investments. The groundwater balance constraint is imposed in all experiments except P, which implies that the value of the dual variable corresponding to this constraint (that is, shadow price) is the implicit tax or subsidy required to induce farmers to pump their tubewells at the level needed for aquifer equilibrium. This tax or subsidy must actually be imposed, at least indirectly, for the model solution to be valid. Similarly, in the saline groundwater areas the objective function of the polygonal submodels includes a term for drainage costs. The interpretation is that farmers in these areas demand the amount of canal water diversions that maximizes their utility, given public drainage costs for which they expect to pay in the form of some kind of tax. Since the government controls the canal water diversions, it is not actually necessary that this tax be imposed to induce the desired level of agricultural production. It is necessary, however, that drainage costs be taken into account in determining the efficient water allocation. The model therefore assumes that the government has social welfare objectives which are completely consistent with maximizing farm output subject to farmer preferences. In this fashion the two-level programming problem discussed by Bisschop and others (1982) can be solved in a linear programming model; that is, the physical externality imposed by the groundwater aquifer can be internalized by a tax or subsidy.

Production and Factor Utilization

Production is measured in value added, and since our concern is efficiency, or relative performance, production is measured relative to experiment D, which approximates actual post-Tarbela conditions. To confine our discussion to policies directed toward changes in water distribution, results from experiments K and L are not presented here. The effect of the several such policies specified in Table 9-4 on agricultural value added and employment is given in Table 9-5. Note that while the overall gains range from 17 percent to 20 percent above the post-Tarbela level for value added and from 14 percent to 16 percent for employment, the gains are markedly different between fresh groundwater (FGW) areas and saline groundwater (SGW) areas. The former show increases of only 2 percent to 4 percent for both measures, while the latter have gains of 55 percent to 65 percent in value added and from 45 percent to 54 percent in employment.

The striking difference between the output responses of FGW and SGW areas clearly signals similar shifts in resource use, and Table 9-6 presents some results on the intensity of labor and land use. Note that the data from the post-Tarbela simulation (experiment D) show divergent levels of input intensity for both inputs between FGW and SGW areas. At the level of the entire basin the effect of alterna-

Table 9-5. *Real Agricultural Value Added and Employment, Indus Basin*

Experiment	Total	Fresh groundwater areas	Saline groundwater areas
	Value added		
D	100.0	100.0	100.0
J	116.6	101.9	155.1
M	119.7	103.3	162.1
O	119.5	103.1	162.2
P	120.1	103.1	164.5
Q	118.4	102.4	159.9
R	120.5	104.4	162.4
S	119.3	102.9	162.1
T	119.5	103.1	162.2
	Employment		
D	100.0	100.0	100.0
J	113.6	103.2	144.6
M	115.1	103.7	149.4
O	115.1	103.7	149.4
P	116.1	103.6	153.8
Q	114.3	102.9	148.6
R	115.9	104.5	150.0
S	115.1	103.6	149.3
T	115.1	103.7	149.4

Note: Measurements are percentages of post-Tarbela levels for value added and employment.

Table 9-6. *Labor and Land Input Intensity by Groundwater Quality Area*

Experiment	Labor intensity[a]		Land intensity[b]	
	FGW areas	SGW areas	FGW areas	SGW areas
	Indus Basin			
D	431	276	138	101
J	445	399	141	138
M	447	413	140	143
O	447	413	140	144
P	446	425	140	144
Q	443	410	140	142
R	450	414	142	144
S	446	413	140	144
T	447	413	140	144
	Punjab			
D	427	235	135	80
J	441	347	137	118
M	444	375	138	125
O	444	375	138	125
P	445	404	138	130
Q	439	378	137	126
R	447	368	138	126
S	443	374	138	125
T	444	375	138	125
	Sind			
D	396	308	154	117
J	415	440	158	154
M	405	442	149	158
O	407	442	149	158
P	392	441	148	154
Q	418	435	161	154
R	418	450	160	158
S	405	442	149	158
T	407	442	149	158

a. Man-hours of labor input per acre.

b. Cropping intensity or cropped acres per acre of irrigated land, normalized to a percentage scale, where 100 denotes each irrigated acre is cropped once a year.

tive system management policies is to draw the levels of input intensity much closer to equality between the two groundwater quality regions, which is accomplished by large increases in intensities for the SGW areas and relatively small increases in the FGW areas. When these data are disaggregated by major provinces, however, the picture is somewhat different. In Punjab, while the trend remains the same, the SGW areas lag significantly behind the FGW areas in input intensity under all policies. In Sind the SGW areas show greater input intensities under most policies. Thus the SGW areas of Sind show much higher input intensities than do the SGW areas of Punjab.

The reason for these divergent patterns becomes clearer when relative water supplies are taken into account, and these data are shown in Table 9-7. As might be expected, water supply per acre also shows a pattern of large increases in the SGW areas and small increases in the FGW areas under all water allocation policies. However, the quantity available per acre in Sind SGW areas is almost twice as large as in Punjab SGW areas. In fact, both the FGW and SGW areas in Sind show significantly greater total water supplies than the corresponding areas in Punjab under all policies—a surprising outcome since it is often argued that the interprovincial distribution of surface water is skewed toward Sind because of political factors. Yet these experiments, some of which release all distributional or equity constraints, show the same pattern of relatively greater distribution of surface water in Sind. It might be argued that the physical capacities of the irrigation system for diversion at various points have constrained the model solutions to this outcome. However, when the shadow prices on capacities of link and main canals are examined, this turns out not to be true to a significant degree. Part of the difference in available annual supplies is due to significantly greater subirrigation in Sind, which is an uncontrollable (by farmers) source that peaks in months when farmers have little land under crops. Yet canal diversions to SGW areas are significantly greater in Sind, and this circumstance argues for greater marginal productivity of water there. One factor contributing to high productivity in this region is Sind's comparative advantage in rice cultivation, which has benefited from the introduction of the high-yielding new varieties. The high productivity of inputs in the water-intensive rice crop has thus resulted in significantly higher optimal diversions to the SGW areas of Sind.

Table 9-7. *Water Supply per Hectare at Root Zone*

Experiment	SGW areas: Public	SGW areas: Total	FGW areas: Public	FGW areas: Private (tubewells)	FGW areas: Total
		Indus Basin			
D	0.579	0.762	0.378	0.348	0.857
J	0.786	1.046	0.381	0.366	0.881
M	0.774	1.039	0.372	0.351	0.860
O	0.759	1.024	0.372	0.348	0.857
P	0.774	1.042	0.390	0.326	0.857
Q	0.753	1.024	0.390	0.354	0.881
R	0.860	1.131	0.418	0.332	0.893
S	0.786	1.055	0.375	0.351	0.863
T	0.759	1.024	0.375	0.351	0.863
		Punjab			
D	0.357	0.451	0.326	0.372	0.814
J	0.494	0.631	0.326	0.390	0.838
M	0.537	0.686	0.332	0.381	0.835
O	0.537	0.686	0.329	0.378	0.832
P	0.561	0.719	0.338	0.372	0.835
Q	0.537	0.689	0.332	0.354	0.841
R	0.533	0.692	0.360	0.363	0.850
S	0.537	0.686	0.329	0.381	0.832
T	0.537	0.686	0.329	0.381	0.832
		Sind			
D	0.750	1.006	0.744	0.235	1.192
J	1.012	1.366	0.780	0.226	1.222
M	0.957	1.314	0.710	0.171	1.091
O	1.006	1.286	0.710	0.171	1.094
P	0.942	1.292	0.820	0.046	1.055
Q	0.924	1.280	0.796	0.186	1.201
R	1.110	1.472	0.881	0.143	1.247
S	0.982	1.341	0.710	0.171	1.091
T	0.930	1.286	0.710	0.171	1.094

Note: Water is measured in meters. The number given is the delta, or height of total water applied per unit of level land. Thus a delta of one implies an application of 10,000 cubic meters per hectare^{-1} (measured at the root zone).

Farm Incomes, Control Costs, and Resource Prices

Since per capita income is a useful and frequently used measure of economic development, albeit an imperfect one, a review of the effect of the system management experiments on per capita farm incomes is pertinent. However, the necessary existence of transfer payments owing to the implicit tax or subsidy on tubewell pumping somewhat complicates the concept of farm income. A distinction must be made between income from farm operations proper—unadjusted income—and income that allows for the impact of the tax or subsidy—adjusted income. The annual public costs of aquifer control for each of the system management experiments are given in Table 9-8. These costs, which are largely subsidy costs, range from 800 million to 1.2 billion rupees per year. Adjusted per capita incomes by provinces and groundwater quality zones are given in Table 9-9, which shows that Sind incomes are greater than Punjab incomes for both groundwater quality zones and under all policies. Basinwide adjusted per capita incomes show increases of 19 to 22 percent with increases ranging from 2 to 13 percent in the FGW areas and from 46 to 84 percent in the SGW areas. Unlike the current post-Tarbela situation, per capita incomes are greatest in Sind SGW areas under all policies, and incomes in Punjab SGW areas are slightly greater than in Punjab FGW areas under most policies. Thus, although the objective was more efficient resource use, the alternative policies have significant implications for income distribution by water quality zone, with something close to equality between zones achieved in Punjab and a reversal of the present situation in Sind. In absolute terms, however, every zone gains in per capita income. In addition, the public costs of aquifer control per capita of farm population, which range from 35 to 50 rupees, are small in relation to the gains in per capita incomes.

Of course, the alternative policies may adversely affect the incomes of the several groups within a region. Some information on income distributional impacts by class is available in the form of the shadow prices on the land and water constraints. These data are summarized in Table 9-10. Note the very large increases in the implicit land rents in the SGW areas under all of the alternative policies and the corresponding sharp decreases in water prices in these areas. A similar pattern occurs in the FGW areas, but as expected the magnitudes of the changes are much smaller. Clearly, landowners, especially in the SGW areas, are well positioned to capture much of the increase in farm

Table 9-8. *Public Costs of Aquifer Control under Groundwater Balance*
(millions of 1977 rupees)

Experiment	Private tubewell subsidy	Annual cost of drainage	Total public cost	Cost per capita of farm population
J	828.4	374.0	1,202.4	50.3
M	612.6	417.2	1,029.8	43.1
O	665.7	418.0	1,083.7	45.3
Q	425.7	411.6	837.2	35.0
R	807.0	290.2	1,097.2	45.9
S	687.9	415.8	1,103.7	46.2
T	684.4	418.1	1,102.5	46.1

Table 9-9. *Adjusted per Capita Income*
(1977 rupees)

		Punjab		Sind	
Experiment	Indus Basin	FGW areas	SGW areas	FGW areas	SGW areas
D	1,081	1,125	708	1,473	1,222
J	1,284	1,188	1,129	1,623	1,780
M	1,307	1,196	1,230	1,613	1,816
O	1,309	1,197	1,231	1,620	1,816
P	1,288	1,151	1,305	1,585	1,800
Q	1,323	1,216	1,209	1,664	1,837
S	1,308	1,195	1,230	1,643	1,816
T	1,309	1,196	1,231	1,650	1,816

incomes. The extent to which tenants and landless laborers benefit from the increase in aggregate farm production depends on the relative change in the demand for, and supply of, farm labor.

The results from experiments K and L, which were run using 1980–81 prices, were very similar to those from experiment J, which differs from K and L only in the price parameters. The important change in relative prices (and only these matter for the model) in these experiments is a sharp increase in the prices of fuels and fertilizer. These changes in costs reduce value added and farm income slightly for experiment K and significantly more for experiment L, which raises energy prices by an additional 50 percent. Since these changes directly affect private tubewell operating costs, the resulting increase in the optimal pumping subsidy caused the public costs of aquifer control to increase by factors of 2 and 4 in K and L, respectively.

To test the sensitivity of results from the system management experiments to the model's assumptions about drainage costs, experiment J was repeated several times with drainage costs increased to 1.5, 2, 3, 5, and 10 times the original level. In these experiments, public costs of aquifer control increased (in the above sequence) by 10, 29, 54, 103, and 168 percent, and per capita incomes in the SGW areas decreased by 0.3, 1, 2.5, 4.5, and 9 percent (in the same sequence). Overall farm income remained virtually constant throughout this sequence of experiments, with increases in the FGW areas offsetting decreases in the SGW areas. Thus this sequence of experiments has demonstrated that the model's results are robust for drainage cost estimates.

Table 9-10. *Annual Shadow Prices of Land and Water*

		Punjab		Sind	
Experiment	Indus Basin	FGW areas	SGW areas	FGW areas	SGW areas
	Land (1977 rupees per hectare)				
D	578	781	30	916	143
J	892	904	289	1,213	1,065
M	993	892	588	1,233	1,405
O	983	877	588	1,230	1,403
P	904	810	590	1,072	1,225
Q	931	865	462	1,255	1,242
R	993	926	420	1,272	1,400
S	983	874	583	1,230	1,400
T	986	877	585	1,230	1,400
	Water (1977 rupees per cubic meter)				
D	1.671	1.233	3.299	0.895	2.583
J	0.997	0.952	2.178	0.525	0.874
M	0.884	1.030	1.543	0.605	0.448
O	0.870	0.993	1.544	0.609	0.457
P	0.999	1.112	1.403	0.822	0.701
Q	0.953	1.033	1.802	0.457	0.683
R	0.799	0.873	1.748	0.438	0.483
S	0.861	1.007	1,553	0.612	0.454
T	0.860	0.993	1.544	0.610	0.459

Note: Annual shadow prices are sums of shadow prices of respective monthly constraints of polygonal submodels aggregated into weighted averages for the areas shown.

Summary

Gains of 17–20 percent in agricultural production and 14–16 percent in employment are possible in the Indus Basin, given more efficient allocation and management of surface and groundwater. These large gains were estimated by holding everything else constant and by making conservative assumptions for water supply. However, these gains require optimally coordinated use of surface and groundwaters. The necessary steps for achieving efficient conjunctive use require investments for drainage in saline groundwater areas and eventual public control of private tubewell withdrawals in the fresh groundwater areas by some combination of taxes, subsidies, quotas, fees, and prices. These steps can be regarded as adjustment costs in a transition toward more efficient resource use. The subsidy costs (shown in Table 9-8) might be unnecessary in practice, since simulation experiments have shown that improved agricultural technology shifts the optimal control from a subsidy to a tax. However, large drainage investments will be required to achieve the gains shown, since these depend on large increases in canal diversions to saline groundwater areas. Until such drainage investments are in place, only limited gains are possible from increases in canal diversions. Control over private tubewell pumping is not needed at present since subsidies to encourage greater pumping are unnecessary until significant increases in drainage capacity exist, and with present water-table depths of less than 20 feet (6 meters) in almost all fresh groundwater areas, there is so far no need for taxation to discourage excessive withdrawals.

Note

This chapter originally appeared in a somewhat different form in *Water Resources Research* vol. 20, no. 11 (November 1984), pp. 1,489–98; copyright by the American Geophysical Union.

References

Agricultural Census Organization. 1972. *Pakistan: Census of Agriculture.* 5 vols. Lahore: Government of Pakistan.

Bachmat, Yehuda, J. D. Bredehoeft, Barbara Andrews, David Holtz, and Scott Sebastian. 1980. *Groundwater Management: The Use of Numerical Models.* Water Resources Monograph Service, vol. 5. Washington, D.C.: American Geophysical Union.

Bisschop, Johannes, Wilfred V. Candler, J. H. Duloy, and G. T. O'Mara. 1982. "The Indus Basin Model: A Special Application of Two-Level Linear Programming." *Mathematical Programming* 20:30–38.

Bredehoeft, J. D., and R. A. Young. 1970. "The Temporal Allocation of Groundwater: A Simulation Approach." *Water Resources Research* 6(1): 3–21.

Brown, Jr., Gardner, and Robert Deacon. 1972. "Economic Optimization of a Single-Cell Aquifer. *Water Resources Research* 8:557–64.

Buras, Nathan. 1963. "Conjunctive Operation of Dams and Aquifers." *Journal, Hydraulics Division, American Society of Civil Engineering* 89 (HY6): 111–29.

Burt, O. R. 1964. "The Economics and Conjunctive Use of Ground and Surface Water." *Hilgardia* 32(2): 31–111.

———. 1966. "The Economic Control of Groundwater Reserves." *Journal of Farm Economics* 48:632–47.

———. 1967. "Temporal Allocation of Groundwater." *Water Resources Research* 3:45–56.

Chaudry, M. T., J. W. Labadie, W. A. Hall, and M. L. Albertson. 1974. "Optimal Conjunctive Use Model for Indus Basin." *Journal, Hydraulics Division, American Society of Civil Engineers* 100 (HY5): 667–87.

Feinerman, Eli, and K. C. Knapp. 1983. "Benefits from Groundwater Management: Magnitude, Sensitivity and Distribution." *American Journal of Agricultural Economics* 65(4): 703–10.

Fiering, M. B. 1965. "Revitalizing a Fertile Plain: A Case Study in Simulation and Systems Analysis of Saline and Waterlogged Areas." *Water Resources Research* 1(1): 41–61.

Greenman, D. W., V. W. Swarzenski, and G. D. Bennett. 1967. "Groundwater Hydrology of the Punjab, West Pakistan, with Emphasis on Problems Caused by Canal Irrigation." U.S. Geological Survey Water Supply Paper 1608-H. Washington, D.C.

IACA. Irrigation and Agricultural Consultants Association. 1966. *Programme for the Development of Irrigation and Agriculture in West Pakistan.* 23 vols. Lahore.

Lieftinck, Pieter, A. Robert Sadove, and Thomas C. Creyke. 1968. *Water and Power Resources of West Pakistan: A Study in Sector Planning.* 3 vols. Baltimore, Md.: Johns Hopkins Press.

Maddock, Thomas, III, and Y. Y. Haimes. 1975. "A Tax System for Groundwater Management." *Water Resources Research* 11(1): 7–14.

Noel, J., B. D. Gardner, and C. V. Moore. 1980. "Optimal Regional Conjunctive Water Management." *American Journal of Agricultural Economics* 62:489–98.

Rogers, Peter, and D. V. Smith. 1970. "The Integrated Use of Ground and Surface Water in Irrigation Project Planning." *American Journal of Agricultural Economics* 52:13–25.

Tipton and Kalmbach, Inc. 1967. "Regional Plans: North Indus Plains." Lahore: WAPDA.

WAPDA. Water and Power Development Authority, Master Planning and Review Division. 1979. *Revised Action Programme for Irrigated Agriculture.* 3 vols. Lahore: WAPDA Press.

WAPDA (Water and Power Development Authority)–Harza Engineering Co., International. 1963. "A Program for Water and Power Development in the Indus Basin in West Pakistan, 1963–1975: An Appraisal of Resources and Potential Development." Lahore.

White House—Department of the Interior Panel on Waterlogging and Salinity in West Pakistan. 1964. *Report on Land and Water Development in the Indus Plain.* Washington, D.C.: U.S. Government Printing Office.

Yu, Wanyoung, and Y. Y. Haimes. 1974. "Multilevel Optimization for Conjunctive Use of Ground and Surface Water." *Water Resources Research* 10(4): 625–36.

Comment

Robert Picciotto

The missing professional ingredient in water resource use is the managerial dimension. In the Indus Basin as in much of the developing world, the scarcest resource—water—is not managed. Addressing a group of World Bank operations managers in 1983, Peter Drucker said that one of the few generalizations which can be made about development is that large projects do not work. Perhaps the same hypothesis can be advanced about large econometric models.

To be more precise, large dams as well as large models can be made to work, but with heavy costs and uncertain benefits—uncertain because what justifies these elaborate structures are simple assumptions about the behavior of the rest of the world. The numerous models on the Indus Basin have taught us that the rest of the world is not docile. Assumptions explain away the very problems which must be addressed and therefore serve as elaborate safety valves for technocrats and professionals who are frustrated by constraints which are, or appear to be, intractable.

The publication of this volume is an acknowledgment of this state of affairs. The consensus is that externalities are really the crux of water resource management and that progress in internalizing these externalities is an urgent challenge for development.

Interdependence is built into the water development measures of the Indus Basin system. O'Mara and Duloy have done a good job of illustrating the system's interrelationships:

- Fresh groundwater underlies only about half of the basin; Pareto optimality demands that surface water should be transferred to areas of saline groundwater, especially in the dry season when sweet groundwater requirements can be met by pumping.
- Fifteen percent of the basin requires groundwater diluted by surface water to ensure that the salinity level does not exceed the tolerance of plants; high returns therefore result from conjunctive use designed for appropriate mixing of saline groundwater.
- Given the high energy cost of pumping, surface water deliveries and storage releases must be integrated with tubewell pumping patterns to help minimize energy demand.
- Surface water deliveries vary considerably from year to year, highlighting the need for flexibility in tubewell operations and intercommand allocations to maximize the benefit of groundwater storage. Incremental surface water supplies give rise to increased aquifer recharge, thus raising the output of tubewell fields and increasing the efficiency of conveying surface water.
- The sequence of actions in investment decisionmaking is critical since, for example, increased canal deliveries can be damaging without adequate subsurface drainage. And using surface and groundwater storage to meet peak requirements must take priority over expensive investments in canal remodeling.

Because reliable information is lacking, the model necessarily does not deal with some phenomena, which are less well understood. Such phenomena include the economic implications of floods and the investments necessary to minimize them and the economics of loss reduction in conveying water through the main canal or through the distributaries. Lack of data is a fundamental problem which often reflects a conflict of interest. Those who control the data have a vested interest in distorting it—groundwater data in the Punjab, for example, and salinization data in Sind.

There are other even more fundamental externalities:

- *Support services.* Externalities contributing to suboptimal returns for water use include research which is not relevant to field problems, extension systems which do not deliver relevant advice or do so at the wrong place or the wrong time, and credit systems which distribute scarce capital to those who need it least.
- *Water rights legislation and water administration.* These institutional constraints tend to favor the "haves"—those at the head of the canal—over the "have nots," whether between provinces or commands or within a command area.
- *Water and power pricing.* Pricing which does not reflect relative scarcities discourages efficient use of single-source water, not to mention conjunctive use.
- *Engineering design standards.* Standards emphasize the pennywise, pound-foolish syndrome if too little is spent for maintenance, measuring devices, drainage, and efficient distribution and loss reduction systems.
- *Administrative systems.* Excessively centralized and inflexible systems are unresponsive to farmers' needs (except for the powerful few) and lack appropriate coordination mechanisms at the local level.

For purposes of analysis, the problem of externality is defined by O'Mara and Duloy as one in which private agents behave as short-term profit maximizers while long-term social concerns are viewed as the province of a single agent called government. Here again, the solution is defined out of existence. In reality, local administrative systems and social groupings are important to rural life, and at this level the social good can perhaps be made an objective.

How to internalize some of the externalities is the challenge posed to the practitioner. There is a need for smaller "real life" models and greater influence on policy from those on the front lines of the development battle.

O'Mara and Duloy capture the key physical interdependencies of the extraordinarily complex Indus Basin system. Their model confirms the importance of a programming approach to investment selection and admirably sets forth the priority of drainage, Sind's rightful claim for additional resources, and the need for change in water rights legislation and administration. It demonstrates that the traditional project approach is inadequate, because of physical interrelationships between and within commands and because growth and equity optimization requires that water and capital allocation be considered in the context of the entire basin.

But there remains a strong case for the project approach which has to do with the management aspects of the problem. According to Drucker, complexity can be handled only by process; it is amenable to organizational solutions. Organizationally there is a sophisticated structure in place within Pakistan. A complex network of institutions is made up of traditional line departments, autonomous public agencies, and a host of coordinating committees with their provincial and local counterparts.

Informal local networks for decisionmaking, including a vigorous private sector, underlie this formal structure. The result is not always harmonious, to say the least. The system is not working for three main reasons:

- The legislative framework for water allocation (that is, the primacy of established water rights) is not conducive to sound decisions regarding water distribution.
- Public control is too far-reaching for the administrative and financial resources available.
- Public agencies have archaic operating procedures for budget, manpower, and monitoring water allocation.

Outside assistance is badly needed in this "software" area. The project approach therefore continues to be relevant if it introduces modern management practices to the allocation and use of Pakistan's scarcest resource—water.

10

Estimating the Externalities of Groundwater Use in Western Argentina

Juan Antonio Zapata

Using groundwater is an attractive way to achieve a modern irrigation system, because it avoids the institutional rigidities that impede the efficiency of surface water systems. Groundwater use reliably matches crop demand for irrigation and is amenable to step-by-step development. It may also play a very important role in areas served by surface water systems, since conjunctive use provides flexibility. Modernization results from better timing of existing irrigation and from its expansion by "stretching" surplus surface flows, which are wasted under a surface irrigation system alone. Flexibility may be enhanced when supply as well as demand is managed by a long-run plan for conjunctive use.

Since groundwater storage is not physically perceptible, however, it is poorly understood, and its potential as a part of current schemes for the development and management of water resources is generally overlooked. For the same reasons, effective policies are not implemented to deal with the externalities of a lower water table which may be observed in many areas where water pumping exceeds the recharge of the aquifer.

Problems of a declining water table may become relevant because one of the important variables in the cost of groundwater use is the depth of the water table. The water table level affects the cost of both investment and operation. On the investment side, it is obvious that the deeper the water table, the more expensive the costs of the drilling, tubes, pumps, and engine. On the operation side, the deeper the water table, the more energy required to pump a given volume of water and the greater the risk of deterioration in the aquifer from seawater intrusion in the coastal areas. Even in noncoastal areas, some saline contamination of the water may occur, since continued pumping lowers the water table and reverses the general direction of groundwater flow. When this flow is reversed, water which normally percolates toward a saline aquifer—and which contains a progressively higher concentration of salts leached from the soil as it approaches this saline aquifer—is drawn backward by the pumping located in a freshwater area. With continued pumping, the quality of the freshwater aquifer therefore changes from good to poor.

The water table level is not an exogenous variable; it depends on the aquifer's variations and its stock of accumulated water. This stock in turn is a function of the recharge, withdrawals, and losses.

Depending on the nature and size of the recharge, pumping may change the water table level either temporarily or permanently, depending on whether long-run withdrawals exceed long-run net recharge (recharge minus losses). Changes in the water table level are considered permanent when the length of time required to replenish the stock extends well beyond the life span of the human agents who caused the change—that is, when the length of time may be calculated in geological rather than human terms.

Groundwater is a common property resource when any individual can withdraw water and when no individual has property rights over the water stored in the aquifer. When the water table is affected by withdrawals, the one who pumps imposes an externality on the rest of the users, who then require more inputs for pumping because of the lowered water table.

The next section presents a model that estimates the externality of energy requirements resulting from a lower water table. This model, which was developed and estimated in Zapata (1969), applies to an area in Mendoza, a province in western Argentina. The empirical results are presented in the succeeding section, followed by the summary and conclusions and an appendix on conjunctive use and management.

The Theoretical Model

The withdrawal of groundwater from a common pool by several users who have no property rights over the water in the aquifer can be expected to result in the well-known misallocative effects of common-property resource exploitation, unless marginal withdrawals do not affect the level of the water table. In cases where the level of the water table depends upon the rate of withdrawal, the marginal cost of pumping includes both the inputs used to lift the marginal amounts of water and the additional cost required to maintain the previous rate of withdrawal at the new depth of the water table. Under these circumstances, the prospective user captures all of the benefits but does not pay the full costs of pumping. For this reason, what appears to be an acceptable project from a private point of view may be undesirable from a social point of view.

The problem of common property resources can be analyzed by either of two equivalent approaches. The first is to define a unit of output which remains invariant. In this case the output would be a given amount of water per well per time period. The cost of production will depend upon the level of the water table, which in turn may depend upon the number of wells in production. The second approach is to define a unit of input which remains invariant—an approach frequently used in analyzing a fishery, where the input is defined as a fishing boat with a given configuration. The output for a unit of input will vary with the number of boats. In the first approach, the problem reduces to a difference between marginal social cost and marginal private cost; in the second approach, the problem reduces to a difference between marginal social product and marginal private product. The theoretical model presented in this chapter uses the first approach, taking into account the availability of information in the arid region of northern Mendoza in Argentina.

Development of new sources of water supply since the early 1950s has been an important aspect of Mendoza's agricultural growth. The supply of surface water has limited this growth process, since no farming is possible without irrigation. Without reservoirs, the supply has been limited by a summer river flow insufficient to irrigate the area currently being cultivated. Since the mid-1950s, groundwater has been used extensively to supplement the river flow because of private investments in tubewells and other equipment to lift water from a common pool. Withdrawals have lowered the water table, and pumping inputs per unit of water withdrawn have increased over time.

This decline in the water table presents a clear case of technical external diseconomies. Water withdrawals by an individual producer increase the cost of obtaining water for all other producers in the same region. In Mendoza's current system of groundwater use, water in the aquifer is essentially a common property resource—that is, any individual can withdraw water, but no individual has property rights that are valid for the future. Moreover, there is no basinwide management of groundwater; thus, economic forces will not lead to optimum use of the region's water resources.

The first step toward improving the allocation of water resources is to estimate the discrepancy between the private and social costs of obtaining water in Mendoza by tubewells. This study develops and estimates a model to measure this discrepancy. In the model, marginal private costs, denoted by *MPC*, are defined as the increase in total pumping costs when a given individual increases the rate of withdrawal by one unit for one time period. Marginal social costs, denoted by *MSC*, are defined as the increase in total pumping costs for the region as a whole when the rate of withdrawal is increased by one unit.

The rate of total water withdrawal by the region in period t is W_t, which is determined by the demand for, and supply of, water. Assume that one individual decides to increase the rate of withdrawal during the first period so that the total withdrawal in that period rises from W_1 to W_1^*. The additional withdrawal, $\Delta W_1 = W_1^* - W_1$, may cause the level of the water table to be lower than it would have been in some or all future periods and therefore may increase the amount of pumping inputs during that period.

The marginal social cost of the incremental withdrawal consists of the direct marginal cost of withdrawing ΔW_1, plus the present value of the incremental costs of all future withdrawals, plus the excess of the potential value of any withdrawals forgone (owing to the increase in the pumping costs) over the pumping costs that those withdrawals would have entailed. This triangular relationship will be ignored in the subsequent analysis; hence the estimates will slightly understate the true value of the marginal social cost as well as the externality.

The individual responsible for the additional withdrawal of ΔW_1 can be expected to consider only his private marginal costs. It is assumed that this individual is aware of how his own activity affects the level of the water table, but considers only how changes in the water table level affect his own pumping costs. Therefore the excess of the marginal social cost over the marginal private cost will be approximately

$$\text{(10-1)} \qquad \Delta W_1 \, (MSC - MPC) = \sum_{t=1}^{\infty} \frac{\Delta MPC_t}{(1+r)^{t-1}} W_t'$$

where r is the discount rate per period and W_t' is the amount of water withdrawn during period t by other individuals. Dividing by ΔW_1 yields

$$\text{(10-2)} \qquad MSC - MPC = \sum_{t=1}^{\infty} \frac{\Delta MPC_t}{\Delta W_1} \, \frac{W_t'}{(1+r)^{t-1}} .$$

The amount of energy required to lift an amount of water a given number of meters has been estimated as *HP* hours $= aD/273n$, where *HP* is horsepower, a is discharge in cubic meters, D is vertical lift in meters, and n is pump efficiency as a percentage. If p is the price of energy consumed per *HP* hour, the cost of lifting a unit of water is $pD/273n$, independent of the quantity pumped. If c is defined as $p/273n$, and p and n are considered to be constants, then marginal private cost is the product of c and D. (For the source of this physical relationship, see Israelsen and Hansen 1962, pp. 52–74.) The cost of lifting the water depends, of course, on the price of energy.

Denoting D as the distance to the water table measured in meters and c as the cost of lifting a unit of water one meter, marginal private cost may be written as $MPC = cD$. Take input prices as constants and let $\Delta D_i/\Delta D_{i-1}$ represent not necessarily the actual ratio of changes in depth but rather the change induced by the earlier W_1; then

$$\text{(10-3)} \qquad \frac{\Delta MPC_t}{\Delta W_1} = c\,\frac{\Delta D_1}{\Delta W_1}\prod_{i=1}^{t-1}(\Delta D_{i+1}/\Delta D_i).$$

It is further assumed, for reasons to be given later, that

$$k = \frac{\Delta D_{i+1}}{\Delta D_i} = \frac{\Delta D_{i+2}}{\Delta D_{i+1}}$$

for all i, permitting equation 10-3 to be written as

$$\text{(10-3a)} \qquad \frac{\Delta MPC_t}{\Delta W_1} = c\,\frac{\Delta D_1}{\Delta W_1}\,k^{t-1}.$$

By substituting equation 10-3 into equation 10-2 and dividing by *MPC*, the final expression for the excess of social over private cost, expressed as a fraction of private cost, is obtained:

$$\text{(10-4)} \qquad \frac{MSC - MPC}{MPC} = \frac{1}{D_1}\,\frac{\Delta D_1}{\Delta W_1}\sum_{t=1}^{\infty}\left(\frac{k}{1+r}\right)^{t-1} W_t'.$$

As this model is used in the case of Mendoza, Argentina, further development of some relations are required, given the state of knowledge about this particular aquifer. It is necessary first to estimate withdrawals or to select surrogate variables for withdrawals and then to develop the relation between water table depth and withdrawals.

Estimation of Withdrawals

Without information on total withdrawals, one must specify a function that may be estimated in order to use the above model. Define a_{ij}^k as the amount of withdrawal of water by the kth well during the ith month of year j. The amount of water produced by the kth well will depend upon seasonal effects relating to plant development, such as temperature and humidity, as well as upon the availability of water from other sources and the cost of pumping. (Output and input prices are assumed to be constant.) The value of the water changes from month to month during the growing season; to capture these fluctuations, a dummy variable is introduced for five of the six months of the irrigation season. The effect of the availability of surface water on the rate of withdrawal of underground water is influenced by the fact that surface water is not price-rationed; rather, a system of water rights is used in Mendoza which allows holders of water rights to use a predetermined share of the streamflow. Surface water allotments thus vary directly with the flow of water in the river. The quantity of water demanded, given the above considerations, will depend upon the cost of additional water, which, given the water-rights system, can be obtained only by pumping. The pumping cost, as indicated above, depends upon the level of the water table. If x_{ij}^k is defined as the amount of surface water available to the area supplemented by the kth well, and D_{ij} as the distance to the water table in that area during the ith month of the jth year, the total water demand for that area can be specified as

$$\text{(10-5)} \qquad a_{ij}^k + x_{ij}^k = b_{1i}^k - b_3^k D_{ij}.$$

Define H as the total number of hectares irrigated by surface water in the entire river basin, n_k as the number of hectares supplemented by the kth well, and A_{ij} as the accumulated river flow during the ith month of the jth year; the following identity is obtained:

$$x_{ij}^k \equiv (A_{ij}/H)\,n_k.$$

To obtain the aggregate demand for underground water, define the water field and then aggregate demand over the number of wells in that field, so that

$$\text{(10-6)} \qquad X_{ij} = \sum_k x_{ij}^k = (A_{ij}/H)\sum_k n_k.$$

Assume that each well supplements the same number of hectares; then $X_{ij} = (n/H)N_{ij}A_{ij}$, where N_{ij} is the number of wells producing in the field during the ith month of the jth year. Substituting equation 10-5 into equation 10-6 yields

$$\text{(10-7)} \qquad W_{ij} = \sum_k a_{ij}^k = b_{1i}N_{ij} - b_2(A_{ij}N_{ij}) - b_3 D_{ij}$$

where b_{1i} is obtained by assuming that all b_{1i}^k are equal over k, $b_2 = (n/H)$, and $b_3 = \sum_k b_3^k$.

Effects of Withdrawals on the Water Table

Once aggregate withdrawals have been specified, as in equation 10-7, the next step is to specify the relation between withdrawal rates and the level of the water table. Let R_{ij} be the groundwater recharge in month i of year j, L_{ij} the outflow (loss) from the groundwater stock during month i of year j, and S_{ij} the stock of groundwater at the end of month i of year j, the following identity is obtained:

$$S_{ij} \equiv S_{i-1,j} + R_{ij} - W_{ij} - L_{ij}.$$

It is possible to establish an inverse relationship between the distance to the water table and the stock of groundwater; the larger the stock is, the higher the water table and the smaller the distance, so that: $D_{ij} = \alpha - BS_{ij}$. Substituting S_{ij}, as derived from the inversion of this equation, into the above equation yields

$$D_{ij} = D_{i-1,j} - BR_{ij} + BW_{ij} + BL_{ij}. \quad (10\text{-}8)$$

As R_{ij}, W_{ij}, and L_{ij} are measured in volume units per unit of time and D_{ij} is measured in meters, the coefficient B is required to transform volume into distance to the water table.

In general, losses from the aquifer L_{ij} are difficult to estimate. Given the present state of hydrological studies of aquifers in Mendoza, it is quite impossible to estimate them. However, a practice sometimes used is to specify losses as a function of the stock of water stored in the aquifer (Heath and Trainer 1968). The greater the stock S_{ij}, the more pressure and hence the larger the loss:

$$L_{ij} = f_0 + f_1 S_{ij}$$

where $f_1 > 0$.

Given the relationship between the stock S_{ij} and distance D_{ij} stated above, the loss may be expressed as a function of D_{ij}:

$$L_{ij} = f_0 + (\alpha / B) f_1 - (f_1 / B) D_{ij} \quad (10\text{-}9)$$

where $f_1/B > 0$. This relationship is consistent with impressions developed by hydrologists that indicate that the distance to the water table has increased in aquifers to the east of Mendoza. These aquifers are believed to be recharged by outflows from the particular pool of underground water considered in this study. As a result of the increased pumping, these outflows have apparently diminished over time, increasing the distance to the water table in downstream aquifers.

The recharge R_{ij} depends upon seasonal melting of winter snows in the Andes to the west of Mendoza, which in turn is reflected in river streamflow. It may also depend upon the stock of groundwater stored in the aquifer (Heath and Trainer 1968, pp. 241–42). Since A_{ij} is the accumulated volume of river flow during the ith month of the jth year, then

$$R_{ij} = g_0 + g_1 A_{ij} - g_2 S_{ij}$$

where $g_1, g_2 > 0$. Given the relationship between stock S_{ij} and distance D_{ij} stated above, then

$$R_{ij} = g_0 + g_1 A_{ij} - g_2(\alpha / B) + (g_2 / B) D_{ij}. \quad (10\text{-}10)$$

Substituting equations 10-7, 10-9, and 10-10 into equation 10-8 yields

$$\begin{aligned} D_{ij} = h_0 &+ h_{1i} N_{ij} - h_2 N_{ij} A_{ij} \\ &+ h_3 D_{i-1,j} - h_4 A_{ij} \end{aligned} \quad (10\text{-}11)$$

where

$$\begin{aligned} h_0 &= [B(f_0 + g_0) + \alpha(f_1 + g_2)] / (1 + f_1 + g_2 + Bb_3) \\ h_{1i} &= Bb_{1i} / (1 + f_1 + g_2 + Bb_3) > 0 \\ h_2 &= Bb_2 / (1 + f_1 + g_2 + Bb_3) > 0 \\ 0 < h_3 &= 1 / (1 + f_1 + g_2 + Bb_3) < 1 \\ h_4 &= Bg_1 / (1 + f_1 + g_2 + Bb_3) > 0. \end{aligned}$$

Equation 10-11 relates distance D_{ij} to the number of wells N_{ij} in such a way that the coefficients may be estimated empirically. The coefficient h_{1i} changes from month to month to reflect seasonal shifts in the demand for water, h_2 reflects the substitution of surface for groundwater, and h_3 relates current level to lagged level. The greater the h_{1i} and h_3, the greater will be the decline in the level of the water table during the ith and subsequent months by the operation of an additional well during the ith month. The larger the f_1 or g_2, the smaller the effects of pumping on the water table because a greater share of water pumped will come from water that would otherwise be lost from, or would not have entered, the underground pool.

The effect of an additional well on the level of the water table will be greater the more inelastic the demand for water; the reason is that the more inelastic the demand, the smaller the decrease in withdrawal per well as a result of the decline in the level of the water table. Finally, it should be noted that the effect of an additional well on the level of the water table is given by $(h_{1i} - h_2 A_{ij})$, which is assumed to be positive.

An alternative hydrological model is one in which the rate of recharge is unaffected by the existing stock, because of lags between the time the snow melts and the time the resulting water reaches the underground pool. In this case, recharge is specified as

$$R_{ij} = \Psi_0 + \Psi_1 A_{i-m,j}, \ \Psi_1 > 0. \quad (10\text{-}10a)$$

Substituting equations 10-7, 10-9, and 10-10a into equation 10-8 yields

$$(10\text{-}11a) \qquad D_{ij} = h_0 + h_{1i} N_{ij} - h_2 N_{ij} A_{ij} + h_3 D_{ij} - h_4 A_{i-m,j}$$

where

$$h_0 = [B(f_0 - \Psi_0) + \alpha f_1] \,/\, (1 + f_1 + Bb_3)$$
$$h_{1i} = (Bb_{1i}) \,/\, (1 + f_1 + Bb_3) > 0$$
$$h_2 = (Bb_2) \,/\, (1 + f_1 + Bb_3) > 0$$
$$0 < h_3 = 1 \,/\, (1 + f_1 + Bb_3) < 1$$
$$h_4 = (B\Psi_1) \,/\, (1 + f_1 + Bb_3) > 0.$$

The difference between equation 10-11 and 10-11a is in the denominator of all coefficients and in the numerator of coefficients h_0 and h_4. The data for the Mendoza region are sufficiently refined to permit us to distinguish between the two models. Indeed, the estimate of the coefficient h_4 was consistently insignificant, and that variable was dropped from the regressions reported below in the section on empirical results.

Externalities Caused by the Additional Well

The model states the number of wells rather than the number of water withdrawals. Data is lacking for withdrawals per well in the period under consideration, which is not unduly restrictive because the analysis is based on how the operation of a typical additional well for one month affects the cost of pumping in that and future months.

In the derivation of equation 10-11, it has been assumed that the rate of pumping is the same for every well. Time periods are defined on a monthly basis. The irrigation season is six months in length (October–March), and it is assumed that withdrawals are evenly distributed over any given month, so that monthly withdrawals are pumped from the average monthly depth.

Let D_{ij} be defined explicitly as the distance to the water table at the middle of month i of year j; the aquifer then has a given time profile of distance to the water table: D_{11}, $D_{21}, D_{31}, \ldots, D_{61}, D_{121}$, and so on. The index i is modulus 6, since the level of the water table during the off-pumping season is irrelevant.

In equation 10-11, change the number of wells in month 1 of year 1, but hold all other things constant; then

$$\Delta D_{11} = h_{11}\, \Delta N_{11} - h_2 A_{11}\, \Delta N_{11}$$

or

$$(10\text{-}12) \qquad \frac{\Delta D_{11}}{\Delta N_{11}} = h_{11} - h_2 A_{11}.$$

Also in equation 10-11, holding N_{ij} and A_{ij} constant yields

$$\frac{\Delta D_{i+1,j}}{\Delta D_{ij}} = h_3$$

so that in general

$$\frac{\Delta D_{ij}}{\Delta D_{11}} = h_3^{\lambda}$$

where $\lambda = 12(j-1) + (i-1)$; therefore

$$(10\text{-}13) \qquad \frac{\Delta D_{ij}}{\Delta N_{11}} = (h_{11} - h_2 A_{11})\, h_3^{\lambda}.$$

It should be noted that the concept of marginal private cost used earlier in this chapter referred to the withdrawal of an additional unit of water, which is now defined as the operation of an additional well for one month. This change is necessary since the operation of wells can be observed over time, but there are no data concerning actual withdrawals.

If each well withdraws a_{ij} units of water during the ith month of the jth year, the marginal private cost of operating a well for that month is

$$(10\text{-}14) \qquad MPC_{ij} = cD_{ij}\, a_{ij}\,.$$

The excess of marginal social cost over marginal private cost is the present value of the increase in the cost of operating the existing wells, which can be written as

$$(10\text{-}15) \qquad (MSC - MPC)_{11} = N_{11} \sum_{j=1}^{\infty} \sum_{i=1}^{6} \left(\frac{1}{1+r}\right)^{\lambda} \frac{\Delta MPC_{ij}}{\Delta N_{11}}$$

where $\lambda = [12\,(j-1) + (i-1)]$. From equation 10-14,

$$(10\text{-}16) \qquad \frac{\Delta MPC_{ij}}{\Delta N_{11}} = ca'_{ij} \left(\frac{\Delta D_{ij}}{\Delta N_{11}}\right)$$

where a'_{ij} differs from a_{ij} because withdrawals per month per well are themselves a function of D_{ij}. Using a'_{ij} rather than a_{ij} results in a slight understatement of the difference between marginal social cost and marginal private cost, which is equal to about one-half the change in distance $(D'_{ij} - D_{ij})$ times the change in withdrawals $(a_{ij} - a'_{ij})$.

If equation 10-13 is substituted into 10-16 and the resulting equation substituted into 10-15, equation 10-17 is obtained:

$$(10\text{-}17)\quad (MSC - MPC)_{11} = cN_{11}\,(h_{11} - h_2\,A_{11}) \sum_{j=1}^{\infty}\sum_{i=1}^{6} a'_{ij}\left[\frac{h_3}{(1+r)}\right]^{\lambda}.$$

If the excess of marginal social cost over marginal private cost is expressed as a fraction of marginal private cost, the resulting final expression involves neither the coefficient c nor the value of a'_{ij}, but does include the ratio of a'_{ij} *to* a_{11}:

$$(10\text{-}18)\quad \frac{(MSC - MPC)_{11}}{MPC_{11}} = \frac{N_{11}}{D_{11}}(h_{11} - h_2\,A_{11}) \sum_{j=1}^{\infty}\sum_{i=1}^{6} \frac{a'_{ij}}{a_{11}}\left[\frac{h_3}{(1+r)}\right]^{\lambda}$$

where $\lambda = 12(j - 1) + (i - 1)$.

Equation 10-18 has an interpretation similar to the measure of the externality derived earlier for the general model. The term $(N_{11}/D_{11})\ (h_{11} - h_2A_{11})$ is the single-period elasticity of depth for withdrawals.

The term h_3^{λ} translates the change in the level of the water table from period 11 to period 12 $(j - 1) + (i - 1)$—that is, h_3^{λ} takes into account the intertemporal dependencies discussed earlier. The term a'_{ij}/a_{11} reflects a downward-sloping demand for water which was not explicitly introduced into the earlier general model and which shifts from month to month in response to seasonal factors.

The value of h_3 is determined by hydrological factors. At one extreme, h_3 would be unity—in the case of an underground lake which has neither inflow nor outflow. In this case the externality is maximized because additional withdrawals during the current period cause the level of the water table to be permanently lower than it would have been without those additional withdrawals.

At the opposite extreme is the case of an underground river with immediate recharge. In this case h_3 is equal to zero; by neglecting the difference between a'_{11} and a_{11}, the right-hand side of equation 10-18 becomes.

$$\frac{N_{11}}{D_{11}}\,(h_{11} - h_2\,A_{11}).$$

Equation 10-18 will be used to obtain estimates of the excess of social over private marginal cost for tubewell irrigation in the Mendoza region.

Empirical Results

This empirical study of the relation between the level of the water table and the other variables specified above is based upon data collected from the files of the Departamento General de Irrigación, the government agency responsible for administration of irrigation water in Mendoza. This agency requires the registration of each well at the time that it is drilled, and the level of the water table is recorded on the registration form. Since no subsequent data are available for any particular well, however, there is only one observation on the level of the water table for each well. Our time series data on depth is therefore based upon a different well or set of wells for each observation. To minimize errors of measurement of the level of the water table owing to irregularities in the terrain, care was taken to choose two areas that are characterized by highly regular terrain. The definition of limits of these areas is, however, somewhat arbitrary, and considerable effort was made to determine the effects of changing those limits.

Data were collected only for those wells that were drilled during the irrigation season—October through March. This season was then divided into six months; one month is the time period used in the analysis. A shorter time period, although preferable, would have severely limited the number of observations; each observation on level, when employed as a dependent variable, must be accompanied by a corresponding observation on lagged level, which appears as an independent variable.

The number of wells drilled in a month, given the definition of the region, is variable. The consequent heteroscedasticity was corrected for by multiplying the dependent and independent variables by the square root of the number of observations on the level of the water table during the corresponding month.

To take into account monthly differences in the amount of water withdrawn, five dummy variables corresponding to November through March were used. The dummies are multiplied by number of wells, so that the coefficient of the number of wells in a given month, h_{1i}, is equal to the coefficient of the number of wells plus the dummy coefficient corresponding to that month. For purposes of estimation, equation 10-11 can be written as

$$D_{ij} = h_0 + h_{1i}N_{ij} + \sum_{i=2}^{6} e_iX_iN_{ij} - h_2N_{ij}A_{ij} + h_3D_{i-1,j}$$

where

D_{ij} = vertical distance to the water table during the ith month of the jth year

N_{ij} = the number of wells in existence at the beginning of the ith month of the jth year

X_i = a dummy variable whose value is either zero or unity

e_i = the coefficient of the dummy variable X_i

A_{ij} = the accumulated volume of river streamflow during the ith month of the jth year.

The availability of surface water is based on data for the

Table 10-1. ***Unconstrained Least-Squares Estimates of Equation 10-11***

Area for definition of N	R^2	*Constant*	h_3	h_{11}	h_{12}	h_{13}	h_{14}	h_{15}	h_{16}	h_2
A	0.862	0.8992	0.5208	0.3770	0.4206	0.5262	0.2373	0.4118	0.4661	0.0002
		(0.7678)	(0.1613)	(0.1726)	(0.1978)	(0.2521)	(0.2704)	(0.2301)	(0.2022)	(0.0017)
B	0.868	0.7352	0.4707	0.2045	0.2202	0.2667	0.1328	0.2118	0.2464	−0.0001
		(0.7503)	(0.1727)	(0.0942)	(0.1104)	(0.1439)	(0.1539)	(0.1346)	(0.1142)	(0.0007)
C	0.869	0.6496	0.4672	0.1596	0.1706	0.2044	0.1066	0.1649	0.1911	−0.0001
		(0.7506)	(0.1761)	(0.0737)	(0.0864)	(0.1128)	(0.1208)	(0.1057)	(0.0897)	(0.0005)
A′	0.693	−1.5683	0.6035	2.4520	3.0879	4.1100	3.7330	3.7370	4.7810	0.0088
		(−2.1309)	(0.2190)	(1.1230)	(1.2448)	(1.3639)	(1.8746)	(1.7177)	(2.2680)	(0.0056)
B′	0.700	1.6834	0.5392	0.5710	0.7243	0.9475	0.9029	0.8886	1.0051	0.0020
		(2.1545)	(0.2198)	(0.2542)	(0.2853)	(0.3123)	(0.4374)	(0.3962)	(0.4739)	(0.0012)
C′	0.679	−1.2126	0.5762	0.3112	0.4002	0.5354	0.4550	0.4714	0.5626	0.0012
		(2.1787)	(0.2244)	(0.1554)	(0.1757)	(0.1904)	(0.2565)	(0.2395)	(0.3040)	(0.0008)

Note: In A, B, and C the number of observations is thirty-one; in A', B', and C' the number is twenty-three. Standard errors appear in parentheses. The areas are as follows: A = San Martin; $B = A$ + Central Region; $C = B$ + Lavalle; A' = Lavalle; $B' = A'$ + Central Region; $C' = B'$ + San Martin. N = units of 100 wells. Data are for 1958–66.

monthly volume of water in the Mendoza and Tunuyán rivers. Since it is possible to separate the areas irrigated by each river, the computation of the variable $N_{ij}A_{ij}$ is straightforward.

The number of wells in existence, N_{ij}, was computed from the same source, which poses a problem with units of measurement. First, there is the problem of interregional effects of wells—that is, the water level in one region may be affected by the number of wells in neighboring regions. Second, there are two types of wells with different time patterns of withdrawals during the irrigation season—those that supplement surface irrigation and those that do not. From the available information it has been possible to distinguish between these two types of wells. Nonsupplementing wells are used each month during the irrigation season, while those supplementing surface water are used only at those times in the irrigation season when the supply of surface water falls short of the demand.

The number of both types of well has increased sharply over the past decade. The correlation coefficient between the numbers of the two types of well over time is more than 0.80. This multicollinearity makes it impossible to use both variables in the same equation. As an alternative, several linear combinations of the two series on numbers of wells of each type have been tried.

Preliminary estimates of equation 10-11, which are presented in Table 10-1, are based upon data for the period 1958–66. For each of the two major regions, three alternative definitions of the stock of wells have been employed. It can be observed in the estimated equations that the coefficient (h_{1i}) of the stock of wells and the coefficient (h_2) of $N_{ij}A_{ij}$ decrease as the stock is increased by enlarging the limits of the region. This result is to be expected because the variance and the number of wells (N_{ij}) increase as more inclusive definitions are used. The water level variable D_{ij} remains invariant, however, and hence its variance remains invariant.

The capacity of the number of wells to explain changes does not vary significantly as the definition of N_{ij} is varied; it is therefore not clear which definition is the most appropriate for explaining changes in the level of the water table. However, this uncertainty is not a problem for subsequent analysis, since the expression for the discrepancy between social and private costs employs the regression coefficient multiplied by the number of wells (see equation 10-18). This product does not change if the various series of number of wells are proportional over time. In a regression, $y = a + bx$, multiplying the variable x by k implies that the estimated coefficient b is divided by k since $b = \Sigma yx/\Sigma x^2$. Then $b = \Sigma yx/k\Sigma x^2$, and the product bx does not change. Since the various series do not grow at exactly the same rate, there are some minor variations among the estimates of the discrepancy between social and private costs as the stock concept is varied.

The estimate of h_3, the coefficient of lagged level of the water table, is quite stable and always statistically significant at the 2.5 percent level. The estimate of h_2, however, is at best marginally significant. In those cases where the estimate of h_2 exceeds the standard error of that estimate, the estimator is always positive, which is the expected sign of h_2. As the algebraic value of h_2 increases, the estimate of the externality decreases (see equation 10-18). Although a good case could be made statistically for treating h_2 as essentially zero, the estimated values of h_2 will be retained with the effect that the externality will tend to be understated.

The estimates of the h_{1i} coefficients reveal a pattern consistent with the seasonal character of the river flow and

Table 10-2. ***Constrained Least-Squares Estimates of Equation 10-11 for San Martin***

Area for definition of N	R^2	*Constant*	h_3	h_{11}	h_{12}	h_{13}	h_{14}	h_{15}	h_{16}	h_2
	0.730	1.1525	0.6	0.2610	0.3670	0.4718	0.2004	0.3323	0.4406	0.0004
		(0.6358)		(0.1335)	(0.1250)	(0.1478)	(0.1734)	(0.1611)	(0.1295)	(0.0013)
A	0.479	0.7444	0.8	0.1201	0.2173	0.2811	0.0165	0.1947	0.2840	0.0005
		(0.6681)		(0.1403)	(0.1314)	(0.1553)	(0.2285)	(0.1693)	(0.1361)	(0.0014)
	0.289	0.3370	1.0	−0.0209	0.0676	0.0903	−0.1674	0.0570	0.1274	0.0007
		(0.7397)		(0.1553)	(0.1454)	(0.1719)	(0.2018)	(0.1875)	(0.1507)	(0.0015)
	0.737	0.9512	0.6	0.1289	0.1761	0.2220	0.0916	0.1590	0.2127	0.0001
		(0.6886)		(0.0655)	(0.0653)	(0.0808)	(0.0974)	(0.0905)	(0.0687)	(0.0006)
B	0.470	0.5759	0.8	0.0468	0.0883	0.1111	−0.0192	0.0728	0.1211	−0.0000
		(0.6812)		(0.0699)	(0.0697)	(0.0861)	(0.1039)	(0.0965)	(0.0732)	(0.0007)
	0.281	0.2007	1.0	−0.0353	0.0004	0.0001	−0.1301	−0.0135	0.0296	−0.0001
		(0.7555)		(0.0775)	(0.0773)	(0.0955)	(0.1152)	(0.1071)	(0.0812)	(0.0007)
	0.734	0.8884	0.6	0.0995	0.1349	0.1686	0.0716	0.1229	0.1628	0.0000
		(0.648)		(0.0494)	(0.0496)	(0.0617)	(0.0749)	(0.0696)	(0.0523)	(0.0004)
C	0.471	0.5319	0.8	0.0339	0.0641	0.0794	−0.0188	0.0513	0.0890	−0.0000
		(0.6906)		(0.0527)	(0.0529)	(0.0559)	(0.0798)	(0.0741)	(0.0557)	(0.0005)
	0.279	0.1754	1.0	−0.0316	−0.0068	−0.0099	−0.1092	−0.0202	0.0150	−0.0001
		(0.7643)		(0.0583)	(0.0585)	(0.0728)	(0.0883)	(0.0820)	(0.0517)	(0.0005)

Note: In all cases the number of observations is thirty-one. Standard errors appear in parentheses. The areas are as follows: A = San Martin; $B = A$ + Central Region; $C = B$ + Lavalle + West Region, N = units of 100 wells. Data are for 1958–66.

the total irrigation requirements. These estimates consistently rise through December, decline sharply in January, and rise thereafter, reflecting that the demand for irrigation water grows more rapidly than the river flow until December–January when the river flow rises very rapidly. The subsequent increase in the coefficient reflects the decline in the river flow plus the autumn irrigation requirements.

Even though the areas (water fields) were chosen to minimize variations in terrain and errors of measurement in the level variable, some errors probably remain. Because lagged water level does not correspond to the same well as the contemporary water level, these errors take the form of measurement error in an independent variable, introducing a potential bias in the coefficient estimates. Estimating the bias requires outside information on either the ratio of the variance of measurement errors to the variance of the observed lagged level or on the true value of the lagged level coefficient (Zapata 1969).

Redefining the dependent variable in equation 10-11 as $D_{ij} - h_3 D_{i-1,j}$ avoids the bias caused by measurement error in the depth variable. The problem, of course, is that the value of h_3 is not known. The extreme values for this coefficient are 0 and 1; the actual value depends on the nature of the groundwater reservoir. If h_3 is 1, an underground lake with no inflow or outflow is indicated; withdrawals in one month will affect the water level in all succeeding periods. If h_3 is 0, the case of a pure river in which withdrawals do not affect future water levels is indicated. The range of 0 to 1 contains an intermediate case in which the effects on successive periods will be lower (higher) when the coefficient is closer to 0 (1).

Zapata (1969, appendix B) shows that under plausible assumptions the estimator of h_2 will be negatively biased. The estimates of h_3 in Table 10-1 range from 0.467 to 0.603. Because of the negative bias, it will be assumed that the value of h_3 does not lie below 0.6.

Equation 10-11 has been redefined as described above, using three alternative values for h_3, with the resulting dependent variables:

$$y'_{ij} = D_{ij} - 0.6\, D_{i-1,j}$$
$$y''_{ij} = D_{ij} - 0.8\, D_{i-1,j}$$
$$y'''_{ij} = D_{ij} - D_{i-1,j}.$$

These definitions of the dependent variables are assumed to span the range of possibilities for h_3; clearly y'''_{ij} corresponds to the upper limit for h_3.

Tables 10-2 and 10-3 present the results for the constrained estimates of h_{1i} and h_2—that is, the estimates which emerge when the lagged water level variable is shifted to the left-hand side of equation 10-11 and the alternative values of h_3 are imposed. In the tables, the larger the values imposed for h_3, the smaller the estimates

Table 10-3. *Constrained Least-Squares Estimates of Equation 10-11 for Lavalle*

Area for definition of N	R^2	*Constant*	h_3	h_{11}	h_{12}	h_{13}	h_{14}	h_{15}	h_{16}	h_2
	0.490	−0.4811	0.6	1.6040	2.2660	2.8200	2.2068	2.3186	3.3430	0.0056
		(1.9174)		(0.9787)	(1.1415)	(1.3735)	(2.1406)	(1.7620)	(2.5704)	(0.0068)
A′	0.370	−0.8549	0.8	1.1610	1.9040	2.2850	1.6946	1.8879	3.0290	0.0052
		(2.0174)		(1.0300)	(1.1722)	(1.4451)	(2.2522)	(1.8539)	(2.7044)	(0.0071)
	0.260	−1.2286	1.0	0.7179	1.5420	1.7499	1.1823	1.4572	2.7149	0.0048
		(2.2044)		(1.1250)	(1.2809)	(1.5791)	(2.4610)	(2.0258)	(2.9552)	(0.0078)
	0.504	−0.4420	0.6	0.3309	0.4836	0.5947	0.4767	0.5044	0.6275	0.0012
		(1.8996)		(0.2093)	(0.2395)	(0.2959)	(0.4748)	(0.3885)	(0.5115)	(0.0015)
B′	0.378	−0.6978	0.8	0.2275	0.3955	0.4681	0.3426	0.3922	0.5329	0.0011
		(2.0186)		(0.2224)	(0.2545)	(0.3144)	(0.5046)	(0.4128)	(0.5435)	(0.0016)
	0.261	−0.9535	1.0	0.1240	0.3074	0.3413	0.2084	0.2800	0.4382	0.0010
		(2.2136)		(0.2439)	(0.2791)	(0.3448)	(0.5533)	(0.4527)	(0.5543)	(0.0017)
	0.490	−0.3090	0.6	0.1965	0.2877	0.3550	0.2454	0.2804	0.3785	0.0007
		(1.9071)		(0.1319)	(0.1506)	(0.1840)	(0.2811)	(0.2368)	(0.3335)	(0.0010)
C′	0.368	−0.6318	0.8	0.1388	0.2407	0.2862	0.1815	0.2248	0.3331	0.0006
		(2.0137)		(0.1393)	(0.1590)	(0.1943)	(0.2968)	(0.2500)	(0.3521)	(0.0010)
	0.254	−0.9546	1.0	0.0812	0.1937	0.2175	0.1177	0.1693	0.2878	0.0006
		(2.2012)		(0.1522)	(0.1738)	(0.2124)	(0.3245)	(0.2733)	(0.3849)	(0.0011)

Note: In all cases the number of observations is twenty-three. Standard errors appear in parentheses. The areas are as follows: A′ = Lavalle; B′ = A′ + Central Region + West Region; C′ = B′ + San Martin. N = units of 100 wells. Data are for 1958–66.

of the h_{1i} coefficients—that is, the smaller the effect of withdrawals on the level of the water table. This is to be expected. Suppose, for example, that

$$y_i = \alpha_0 + \alpha_1 X_{1i} + \alpha_2 X_{2i} + U_i$$

which is estimated as

$$y_i' = (y_i - \alpha_1' X_{1i}) = \alpha_0' + \alpha_2' X_{2i} + U_i'.$$

The estimator of α_2' is

$$\alpha_2' = (m_{yX_2})/(m_{X_2X_2}) = (m_{yX_2} - \alpha_1' m_{X_1X_2})/(m_{X_2X_2}).$$

Clearly, $d\alpha_2'/d\alpha_1' < 0$ for $m_{X_1X_2} > 0$. In this case X_1 corresponds to lagged levels, X_2 corresponds to number of wells. The data reveal a positive correlation between these two variables. In the case of three independent variables, the above result holds strictly only in the case where X_3 (that is, the variable $N_{ij}A_{ij}$) is only "weakly" correlated with the other two independent variables.

The negative correlation between the imposed value of h_3 and the values of h_{1i} reduces the limits of the estimated externality under alternative values of h_3. This occurs as our estimate of the externality involves the product of h_{11} and various powers of h_3.

The externalities for alternative values of h_3 have been estimated using the coefficients from Tables 10-2 and 10-3. Equation 10-18 requires specification of A_{11}, N_{11}, r, D_{11}, as well as the ratio a_{ij}/a_{11}. For the entire period 1958–66, A_{11} was taken to be the average value of A_{1j}; N_{11} and D_{11} were taken to be the number of wells and distance to the water table of the last observation in the sample. A discount rate of 1 percent a month was used.

In the regressions reported in Tables 10-2 and 10-3, the variable D_{ij} was defined as the static distance to the water table—that is, the distance that exists at a zero rate of withdrawal for the particular well. From the point of view of the cost of withdrawing water, and hence the externality, it is important to take into account the drawdown—that is, the local depression in the water table induced by pumping. Since this drawdown is independent of the distance to the water table, it was neglected in the definition of the variable D_{ij} in the regression equations, which can affect only the constant term. In equation 10-18, however, it was necessary to define D_{11} inclusive of the drawdown. Data furnished by the Departmento General de Irrigación of Mendoza indicate that the drawdown for the average well is 14 meters. Since the final observation on static distance to the water table was about 6 meters, D_{11} was taken to be 20 meters for equation 10-18. For evidence that the drawdown is independent of the static distance to the water table, see Israelsen and Hansen 1962, chapter 5. This local drawdown disappears when the well stops pumping. It is completely different from the decline in water table of the model, which results from decreasing the stock of water.

A more serious difficulty arises with the estimation of the ratio a_{ij}/a_{11}. Equation 10-5 and the definition of x_{ij}^k yield the following expression for the typical well:

$$a_{ij} = b_{1j} - b_3 D_{ij} - \left(\frac{n}{H}\right) A_{ij}.$$

The ratio of a_{ij} to a_{11} therefore becomes

$$\frac{a_{ij}}{a_{11}} = \frac{b_{1j} - b_2 A_{ij} - b_3 D_{ij}}{b_{11} - b_2 A_{11} - b_3 D_{11}}$$

where $b_2 = n/H$, as was defined earlier. Multiplying numerator and denominator by the factor $B/(1 + f_1 + g_2 + Bb_3)$ converts the b coefficients into h coefficients and yields

$$\frac{a_{ij}}{a_{11}} = \frac{h_{1j} - h_2 A_{ij} - Bb_3 h_3 D_{ij}}{h_{11} - h_2 A_{11} - Bb_3 h_3 D_{11}}.$$

From the estimates in Tables 10-2 and 10-3, all of the elements of this ratio can be computed except b_3—the slope of the demand curve for underground water. Neglecting completely the terms involving b_3 yields

$$\left(\frac{a_{ij}}{a_{11}}\right)^* = \frac{h_{1j} - h_2 A_{ij}}{h_{11} - h_2 A_{11}}.$$

Since $Bb_3h_3 > 0$, it follows that

$$\left(\frac{a_{ij}}{a_{11}}\right)^* \begin{matrix} > \\ < \end{matrix} \frac{a_{ij}}{a_{11}} \text{ as } D_{ij} \begin{matrix} > \\ < \end{matrix} D_{11}.$$

Table 10-4. *Excess of Marginal Social over Marginal Private Cost as a Percentage of Marginal Private Cost*

Region	*October*	*November*	*December*	*February*	*March*
San Martin					
A	19.3	20.6	17.8	15.7	12.6
B	20.8	21.9	19.1	17.2	13.3
C	21.0	22.1	19.3	17.4	13.5
Lavalle					
A'	31.2*	30.3	20.1	33.6**	30.8**
B'	29.5*	29.1	19.0	30.3**	25.4**
C'	29.3*	29.0	19.3	29.0**	25.3**

Note: In computing these estimates, the term $(a_{ij}/a_{11})^*$ was set to equal zero in all cases for which the corresponding h_{1j} term was not statistically significant, which will result in an understatement of the externality. Those estimates that appear without asterisks were obtained by setting at zero all values for $(a_{ij}/a_{11})^*$ that were not statistically significant at the 5 percent level; a single asterisk implies the 10 percent level, and a double asterisk implies the 15 percent level. The 10 and 15 percent levels were required in some cases owing to the fact that no values of h_{1j} were significant at lower levels. *A, B, C, A', B',* and *C'* are defined in Tables 10-1, 10-2, and 10-3.

The value chosen for D_{11} corresponds to the end of the period in question, and in general that value exceeds the values of D_{ij} for the corresponding season. It is therefore unlikely on average that D_{11} will fall short of D_{ij}; the opposite is more likely. Employing the ratio $(a_{ij}/a_{11})^*$, rather than the ratio a_{ij}/a_{11}, can be expected to understate the externality computed in equation 10-18.

Estimates of this externality are presented in Table 10-4. These estimates are based on those regression coefficients with the highest level of statistical significance; that is, for h_3 restricted to 0.6. Equation 10-18 was computed for each month of the irrigation season, for each definition of the relevant stock of wells N_{ij}, and for the two regions under consideration. That is, the estimates in Table 10-4 are for the externality associated with the withdrawal of additional water during the months specified in the table. January does not appear because the estimated externality was approximately zero in that month owing to the very high rate of river flow. When larger values of h_3 were used, the level of statistical significance of the estimates of the h_{1i} was generally quite low. Alternative estimates of the externality employing larger values of h_3 are consequently less reliable. The estimates so obtained were in some cases greater and in some cases smaller than those in Table 10-4; the effect of increasing the value of h_3 appears on the average, to reduce slightly the estimate of the externality.

The results in Table 10-4 indicate that there are substantial externalities for pump irrigation in Mendoza, particularly in the Lavalle region. The externality tends to be greatest during the first months of the irrigation season and declines to an insignificant level during January, when the river flow is extremely high, only to rise again during the final months of the season as the river flow declines and as pre- and post-harvest water requirements rise. On the average (excluding January), the externality is approximately 20 percent for the San Martin region and 30 percent for the Lavalle region. The implications of these estimates will be discussed in the next section.

Summary and Conclusion

The supply of irrigation water imposes the fundamental constraint on the agricultural output of Mendoza, Argentina. Rainfall is for all practical purposes negligible, and the supply of surface water, although variable from year to year, is not under the control of the individual farmer. In recent years pump irrigation has been extensively exploited to supplement existing supplies of surface water and to reduce the variability in the annual supply of water. This study has been concerned with external effects of pump irrigation, since the level of the water table depends on the rate at which groundwater is extracted, and since individual farmers extracting the groundwater have no property rights over the water remaining underground. A model has been developed and tested to estimate the size of the external effects.

The model analyzed corresponds to a steady state in which net inflows to the aquifer are a function of the level of the water table, which in turn is influenced by pumping withdrawals. Under the assumption that net inflows are limited only by the amount of water in the aquifer and not by external supply, there exists a unique steady-state level of the water table for each rate of pumping withdrawal.

Alternative versions of the model correspond to alternative hydrological assumptions. At one extreme, in the underground river model, current withdrawals have no influence whatsoever on the future level of the water table; the steady-state solution is realized without lag. At the other extreme, in the underground lake model, current withdrawals permanently lower the level of the water table; since inflows are precluded, no steady-state level of the water table exists as long as pumping continues. The intermediate and general case is the one in which current withdrawals have a transitory effect on the level of the water table; withdrawals lower the water table, but eventually inflows restore the water table to the original level. More precisely, let $W^*(t)$ represent a particular withdrawal program, to which will correspond a given time profile of distance to the water table, $D^*(t)$. Let withdrawals be increased by an amount Δ over the interval from t' to $t' + \delta$. The proposition is that $D(t)$ will return to $D^*(t)$ within a finite time period. This intermediate case appears to be relevant for the Mendoza region.

It is important to recognize that, apart from certain particular cases (Zapata 1969, pp. 6–9), the externality of withdrawing groundwater will have a time element—that is, withdrawing an additional unit of water today will increase today's pumping costs and also make pumping costs in some future periods higher than they otherwise might have been. For this reason, a steady-state model is not sufficient (Zapata 1969, pp. 6–9.) Rather, one must develop the analysis in such a way that the temporary departures from the steady state are explicitly taken into account. To neglect these departures would cause the externality to be understated.

To complete the model, demand must be introduced as well as supply. The demand for groundwater was specified in linear form as a function of price (the cost of extracting water, which depends upon the level of the water table), the amount of surface water available for irrigation purposes, and a seasonal effect reflecting variations in water demand. The seasonal effect was introduced by two methods. The first alternative constructs an index based upon experimental data relating to such water requirements as temperature and humidity. The second—the one used to obtain the empirical results reported earlier—introduces monthly dummies into the regressions. The dummy variables were multiplied by the number of wells to reflect the aggregate effect of seasonal factors. The coefficients of the seasonal dummies are expected to rise through December, fall in January, and later rise again. During the first part of the irrigation season (October–December), the demand for water grows much more rapidly than does the supply of surface water, but in late December and January surface water rises very rapidly as rivers are fed by melting snow in the Andes. The subsequent rise in the coefficients reflects the declining supply of surface water as the snow accumulation is exhausted and as the preharvest demand for irrigation water rises.

The size of the externality is estimated in two steps. The first step is to estimate a reduced-form equation reflecting both the demand and supply forces. This equation (equation 10-11), expresses the current level of the water table as a function of the lagged level, the rate of withdrawal of water (represented by the number of wells), and a final variable to capture the effect of the supply of surface water on the relation between the number of wells and the aggregate rate of withdrawal. Since the equation in question is a reduced form of the structure, the coefficients of that equation are combinations of the structural coefficients. No attempt was made to estimate the structural coefficients because the reduced-form coefficients are sufficient to estimate the externality.

Two of the reduced-form coefficients are of special significance. The coefficient of the number of wells will be zero if there is no relation between the stock of water in the aquifer and the level of the water table, which is the case for an underground river whose depth cannot be depressed by the withdrawal of water and where no externality can exist. A positive coefficient for this variable indicates that the underground stock and the level of the water table are related, and hence an externality will be present whenever pumping takes place. The externality described here is the kind associated with the crowding effect on highways—the externality disappears once the crowding is eliminated.

The second crucial coefficient is that of the lagged level of the water table. If recharge occurs with no significant lag, the level of the water table will depend only upon current withdrawals; the lagged variable, which captures the effect of prior withdrawals, will carry a zero coefficient. In this case the externality is limited to the concept described in the preceding paragraph. If recharge takes a significant amount of time, however, previous withdrawals will be relevant in determining the current level of the water table, and hence the lagged dependent variable will carry a nonzero coefficient. In the limiting case of an underground lake, recharge will be zero; current withdrawals will thus affect only the rate of *change* in the level of the water table. In this case the coefficient of the lagged variable will be unity.

The existence of an externality depends upon a positive coefficient for the withdrawal variable (number of wells), and the duration of the externality depends upon the coefficient of the lagged dependent variable. Measurement errors pose a serious problem in estimating the coefficients, particularly the coefficient of the lagged dependent variable. The data on the level of the water table are obtained by measurements taken as the drilling of each well is completed. This provides just one observation per well; time series data can be built up only by combining measurements on a series of wells. As a consequence, terrain irregularities introduce error into these measurements. Although care was taken to define regions where the terrain was highly regular, some errors must surely be present. Because the lagged dependent variable has a key role as an independent variable, errors in measurement in the dependent variable introduce potential bias in the regression coefficients. It is expected that the bias is negative in the case of the coefficient of the lagged dependent variable. To deal with the bias, the reduced form equation was estimated for two regions, and on the basis of estimates thus obtained for the coefficient of the lagged variable a lower limit was established. The lagged dependent variable was then introduced, with a constrained coefficient, on the left-hand side of the equation for estimation of the remaining coefficients. The empirical results of a number of experiments carried with constrained and unconstrained models are described, all of which suggest that measurement error adds to the size range of the estimated externality, but does not vitiate the conclusion that the size of that externality is substantial.

The second step in measuring the externality is to introduce the estimated coefficients of the reduced-form equation into an equation that presents the externality as a fraction of the marginal private cost of pumping. The available data permitted an estimate not for the externality of withdrawing an additional unit of water but rather for the externality of operating an additional well for one month during the irrigation season. As can be seen from equation 10-18, the externality so measured is an increasing function of the coefficient of the number of wells and the coefficient of the lagged dependent variable. The externality, as measured by this equation, includes the contemporaneous costs imposed upon other pumpers by an incremental withdrawal of water, plus the present value of such future costs as a consequence of the lag in recharge. A discount rate of 1 percent a month was employed to obtain present values.

The external effect of pump irrigation was estimated, by using the model described above, for two regions in the province of Mendoza, Argentina. The empirical results are highly consistent with a hydrological model in which the level of the water table is positively related to the stock of water in the aquifer and in which recharge occurs with a significant lag. The coefficient of the number of wells is positive and statistically significant at the 5 percent level in all but one or two months of the irrigation season, depending upon the region. The months for which statistically significant results were not obtained correspond to a period during which the supply of surface water is abnormally high and the output of each well therefore low. In the unconstrained regressions the coefficient of the lagged dependent variable ranges from greater than 0.4 to 0.6, indicating a significant lag in recharge. This conclusion is strengthened by the expectation that this coefficient is subject to a negative bias owing to error of measurement. If 0.6 is taken as a plausible value for this coefficient, over 20 percent of the initial decline in the level of the water table persists three months after that decline has been induced by additional pumping. The amount of external effect varies with the timing of the additional pumping, since the irrigation season covers only part of the year; a depressed water table during the off-season will not have detrimental effects. In addition, owing to the fact that the externality is measured as the excess of social over private costs arising from the operation of an additional well for one month, the estimate of the externality tends to be greatest at the beginning of the season when groundwater demand is highest, and the lowest during the middle of the season when surface water supply is greatest. The estimates obtained indicate that during the five months of the irrigation season for which an appreciable externality exists, the excess of marginal social over marginal private costs of pump irrigation ranges from more than 10 percent to about 34 percent of the marginal private costs. This range arises from geographic as well as seasonal factors.

The qualitative nature of these results is strengthened by the observation that the behavior of the seasonal effect, as revealed by the coefficients of the dummy variables described earlier, is highly consistent with factual evidence concerning the time pattern of water withdrawals. Furthermore, the measure of the externality is relatively insensitive to the definition of the water field; that is, the size is not influenced by the somewhat arbitrary choice of the geographic area used in defining the number of wells at a point in time. The findings of this study are briefly summarized below:

- The hydrological structure of Mendoza, Argentina, is such that an inverse relationship exists between the rate of withdrawal of groundwater and the level of the water table.
- The influence of withdrawals of groundwater on the level of the water table persists at a significant rate for several months after those withdrawals.
- The size of the effect of withdrawals on the level of the water table critically depends upon the point in the irrigation season at which those withdrawals are made.
- The pattern of property rights in the Mendoza region is such that individual farmers will, in calculating their

private costs consider only a portion of the total costs of withdrawing groundwater.

- As a result of the above, the market equilibrium leads to inefficient use of groundwater, since social cost exceeds private cost by 10 to 34 percent.

In planning for efficient use of water resources, knowledge of the existence of and size of externalities is vital. These external effects are relevant even if the rate of decline in the water table is small or negligible. The results of my study indicate that this rate is currently small. For the San Martin area, the time trend for the level of the water table suggests that the level falls only one meter in four years. If this rate were to be maintained, it would take a rather long time to reach what is today considered an uneconomic level (90 meters). As the number of wells increases, however, the rate of withdrawal and therefore the rate of decline in the water table will accelerate. Finally, the rate of decline in the water table is not a reliable guide to the nature and extent of the externality, since that externality would continue to exist even if the level of the water table were to stabilize. As can be easily seen from the steady-state model, the water table would stabilize at a level below the optimal one.

The case of common-property resources has been thoroughly analyzed (Pigou 1932; Gordon 1954, pp. 124–42; Milliman 1956, pp. 426–37). Taxing the use of such resources is the policy that is most frequently employed to remedy the misallocative effects arising from private exploitation of that resource. One of the major problems in practice is, of course, the estimation of the extent of the externality and hence the appropriate tax rate. In this study we have developed and estimated a model that produces just that type of information.

Perhaps the most compelling policy implication of this study is that extracting groundwater in Mendoza involves externalities sufficiently large to merit—even demand—public concern.

Appendix. Conjunctive Use and Management

The conjunctive use and management of surface and groundwater leads to the most flexible of systems, which can adapt both to the rate of growth of the demand for water and to the limiting boundary of water availability. The second type of flexibility permits a greater expansion of the area based on the storage characteristics of aquifers and dams and the variability of rainfall and runoff.

An analysis of the hydrological balance is needed to determine the most flexible system of using surface and groundwater reservoirs for storage. Storage presents different problems, depending on whether water is stored in reservoirs or in underground aquifers through artificial recharge. Surface reservoirs are limited by their total capacity, although the rate of storage may be very high. Groundwater aquifers may have a very large storage capacity, but the rate at which water can be stored underground may be rather low, because it is subject to the percolation capacity of the recharge facilities and to the capacity of structures conveying the water to them (Wiener 1971, pp. 385–97). With this approach, surface storage serves as a shock absorber between irregular flows and the limited absorptive capacity of recharge installations.

Other energy considerations may be taken into account in the conjunctive management of surface and groundwater storage. In an extremely arid environment the generation of electricity at the reservoir dam for an off-peak load is not competitive with the use of water for irrigation, given the possibility of downstream recharge of the aquifer. And during the pumping season the cost of groundwater can be minimized by the conjunctive use of surface water delivered to the reservoir dam to generate electricity for pumping.

This approach was used in a simulation model developed for the evaluation of a proposed project in Mendoza. The results indicate that the storage capacity for the system—surface reservoir and the aquifer—was substantially increased by a set of management rules including artificial recharge. At the same time conflicts among different purposes (irrigation, energy, and recreation) were minimized, and the size of the proposed dam was reduced. (See the Departamento General de Irrigación's report on the Aprovechamiento Multiple Potrerillos, Mendoza, Argentina, 1980.)

Conjunctive use of surface and groundwater may also affect water distribution within areas of a region, resulting in a saving of water and energy resources. For the upper and lower areas of a valley, for example, certain amounts of surface and groundwater would have to be allocated in such a way that costs can be minimized. In some cases costs might be reduced if surface water were distributed upstream and groundwater pumped downstream. Furthermore, when the upstream area lies above an unconfined aquifer, water conveyance losses can be pumped downstream, which changes the benefits of lining canals. Given pumping costs, it is not certain that improving efficiency upstream is profitable.

References

Heath, Ralph C., and Frank W. Trainer. 1968. *Introduction to Ground Water Hydrology.* New York: Wiley and Sons.

Gordon, H. Scott. 1954. "The Economic Theory of a Common-Property Resource: The Fishery." *Journal of Political Economy* 62:124–42.

Israelsen, Orson W., and Vaughn E. Hansen. 1962. *Irrigation Principles and Practices.* New York: Wiley.

Milliman, Jerome W. 1956. "Commonality, the Price System, and Use of Water Supplies." *Southern Economic Journal* 22 (April): 426–37.

Pigou, A. C. 1932. *The Economics of Welfare.* 4th ed. London: Macmillan.

Wiener, Aaron. 1971. *The Role of Water in Development.* New York: McGraw-Hill.

Zapata, Juan Antonio. 1969. "The Economics of Pump Irrigation: The Case of Mendoza, Argentina." Ph.D. dissertation, University of Chicago.

Comment

D. N. Basu

Zapata treats groundwater use econometrically without initially detailing a distributed parameter model of the aquifer's physical characteristics and its response to farmer pumping and surface canal seepage. Instead, he focuses on pumping over an infinite horizon, where each farmer considers only his own private costs, which are specified solely in pumping costs as a linear function of water table depth. The discrepancy between private costs and social costs is written in terms of changes over time in the depth to the water table.

Zapata first writes the demand for groundwater as a decreasing function of the cost of pumping—that is, as a decreasing function of water table depth. He then writes a material balance identity, transforming water stock variables into equivalent expressions of water table depth and using an inverse relation between aquifer water stock and water table depth. Finally, Zapata substitutes in the material balance the demand for withdrawals and some expressions for underflows in water table depth and surface water flows, which yields a relation between depth and number of wells, surface water flows, and the lagged value of depth.

From this basic relation, Zapata is able to estimate the discrepancy between marginal private costs and marginal social costs as a proportion of marginal private costs for each additional well. This estimate of the externality cost is written as the product of a single-period elasticity of water table depth to withdrawals and an intertemporal factor that mechanically links the change in depth over time with a demand adjustment that depends upon the cost of pumping. The parameters of this relation are derived from the regression estimates of the basic relation already described.

Zapata estimates the difference between marginal private costs and marginal social costs in general physical relationships and in the effects of physical interdependence (that is, externality), which are assumed to summarize increased pumping costs adequately. Within this framework, he uses data on number of wells and surface water flows, as well as rather sparse data on water table depth, to derive lower-bound estimates of the externality cost—that is, the amount of the optimal tax on pumping required to equalize private and social costs.

The model accepts the existing institutional arrangements for water allocation (except for the optimal pumping tax). It neglects dynamic optimization issues such as the optimal level of storage in the aquifer, as well as other environmental effects—for example, the mixing of fresh with saline groundwater, long-term salt disposal by horizontal drainage, and control of potential land subsidence.

Zapata's method thus uses minimum information to estimate the short-run optimal tax on pumping. Although his method guides policy on aquifer management toward more efficient short-term resource use, it should not be used as an excuse to neglect long-term resource management issues.

11

Risk Aversion in Conjunctive Water Use

John D. Bredehoeft and Robert A. Young

The South Platte River system of Colorado uses surface and groundwater conjunctively for irrigation. Yet actual well capacity is approximately sufficient to irrigate the entire area. This apparent overinvestment in wells raises a question. To what extent is groundwater being developed as insurance against periods of low streamflow? Our model couples the hydrology of a conjunctive stream aquifer system to a behavioral-economic model which incorporates farmer behavior. In this chapter we use a simulation model to investigate the economics of an area patterned after a reach of the South Platte Valley.

Two earlier studies addressed the management issues of groundwater development. The first study (Bredehoeft and Young 1970) focused on a groundwater mining situation: the second study (Young and Bredehoeft 1972) addressed the problem of conjunctive surface and groundwater use. These studies represent the ends of the spectrum in groundwater management problems. At one end are developments where all the water is being mined, as in the High Plains of Texas and New Mexico; at the other end is the perpetual development in which the groundwater system is more or less fully renewed each year. The aquifers in the South Platte and Arkansas valleys of Colorado are examples of perpetual conjunctive use systems. Each of our studies examines institutional options for managing the systems.

The situation in Colorado typifies the general problem of groundwater management, especially as it applies to the conjunctive use of surface and groundwater. The legal institutions have evolved in an interesting way. In 1966 the state engineer of Colorado ordered thirty-nine pumpers in the Arkansas Valley to stop pumping. Their pumping was reducing the flow of the river and thereby damaging the surface water rights of more senior downstream users. Under a 1965 Ground Water Management Act, the state engineer has the power to restrict such pumping. A test case—*State of Colorado* v. *Roger Fellhauer,* 1968—eventually went to the Colorado Supreme Court for settlement. In ruling on the case the court upheld the law, which gave the state engineer power to regulate groundwater pumping, but indicated that he had acted, in this instance, in an arbitrary and discriminatory manner (Daubert 1978). Although the *Fellhauer* case is a single example, it illustrates the general problem of groundwater management in the United States. It is difficult to establish a centralized authority which has the power to regulate groundwater development in a manner that truly achieves what system analysts agree is optimal.

To design effective and appropriate institutions one must understand what motivates individual users of the resource—in this instance, irrigators. This investigation couples a hydrologic model with an economic model to examine the objectives of the individual farmers. The analysis is designed to test our hypothesis of what we now believe is the appropriate objective function. Our more general thesis is that the sophisticated management models available to us can be used to explore the implications of various management institutions, thereby helping to select feasible institutions within which one can more effectively manage the resource.

In designing effective institutions for water management, one must understand the interaction between the water supply, in this case both surface and groundwater, and the economic factors which motivate the farmers' actions. Young and Bredehoeft (1972) examined the problems of conjunctive management of surface and groundwater for irrigation. A simulation model was presented which captured many of the essential elements of both the hydrologic system and the economics of allocating irrigation water. This model was used to investigate a hypothetical reach of stream and interconnected aquifer which supported an agricultural economy; this reach was patterned

Figure 11-1. *Optimum Well Capacity*

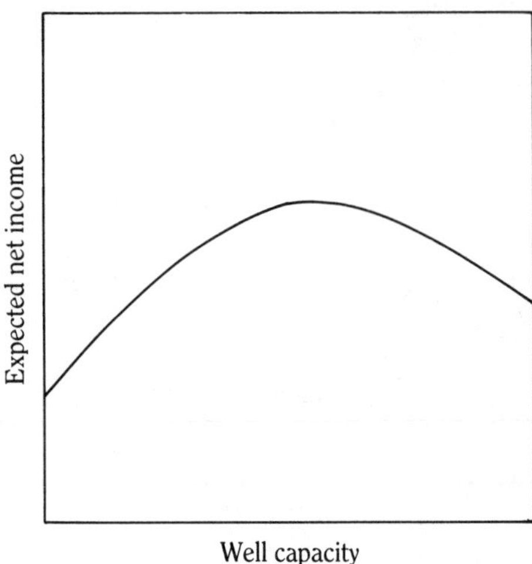

Figure 11-2. *Expected Net Income versus Well Capacity (Based on 1972 Results)*

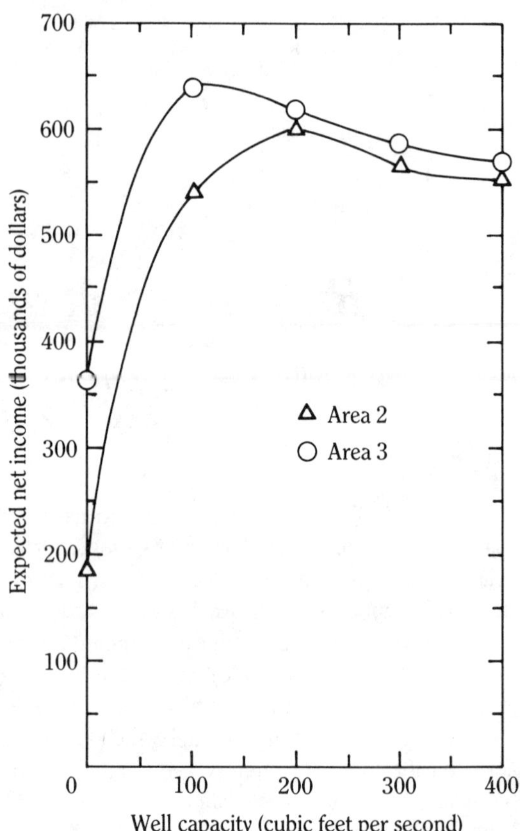

after a reach of the South Platte River in Colorado. The 1972 study also investigated the dynamics of the local farming economy, the institutions within which it operates, and the investment in wells. In this chapter we report our further studies of the South Platte system, which are based upon the 1972 model with the economic parameters updated to 1978 values. The chapter focuses on a question that arose because our earlier model failed to predict accurately the investment in wells. What role does the variation in water supplies play in motivating farmers' investments in groundwater capacity?

Optimal Groundwater Capacity

In the previous study (Young and Bredehoeft 1972) we hypothesized that the farmers invest in a way that maximizes their expected net income. We used a ten-year mean income as a measure of the expected income. This concept assumes that maximizing mean income is the single most important objective. In using this single-valued objective function, we gave no consideration to the possibility that farmers install wells to ensure themselves of a water supply during times of low streamflow. Well capacity was treated as a decision variable, with the simple thesis that there was an optimum well capacity which would maximize the expected net income. This concept is illustrated in Figure 11-1.

Figure 11-2 plots the expected net income versus well capacity constructed from our 1972 results. These results suggested that there was an optimum well capacity for a system like the South Platte. Based upon the analysis, however, the optimum capacity would be approximately one-half of the capacity actually installed in the South Platte system at that time. This, of course, suggests that the assumptions regarding farmers' motivations incorporated into our analysis were in some way incomplete or inaccurate. The role of groundwater supplies in reducing variation in water supplies (providing insurance) suggests that an alternative hypothesis be sought in the theory of choice under imperfect knowledge (Raiffa 1970).

Expected Income versus Variance

In most cases of uncertainty, one is willing to pay something to reduce the variance in one's income. In the conjunctive use of surface and groundwater, it is uncertain that available surface water will meet demands, particularly when the farmer is making a decision to plant at the beginning of the growing season based upon a forecast of flow for the season. Planting decisions are based upon estimates of the runoff to be expected that season from both reservoir and snowpack storage. How climatic conditions will affect the growing seasons is unknown, however, which introduces an additional uncertainty.

In a conjunctive use system, groundwater represents an alternative source of water which is far more certain than surface water. Errors in estimating the runoff—that is, planting decisions which do not reflect the availability of

Figure 11-3. *Tradeoff between Expected Income and Income Variance*

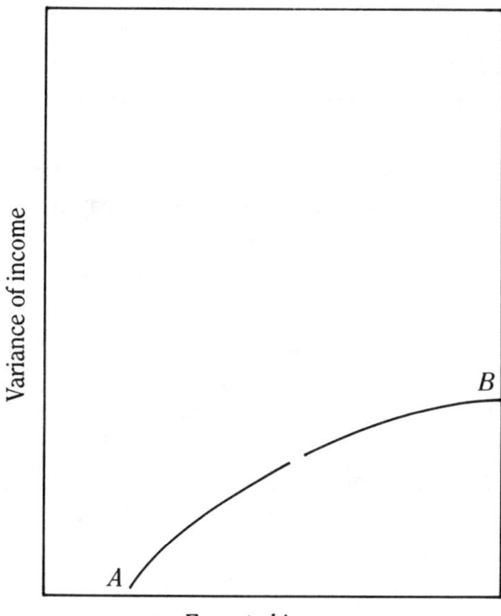

Figure 11-4. *Ideal Conjunctive Groundwater and Surface Water System*

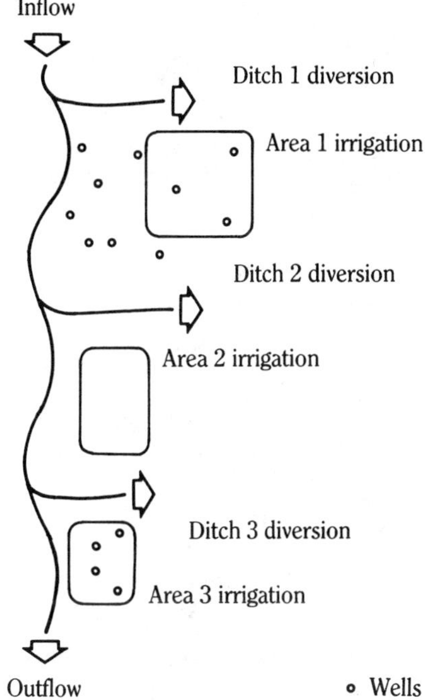

Source: Young and Bredehoeft (1972).

surface water during the growing season—can be partially compensated for by simply turning on the pumps. The problem is, how large should the pumping capacity in the system be? Or, to put it in economic terms, what is the utility-maximizing investment in well capacity?

It is true that pumping groundwater is more expensive than simply diverting surface water. One would expect there to be a tradeoff between expected income and the variance of that income that takes the form illustrated in Figure 11-3. Curve *AB* represents an iso-utility function, which relates expected income versus the variance of income at a constant utility level. Curves such as *AB* are usually referred to as "gambler's indifference curves" (Raiffa 1970) and show that an increase in expected income is required to offset increased variance. Conversely, to reduce the variance, the investor is willing to reduce expected income. Most individuals are to some degree willing to accept a reduction in expected income to achieve reduced income variance—that is, most individuals are "risk averse." In our case, farmers investing in well capacity beyond the point of maximizing expected income can be seen as attempting to reduce the variance in their income by providing additional well capacity. Their behavior can be interpreted as moving down the gambler's indifference curve from *B* toward *A* in Figure 11-3.

This study investigates the relationship between the decision to invest in well capacity, the expected income, and the variance in that income. We still believe that the model developed in 1972 captures most of the essential elements of the problem. The simulation procedure is identical to that used in our 1972 study; however, the variance in expected income owing to the variation in water supply was examined, something which was not done in 1972. (As before, the model assumes water to be the only stochastic variable; prices and technology are assumed to be known with certainty.) In particular, we have simulated the operation of the system through a nine-year period with various levels of installed well capacity. At zero well capacity the system is totally dependent on streamflow for irrigation, and the variation in yearly net income is quite large. As more and more wells are installed, the variation in net income is reduced; at the same time, the expected net income is increased. Finally, sufficient well capacity is installed to make the system capable of total irrigation by groundwater; at that time the variation in annual net income becomes small, nearly zero. To make the results relevant to more recent conditions, we have revised the economic inputs (prices and costs) to 1978 values.

Simulation Model

In our earlier study (Young and Bredehoeft 1972) we argued that our simulation approach represents a conjunctive irrigation system. The key elements of the physical system are illustrated in Figure 11-4. A stream provides water for a set of irrigation districts; the irrigation districts

Figure 11-5. *Flow Chart of the Hydrologic Simulation*

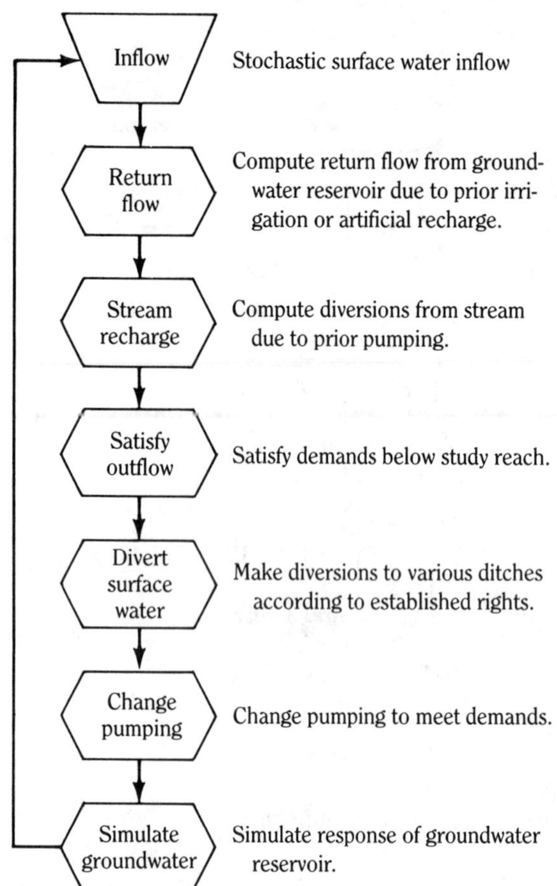
Inflow
Stochastic surface water inflow
Return flow
Compute return flow from ground-water reservoir due to prior irrigation or artificial recharge.
Stream recharge
Compute diversions from stream due to prior pumping.
Satisfy outflow
Satisfy demands below study reach.
Divert surface water
Make diversions to various ditches according to established rights.
Change pumping
Change pumping to meet demands.
Simulate groundwater
Simulate response of groundwater reservoir.

Source: Young and Bredehoeft (1972).

have varying water rights—some senior, some junior, and some intermediate. Groundwater provides an alternative source of water. In a system such as the South Platte in Colorado, much if not most of the groundwater comes from natural or induced recharge to the groundwater system from the stream and from recharge from irrigation itself. The demand for water in the system, as it is elsewhere in much of the western United States, is for crop irrigation.

The farmers operate on an annual cycle. At the beginning of the growing season an estimate of the streamflow is made for the entire growing period. Given that estimate and a judgment as to what an individual's available irrigation water will be during the growing season, the farmer makes a decision to plant certain crops. His available irrigation supply includes his share of irrigation water from the river as well as his installed capacity to pump groundwater. He makes his planting decision in an effort to maximize his net income for the year.

Once a planting decision is made, the farmer must irrigate with the water available at any time during the growing season in a way that continues to maximize his net profits for the year. If his only supply is surface water, and the surface water is less than he planned for, the farmer must decide which crops to irrigate with how much water in order to continue to maximize his income for the season. As the capacity to pump groundwater is increased, a shortage of surface water can be compensated for by turning on the pumps.

Each action related to water—either to divert streamflow or to pump groundwater—affects the available water in the system, especially the flow of the stream. Pumping groundwater has a delayed effect on the river; that time delay varies from place to place in the aquifer.

In our earlier effort we presented a simulation model which coupled both the hydrologic response of the surface and groundwater systems as well as the behavior of the water users in the system. The essence of the hydrologic simulation is presented in Figure 11-5. The basic hydrologic component of the model is a groundwater simulation subelement to which is coupled a stream reach model. To the hydrologic component is coupled a behavioral component represented as a sequence of linear programs. These linear programs make allocation decisions designed to maximize net income.

The general sequence of events in the simulation, listed in Figure 11-5, illustrates the method. At the beginning of the season a linear program takes a planting decision which should maximize net income within a given set of economic conditions (crop prices and costs). The decision uses estimates of the streamflow throughout the growing season. Once the planting decision is made, a second linear program allocates the available water, ground and surface, in the first month of the growing season in a way that maximizes the profits at the end of the growing season. The hydrologic component then takes over and simulates one month of operation of the conjunctive surface and groundwater system. At the beginning of the second month a different quantity of surface water is available, and a second operating decision is made for the second month of the growing season by the linear program which simulates operation. The sequence of simulation proceeds until the entire season is complete.

At the end of the season a net profit is computed. Both fixed and operating costs are included in the net benefit calculations. These costs include both the cost of operating wells as well as the initial cost of investing in wells amortized over the expected life of the well. Crop yields are adjusted during the growing season to reflect the water actually applied in each growing month.

In summary, the procedure we outlined in 1972 consists of what we believe to be the characteristics of a conjunctively operated irrigation system: a linear program which makes a crop planting decision based upon estimates of the water available through the growing season is coupled with a hydrologic model which simulates the response of groundwater and surface water systems, and a sequence of linear programs simulates the operation of the irrigation system through the growing season.

In this chapter, the prototype aquifer is the same as that studied in 1972, the reach along the South Platte from approximately Kersey to Balzac. The same ten-year period of streamflow, 1951–60, which includes the critical period of record for the South Platte, was used for the simulation.

Water Rights and Delivery Systems

The reach was divided into three subareas. Subarea 1 contained 5,500 acres of irrigated land; subareas 2 and 3 contained 30,000 acres each. Each subarea was assumed to be supplied by a single ditch company. Diversions were assumed to occur at the upstream limit of the subarea. These subdivisions are hypothetical and do not reflect real ditch companies along this reach of the South Platte.

A set of priorities and water rights was arbitrarily set up for each of the three subareas in the reach. The rights and priorities are listed in the appendix. This set of rights provides area 1 with senior surface water rights, which meet its demands most of the time; area 2 with junior rights, in which pumping is clearly desirable; and area 3 with intermediate rights.

The downstream demand for streamflow was set at 50 percent of the inflow. This figure represents the downstream users' water rights. In reality, the flow at the Balzac gage, which is near the lower end of the study reach, averages 46 percent of the flow at the upstream Kersey gage.

The study consisted of varying the pumping capacity over the range of interest. Wells are assumed to be randomly distributed in areas 2 and 3. This distribution was intended to simulate the more or less unregulated development of irrigation wells that has taken place in the South Platte Valley during the past thirty years.

Demand Model

All water demands are assumed to be for crop irrigation. The four crops included (corn, sugar beets, beans, and alfalfa) are reasonably representative of conditions in the study area. The data assumed as a basis for the income estimates, including prices, crop yields, and yield adjustment coefficients, are presented in the appendix. The data that form the basis for cost parameters are also presented in the appendix.

Procedures and Results

How did the net benefits and the variance of those benefits change with additional well capacity? Each simulation run of the model produces a net income for each

Figure 11-6. ***Annual Net Benefit versus Logarithm of the Flow during the Critical Month of the Growing Season***

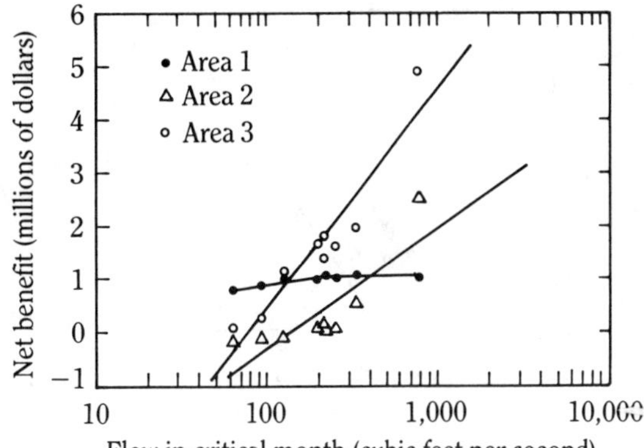

Note: No wells in system.

subarea for each year simulated. These data for a multi-year simulation can be assembled to yield an expected income and a variance in income. As wells are introduced into the system, both the expected income and the variance change. It is this relationship that we examine.

We first examine the system under conditions in which the forecast of monthly flows is quite accurate, so that the forecasted flows are largely realized. This is, in effect, a condition of certainty. Under these conditions, good planting decisions are made. Later, we will examine the effect of less accurate forecasts and therefore less profitable planting decisions.

The simulation was conducted using monthly flows at the upstream gage as the input. During any single growing season, one month is usually more critical than any others; we have defined the lowest monthly flow during each growing season as the critical flow for that season.

We have plotted the net benefits predicted by the model for each year versus the logarithm of the flow during the critical month. Figure 11-6 is such a plot of the predicted annual net benefit, derived with zero well capacity, versus the logarithm of flow during the critical month.

We have examined a number of functions to fit our results. The data fit reasonably well with the relationship

$$\phi_g = a + b \log (q_{cm}) \quad (11\text{-}1)$$

where

ϕ_g = annual net benefit, for a specific well capacity (the net benefit is expressed in 1978 dollars), g

g = installed pumping capacity (cubic feet per second)

q_{cm} = average monthly streamflow in the critical month of the growing season (cubic feet per second)

a = zero intercept of the assumed relationship

b = slope of the semilogarithmic relationship: the slope is a measure of the variability of the annual net income.

Figure 11-7. *Acreage Planted during a Season versus the Critical Flow*

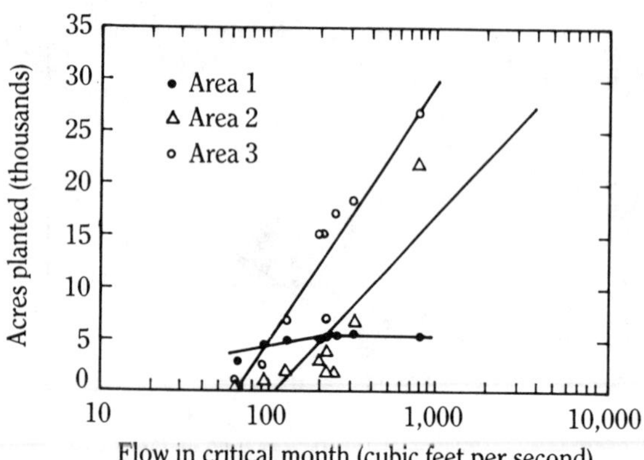

Note: No wells in system.

Figure 11-8. *Net Income for Area 3 versus Inflow*

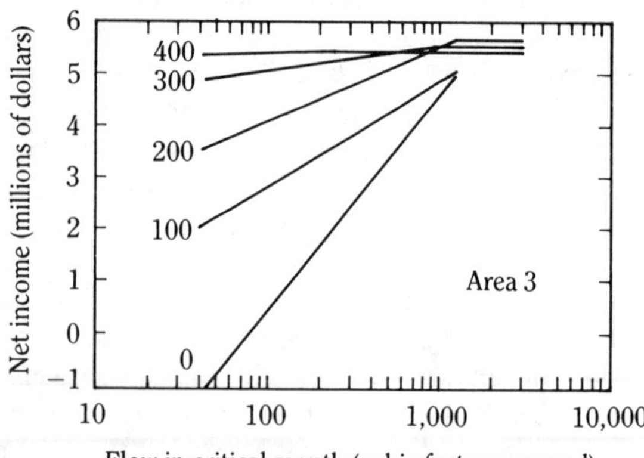

Note: Each line represents a different installed well capacity.

This relationship applies up to the point that all the available land is irrigated. Beyond that point there is no increase in net benefit with an increase in flow, which can be seen clearly by examining the results for area 1 plotted in Figure 11-6. Once all available acreage is irrigated to the maximum water requirement, an increase in water does not increase the net benefit; that is, the available acreage constrains the net benefit. The expected value and the variance of the annual net benefit are given by

$$E[\phi_g] = a + bE[\log(q_{cm})] \quad (11\text{-}2)$$

$$V[\phi_g] = b^2 V[\log(q_{cm})] \quad (11\text{-}3)$$

respectively. Because their applicability is limited, equations 11-1, 11-2, and 11-3 are conditional statements. Equation 11-3 ignores the scatter about the line of best fit.

The benefits are more or less linearly related to how much acreage is planted. The acreage planted should show a relationship to critical flow similar to that of net benefits. Figure 11-7 plots acreage planted versus critical flow.

At flows larger than approximately 200 cubic feet per second, all the available acreage in area 1 is planted, because of the senior surface water rights available to area 1 and because area 1 has only 5,500 acres available. Areas 2 and 3 have 30,000 acres each available for planting. Without wells the total area is not planted until the flow reaches high values. The differences between areas 2 and 3 reflect the differences in surface water rights. Area 2 has poor surface water rights; in dry years it gets inadequate surface water, as can be seen in Figures 11-6 and 11-7. Area 3 has intermediate surface water rights.

Even with substantial pumping capacity in the system, the available acreage is not planted either in areas 2 or 3 until the flow in the critical month exceeds approximately 800 cubic feet per second. As we indicate below, the critical monthly flow has never exceeded 800 cubic feet per second at the Kersey gage—the inflow station—in the seventy-seven years of record. In our case, equations 11-1, 11-2, and 11-3 apply to areas 2 and 3 most of the time.

The Effect of Wells

The simulations were repeated with differing well capacities. Figure 11-8 shows the effects of increasing capacity upon net income. Both the annualized capital costs of investing in wells and the operating costs are included in the net income calculations. Comparison of the lines com-

Figure 11-9. *Acreage Planted in Area 3 versus Inflow*

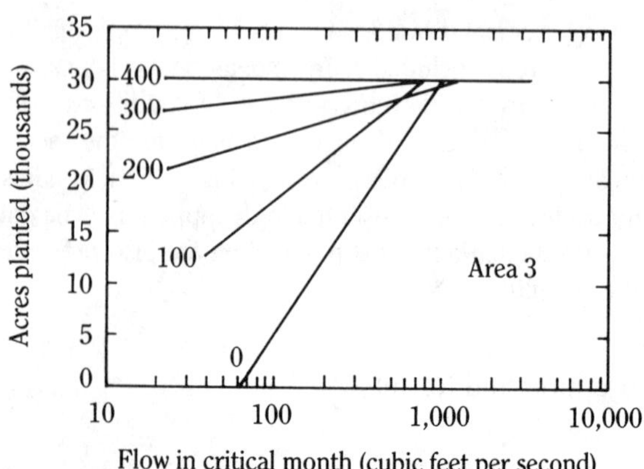

Note: Each line represents a different installed well capacity.

Figure 11-10. *Net Income versus Critical Flow for Differing Well Capacities*

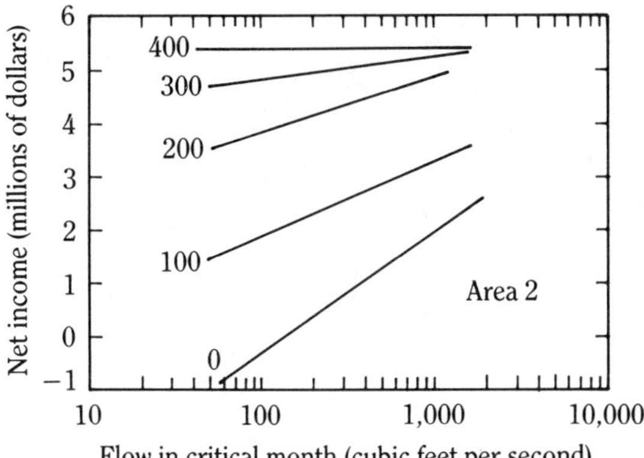

Figure 11-11. *Acreage Planted versus Critical Flow for Differing Well Capacities*

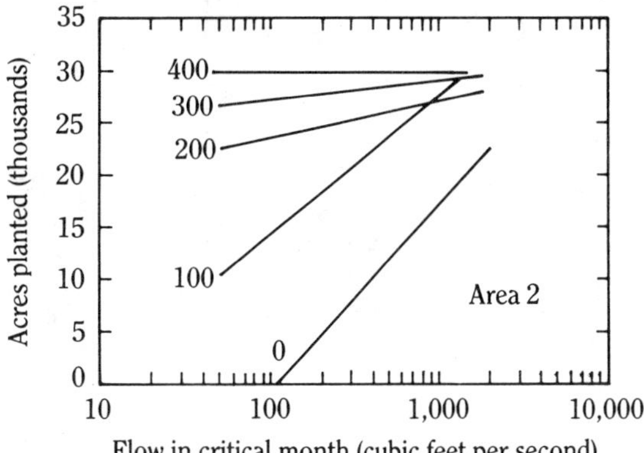

puted for various pumping capacities shows that increasing the capacity to pump groundwater reduces the variance in net income, as expected.

We have also examined the acreage planted as the well capacity is increased. These results are plotted in Figure 11-9. As more well capacity is installed, more of the acreage is planted. At a capacity of approximately 400 cubic feet per second, all the acreage is planted all of the time. A capacity of 400 cubic feet per second for a 100-day growing season can provide an application of approximately 2.6 feet on 30,000 acres. This is sufficient to irrigate fully any of the commonly grown crops in the South Platte Valley.

The results for area 2 are similar to those for area 3. The area 2 results are shown in Figures 11-10 and 11-11. Comparing Figures 11-8, 11-9, 11-10, and 11-11, we see that installing well capacity not only reduces the variance in the net income but also increases the expected income. The median critical flow for the nine-year period simulated is 215 cubic feet per second. The net income associated with this nine-year median critical flow increases as the well capacity is increased until the capacity is such that all available acreage is planted all the time. The variance in income at this capacity is small: our results indicate that it is essentially zero. Groundwater provides a dependable supply that allows full development of the irrigable land.

To examine the long-term effect of installing wells in the system, we have made several assumptions:

- The relationship between critical monthly flow and net benefits, as well as the relationship between critical monthly flow and acreage planted developed from the nine-year simulation, applies to the system over all ranges of flow. (These relationships are presented in Figures 11-8, 11-9, 11-10, and 11-11.)
- The net benefit associated with the median critical monthly flow for the entire period of record at Kersey is a good estimate of expected net benefit.
- The acreage planted associated with the median critical monthly flow for the entire period of record is also a good estimate of expected acreage planted.

The historical streamflow data at the upstream gage at Kersey for the period 1901 (when the station was established) through 1979 were examined. (There are no data for the years 1904 or 1914.) For this period the mean critical monthly flow, by our definition of critical flow, is 251 cubic feet per second; the median critical monthly flow is 190. The flow during the critical month has never exceeded 800 cubic feet per second and has exceeded 700 cubic feet per second in four years only. The critical monthly flow has been below 100 cubic feet per second in thirteen years, below 70 in four years, below 60 once, and never below 50.

We have summarized our results by examining the relationship of well capacity to long-term expected net income, to long-term expected acreage to be planted, and to a measure of the variance of the annual net income b^2. The variance of the annual net income is a function of b^2, where b is the slope of the relationship $\phi_g = a + b \log (q_{cm})$. Plots of these relationships are shown in Figure 11-12. With these results, we can also examine the expected per acre benefit for differing well capacity. The benefit is $E[\phi_A] = E[0_g/A_p]$, where ϕ_A is the annual net benefit per acre and A_p is the annual acreage planted. The expected per acre annual income is plotted as a function of well capacity in Figure 11-13. In making this summary we have used the assumptions stated about the relationship between critical monthly flow and net benefits as well as acreage planted.

At well capacities of 200 cubic feet per second and beyond, the change in per acre net benefits is small. For area 3, the expected per acre benefit decreases at capacities larger than 300 cubic feet per second. However, these changes in expected per acre benefits are offset by the larger acreages which can be planted with increased well

Figure 11-12. ***Well Capacity versus Expected Net Income, Expected Acreage to be Planted, and Income Variance***

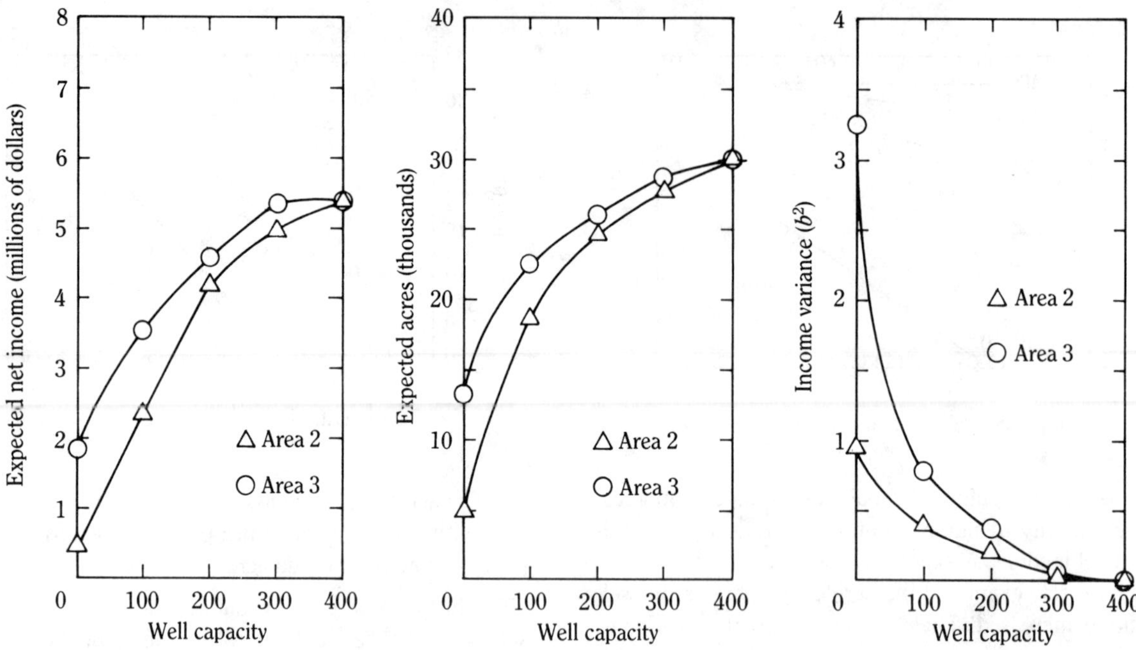

Note: Well capacity in cubic feet per second.

Figure 11-13. ***Expected Income Per Acre versus Well Capacity***

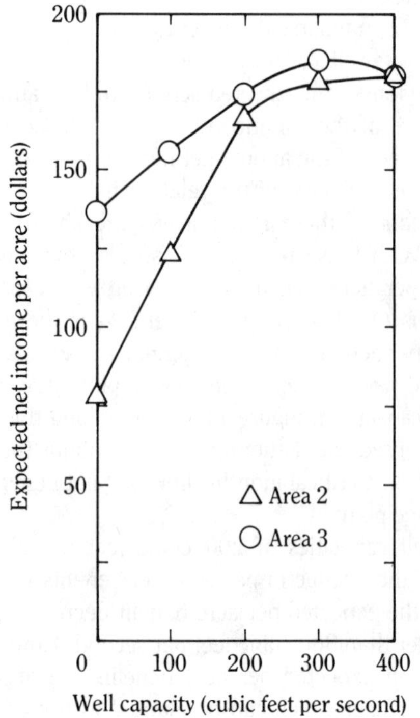

capacity. Total expected net income for the areas increased, even though the changes in per acre income were small (or even negative).

Comparison with 1972 Study

Our 1972 results indicate that although an optimum capacity existed, the net benefits versus well capacity was more like a broad plateau within a wide spectrum of pumping capacities producing approximately the same net benefits. These results are shown in Figure 11-2. The absence of a sharply peaked optimum is typical of many water resources systems, which of course, indicates that one can operate over a wide range of values without diminishing the net benefit significantly.

There was sufficient well capacity in the South Platte system in 1970 to irrigate without surface water. Translating that in terms of Figure 11-2 indicates that the installed well capacity of between 300 and 400 cubic feet per second is beyond that needed to maximize expected net benefits. However, it is not surprising for risk-averse farmers to have installed this capacity. The reductions in expected benefits are not that large and apparently were in a sense viewed as the insurance costs of a dependable water supply. We did not make this point in 1972.

The economics of farming changed during the ten years of our studies in the South Platte Valley. The net benefits increased approximately an order of magnitude above what they were in the 1972 results (compare Figure 11-12 with Figure 11-2). Figure 11-12 indicates that it now makes more sense to invest in sufficient well capacity to irrigate all available acreage—in other words, to act as if no surface water were available. The expected income increases continuously to this point; at the same time, the variance in income correspondingly decreases as pumping capacity increases. The variance can be driven to essentially zero by increasing the well capacity to the point at which all the acreage can be irrigated by wells.

Work along the main stem of the South Platte by the U.S. Geological Survey suggests that the installed well capacity is probably sufficient to provide approximately 0.5 to 0.6 feet of water per month on the irrigated land. Allowing for areas where the water rights are sufficiently senior that wells are unnecessary even during critical months (as illustrated by area 1), this installed capacity is adequate to irrigate all available land. Although we have examined the issue of risk aversion in this investigation, the results suggest that either maximizing one's expected income or minimizing its variance leads to the same well capacity in this instance. This result is somewhat different from our 1972 results. We now believe that risk aversion is an important element governing farmers' investment behavior and should be incorporated into the analysis.

Downstream Effects of Groundwater Withdrawals

Pumping groundwater in a system in which the aquifer is intimately connected with the stream must ultimately reduce streamflow. The institutions which govern groundwater development in Colorado have undergone several iterations: in at least two instances the courts have rendered them ineffective (Daubert 1978). One of these instances, the *Fellhauer* case, was mentioned in the introduction. In the late 1970s Colorado entered into a scheme in which well owners are required to augment streamflow during the growing season. Groundwater users must have the capability to replace streamflow depletions caused by pumping groundwater. The present policy requires that the well owners shall provide an amount of water equal to 5 percent of the amount of groundwater to be pumped. The well owners have the responsibility for designing their own plan of augmentation, which must be approved by either the water clerk of the district or the state engineer before pumping is allowed.

One of the largest augmentation plans is that operated along the South Platte by the Ground Water Users Association of the South Platte (GASP). At present, this group assesses each of its members 25 cents for each acre-foot pumped during the growing season. These funds are used (1) to rent capacity in reservoirs tributary to the South Platte for the storage of surface water during the nongrowing season, which can be used during the growing season to augment surface supplies, and (2) to operate a series of wells which can pump directly into the surface water system to augment surface flow.

This system of augmentation allows the well owners to make their own decisions about the use of wells. It further protects senior surface water users by augmenting the streamflow during critical periods. Is the 5 percent augmentation adequate to meet the surface water requirements during a critical period? We have operated our simulation model to analyze this question.

Our prototype is a reach along the South Platte Valley from the Kersey stream gage to the Balzac stream gage. The flow at the Balzac gage averages 46 percent of the flow at the upstream Kersey gage. We set the downstream demand for water at 50 percent of the inflow into the reach. In the simulation this demand is met before water is diverted within the study reach. In simulations in which no groundwater is pumped, this demand is always met. Pumping groundwater depletes the stream, however, and can leave downstream users without sufficient water in the stream, particularly in periods of low flow.

In Figure 11-14 we have plotted the flow into the reach at Kersey versus the simulated flow out of the reach at Balzac for the forty growing-season months simulated during the ten years studied. Historical flow at Kersey for 1951–60 was used as input; this includes the critical period 1953–56 for the South Platte. Figure 11-14 is plotted for a total pumping capacity of 400 cubic feet per second within the reach, 200 cubic feet per second in area 2, and 200 in area 3. At inflows less than approximately 400 cubic feet per second, there begin to be downstream shortfalls. At inflows less than approximately 200 cubic feet per second these range up to 20 or 30 cubic feet per second, which suggests that 5 percent will provide sufficient augmentation most of the time.

Figure 11-14. ***Inflow versus Simulated Outflow for Forty Months of Growing Season***
(cubic feet per second)

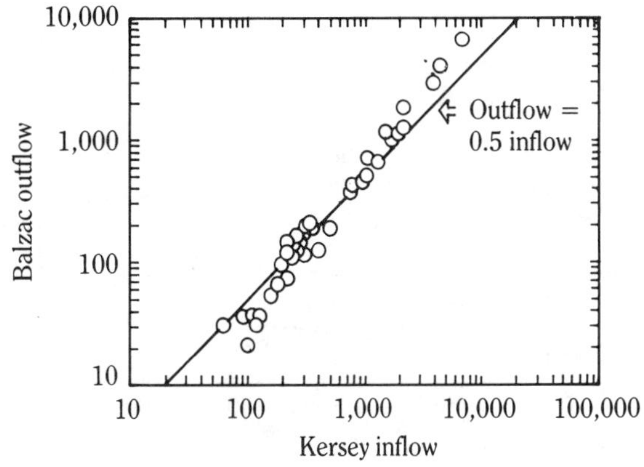

Note: Installed well capacity is 400 cubic feet per second.

Figure 11-15. *Inflow versus Simulated Outflow for Forty Months of Growing Season*
(cubic feet per second)

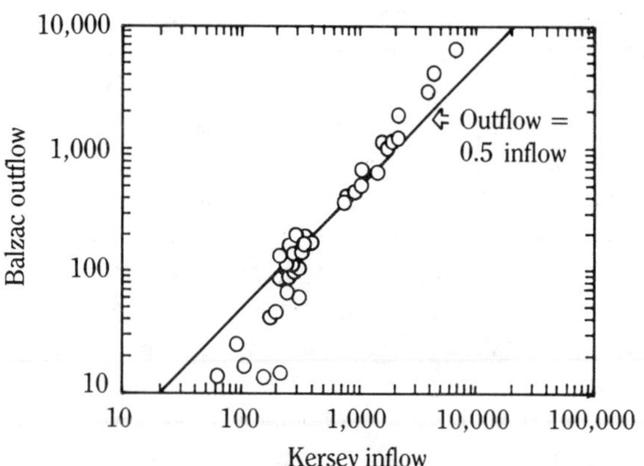

Note: Installed well capacity is 800 cubic feet per second.

Figure 11-15 displays the impact of a total installed well capacity of 800 cubic feet per second in the study reach. At inflows less than approximately 300 cubic feet per second, the downstream shortfalls are significant, and at inflows of approximately 100 cubic feet per second, the stream is dried up during three months. At inflows less than approximately 200 cubic feet per second, the shortfalls range up to 95. This would suggest that as much as 10 percent augmentation would be required in dry years with the larger pumping capacity.

Figure 11-16. *Flow Estimates versus Expected Net Income*

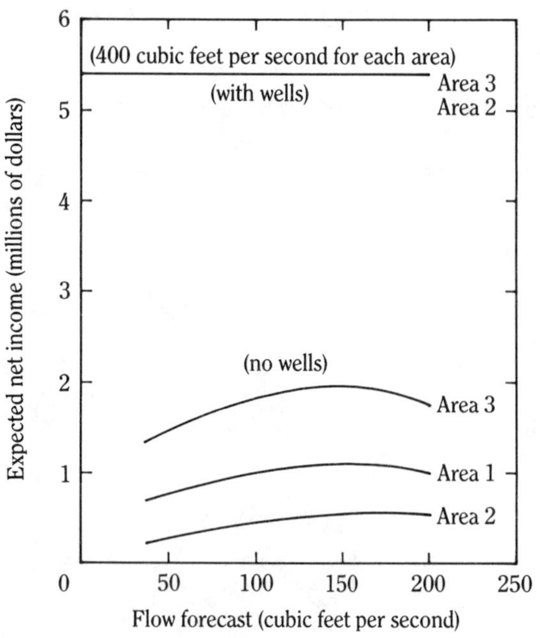

Note: Flow estimates at 100 percent represent certainty.

Table 11-1. *Expected per Acre Benefit with Varying Accuracies of Flow Forecasts*

	Inflow forecasts as a percentage of actual flow		
Area	*50 percent*	*100 percent*	*200 percent*
1	150	192	192
2	57	79	106
3	110	138	135

Note: No wells in system.

Value of Improved Flow Forecast

Up to this point we have treated the system under conditions of certainty—that is, flow forecasts are realized exactly. In reality, the forecasts are in error. An inaccurate forecast of river flow for the season leads to a less than ideal planting decision. Of interest is the cost of a poor forecast or, conversely, the value of an improved forecast in a conjunctive use system. The forecast is one of surface flow; the availability of groundwater is determined by the pumping capacity of the system.

In an effort to investigate this question, we simulated the system with accurate estimates of flow (a condition of certainty), estimates which were 50 percent of the flow, and

Table 11-2. *Water Rights and Their Priorities*

	Diversion right (cubic feet per second)		
Priority	*Area 1*	*Area 2*	*Area 3*
1	25		
2			20
3	15		
4			20
5	25		
6			20
7		20	
8	10		
9			50
10		20	
11			50
12		50	
13			100
14		50	
15			150
16		50	
17			200
18		200	
19			200
20		200	

Table 11-3. *Monthly Mean Flow for South Platte River at Kersey, Colorado*
(cubic feet per second)

Water year	October	November	December	January	February	March	April	May	June	July	August	September	Yearly average
	Monthly mean discharge												
1951	171	420	410	442	545	324	305	302	1,039	283	1,304	306	487
1952	617	640	613	614	551	542	878	2,189	1,936	260	246	175	772
1953	385	442	529	527	432	399	406	217	386	264	249	143	365
1954	248	504	483	466	363	308	111	63.2	109	159	120	76.9	250
1955	129	230	293	332	327	309	187	91.9	182	100	307	154	220
1956	267	367	344	332	287	268	179	339	289	127	216	91.3	259
1957	149	362	355	369	418	368	567	4,443	3,942	1,758	775	467	1,169
1958	710	975	830	689	626	542	1,034	6,824	2,133	332	326	190	1,275
1959	459	477	477	500	556	601	936	1,046	750	214	306	360	556
1960	677	634	595	529	623	1,101	815	1,553	917	265	193	181	674

Source: Data from U.S. Geological Survey (1964).

Table 11-4. *Yield Adjustment Coefficient*

Number of irrigations	Period 1	2	3	4
	Corn			
0	0.8	0.8	0.8	0.7
1	1.0	1.0	0.9	0.8
2	1.0	1.0	1.0	1.0
3	1.0	1.2	1.2	1.2
	Alfalfa			
0	0.9	0.9	0.9	0.9
1	1.0	1.0	1.0	1.0
2	1.0	1.0	1.0	1.0
3	1.0	1.2	1.2	1.2
	Beans			
0	1.0	0.8	0.7	0.8
1	1.0	0.9	0.8	0.9
2	1.0	1.0	1.0	1.0
3	1.0	1.0	1.0	1.0
	Beets			
0	0.7	0.9	0.7	0.7
1	0.9	1.0	0.9	0.8
2	1.0	1.0	1.0	0.9
3	1.0	1.0	1.0	1.0

Table 11-5. *Crop Yields and Prices*

Crop	Yield per acre (before adjustments)	Unit price (dollars)
Corn	108 bushels	2.75
Alfalfa	3.2 tons	61.20
Beans	1,740 pounds	22.60
Beets	17.8 tons	37.80

Table 11-6. *Periodic Variable Costs*
(costs in dollars per acre planted)

Crop	Preseason	Period 1	2	3	4	Total
Corn	47	28	19	7	39	140
Alfalfa	43	15	15	15	14	104
Beans	18	12	36	8	45	119
Beets	58	41	60	11	91	261

Table 11-7. *Annual Operating Costs and Fixed Costs for Water*

Costs	Unit price
Operating costs (dollars per acre-inch)	
Surface water	0.25
Groundwater	0.46
Water application	0.30
Fixed costs (dollars per acre of cropland)	35.0

estimates which were 200 percent of the flow into the reach. The results are shown in Figure 11-16.

Without wells, underestimating the flow decreases the expected income. It is also useful to examine this decrease in terms of per acre income. This result is shown in Table 11-1.

The cost of overestimating the flow is much less than underestimating. This is because an underestimate at the high flow results in less acreage planted than would be optimum. But overestimating the flow results in too much acreage being planted; however, the costs associated with planting too much acreage are not as high as those associated with planting too little acreage.

As pumping capacity is installed in the system, the value of an improved flow forecast diminishes. As the capacity approaches that which is capable of irrigating all the available acreage, the value of an improved flow forecast is negligible. At the higher pumping capacities, poor flow forecasts are compensated for by pumping groundwater.

Summary and Conclusion

To be useful, simulation must almost always be based upon some prototype. The results are closely tied to a particular real system. This problem is inherent in using simulation as a tool for optimization.

Table 11-8. ***Tableau for the Planning Stage Linear Program***

	Water supplies								Corn irrigation								
Constraints	*SW1*	*SW2*	*SW3*	*SW4*	*GW1*	*GW2*	*GW3*	*GW4*	*C1*	*C2*	*C3*	*C4*	*C5*	*C6*	*C7*	*C8*	*C9*
Water 1	−1				−1				0	0	0	6	6	6	6	6	6
Water 2		−1				−1			12	12	0	6	0	6	0	0	6
Water 3			−1				−1		6	0	0	6	12	6	0	12	0
Water 4				−1				−1	12	18	18	12	12	6	18	6	12
SW1	1																
SW2		1															
SW3			1														
SW4				1													
GW1					1												
GW2						1											
GW3							1										
GW4								1									
Land									1	1	1	1	1	1	1	1	1
Corn land									1	1	1	1	1	1	1	1	1
Alfalfa land																	
Bean land																	
Beet land																	
Net revenue (dollars)	−0.55	−0.55	−0.55	−0.55	−0.76	−0.76	−0.76	−0.76	147	136	88	159	145	147	122	133	145

We believe that the results of our analysis of the South Platte system provide an understanding of the economics of other similar systems; however, one can only infer this from our results. Nevertheless, the methodology presented has wide application.

As indicated in earlier studies, using the groundwater aquifer as a reservoir in a conjunctive water system such as that typified by the South Platte in Colorado greatly increases the economic benefits to be derived from the system. As pumping capacity is installed in the system, the expected net benefit is increased greatly depending upon availability of surface water—that is, surface water rights. For an area typified by our area 2, the expected net income is increased thirteen times from the situation in which it depends only on surface water to that in which it can pump sufficient groundwater to meet its need (see Figure 11-11).

Our results suggest that under the current economic condition existing in the South Platte Valley in Colorado, the most reasonable groundwater capacity is a total capacity capable of irrigating all the available acreage with groundwater. Our results suggest that the additional costs of pumping groundwater are not significant. The best strategy is to discount surface water entirely when considering what pumping capacity to install—if, of course, one can augment streamflow to the downstream users in periods of low flow.

Installing sufficient pumping capacity to irrigate all available acreage has two benefits: it maximizes the expected net benefit, and it reduces the variance in expected income to nearly zero. Given these results, it is understandable that the actual installed capacity in the South Platte system is capable of supplying approximately 2.5 feet of water to the available acreage.

Our results suggest that the present Colorado plan of 5 percent flow augmentation may be small in really dry years. We expect that an augmentation capacity approaching 10 percent may be necessary in such dry years as occurred from 1953 to 1956. An augmentation of 5 percent is, however, adequate much of the time.

As pumping capacity is installed in a conjunctive use system, the value of flow forecasts diminishes. Poor forecasts are compensated for by pumping from the wells. It is questionable in systems such as the South Platte with a large installed well capacity for most of the area that one should spend much effort on improving flow forecasts for water supply purposes.

Appendix

The pertinent hydrologic and economic data are presented in Tables 11-2 to 11-8.

Note

This chapter originally appeared in a somewhat different form in *Water Resources Research,* vol. 19, no. 5 (October 1983), pp. 1,111–21; copyright by the American Geophysical Union.

References

Bredehoeft, J. D., and R. A. Young, 1970. "The Temporal Allocation of Groundwater: A Simulation Approach." *Water Resources Research* 6(1): 3–21.

(acre-inches)				*Alfalfa (acre-inches)*								*Beets (acre-inches)*						
C10	*C11*	*C12*	*C13*	*A1*	*A2*	*A3*	*A4*	*A5*	*A6*	*A7*	*Beans*	*B1*	*B2*	*B3*	*B4*	*B5*	*Fallows*	*Constraints*
6	6	6	6	0	0	4	4	4	4	4	0	4	4	4	4	4	0	Water 1
6	0	6	0	12	4	12	4	12	12	4	8	4	4	0	4	0	0	Water 2
6	6	0	0	0	4	4	4	0	4	0	8	8	0	8	4	0	0	Water 3
0	6	6	6	4	4	4	4	4	0	4	8	12	8	12	4	4	0	Water 4
																		SW1
																		SW2
																		SW3
																		SW4
																		GW1
																		GW2
																		GW3
																		GW4
1	1	1	1	1	1	1	1	1	1	1	1	1	1	1	1	1	1	Land
1	1	1	1															Corn land
				1	1	1	1	1	1	1								Alfalfa land
											1							Bean land
												1	1	1	1	1		Beet land
126	105	119	77	89	62	92	89	92	92	79	274	411	381	405	381	334	−6	Net revenue (dollars)

Daubert, J.T. 1978. "Conjunctive Ground and Surface Water Allocations: The Economics of a Quasi-Market Solution." Ph.D. dissertation, Colorado State University, Fort Collins.

Raiffa, Howard. 1970. *Decision Analysis: Introductory Lectures on Choices under Uncertainty.* Reading, Mass: Addison-Wesley.

State of Colorado v. *Roger Fellhauer,* Fellhauer v. the people. State of Colorado, no. 167 Colo. 320, 447 P. 2d 986, 1968.

U.S. Geological Survey. 1964. "Compilation of Records of Surface Waters of the United States, October 1950 to 1960, 6-A Missouri River Basin above Sioux City, Iowa." *U.S. Geological Survey Water Supply Paper* 1730.

Young, R. A., and J. D. Bredehoeft. 1972. Digital Computer Simulation for Solving Management Problems of Conjunctive Groundwater and Surface Water Systems. *Water Resources Research* 8: 533–56.

Comment

Gerald T. O'Mara

Bredehoeft and Young focus on the incentives that motivate individual irrigators to augment their surface water supplies by investing in tubewells. Although their immediate objective was to explain an apparent overinvestment in tubewell capacity, their ultimate objective was to understand farmer behavior in order to develop institutional options for more efficient conjunctive use. Bredehoeft and Young display technical ingenuity in simulating the physical interdependence between well-pumping farmers and return flows from the aquifer to the stream (which in turn were surface supplies to downstream users). But the results of their experiments apply only to the system modeled—a reach of the South Platte River in Colorado.

The authors found that farmers benefit from investing in tubewells to the point at which groundwater is capable of providing their total water supply; they thereby avoid relying on a highly variable surface water supply. The authors believe that their findings transfer to similar physical contexts in developing countries. Developing countries are different, however, since they have lower resource productivity and a higher relative price of pumping equipment. It is true, however, that groundwater is widely used to stabilize irrigation water supply in developing countries. The motivation in this situation is clearly quite different, since well-owners will substitute the cheaper (that is, subsidized) surface water for groundwater whenever a timely surface water supply is available.

The main institutional option discussed was that well users be legally required to supply replacement water to downstream users. This option is feasible only if replacement water supplies are available—a happy circumstance that does not always exist. It is doubtful what the Colorado judiciary would do when confronted with a situation in which alternative supply was not available.

Groundwater as a Constraint to Irrigation

Robert G. Thomas

This chapter analyzes two arid countries—Qatar and Libya—to illustrate the complex economic and social effects of using groundwater for irrigation. These effects intensify as water becomes either more scarce or more costly or both. Like other agricultural inputs, water becomes a constraint when its use exceeds available amounts. With groundwater, however, the nature of the use often affects the level of availability. And like soil quality, the water quality in many areas may be affected by both natural and man-induced causes.

Another dimension of water resources and irrigation planning is time. Groundwater in a very small basin may be completely replenished by each rainstorm; at the other extreme, thousands of years may be required in large basins to reach equilibrium of water quality after big changes of input or extraction.

In large basins, the constraint to irrigation may be economical extraction, not recharge. The mining of groundwater—whereby the recharge rate is exceeded—is another complex issue. Groundwater may be managed, but the institutional and legal instruments required tend to be either costly or ineffective or both. As a result, all groundwater basins are used, but few indeed are consciously managed for the optimal public and private good. Those that are managed have sources of imported water. When imported water is not or cannot be made economically available, the use of groundwater will simply slow down as it is depleted or as its quality is degraded.

Groundwater experts commonly resist making quantitative estimates (geologists) or presenting quantitative estimates to several significant figures (engineers). Of course, correct figures are essential, but the size of various groundwater parameters can at best be estimated only. Although this situation may be improved somewhat by various statistical devices, uncertainties will always remain—especially as studies become more detailed for different purposes. The technically most sophisticated problems involve not only water quantity but also water quality, wastewater properties, and aquifer thermal properties. Fortunately, for most irrigation in developing countries, groundwater analysis is usually not so complex.

Even in relatively simple areas, the social and economic effects of irrigation may not be easy to understand. To illustrate, I will present two brief case histories in which water use has clearly exceeded natural supply and in which money is not necessarily a constraint. Both of these studies have resulted from many years of data collection and analysis, and both provide a realistic basis for decisions in the management of groundwater resources.

Qatar

Since Qatar began producing oil in 1948, output has risen to about 500,000 barrels a day in 1980, which provides an annual income of about $5.5 million. Indigenous population has grown from 25,000 in 1948 to about 60,000 in 1980, but the total population in 1980 was about 265,000.

Qatar is located in the Persian Gulf on a peninsula adjacent to the Arabian peninsula. With a land area of about 11,610 square kilometers, its length is about 180 kilometers and its maximum width about 85 kilometers.

Roughly 9,000 square kilometers of Qatar's domed-shaped area has about 800 depressions with no external drainage—an important factor in water resource use. The highest point is only about 80 meters above sea level, so there is little orographic effect. Rainfall occurs in brief but intense convective showers between November and March. Average rainfall figures are extremely misleading, however; in some years there may be only one rainfall for a given area.

Table 12-1. *Qatar's Average Water Balance, 1962–63 to 1979–80*

Item	Millions of cubic meters per year
Natural recharge	40
Irrigation return	8
Total pumping	38
Outflow	27
Change in storage	−2
Change in freshwater reserves	−15

Source: FAO.

Fresh groundwater occurs in an aquifer of fractured Tertiary marine limestone. The lens of freshwater floats on saline water. Increased extractions have lowered the water table, but the displacement of freshwater by saltwater intrusion has been much more significant than the change in the amount stored. This situation has also delayed the deterioration of water quality.

Intense rains form small lakes and ponds which last only a few minutes to a few days in the internal drainage area, and some of the water goes into the rocks below. With an average rainfall of 69 millimeters a year, the average annual recharge for 1962–63 through 1979–80 is 27 millimeters, which means a total average recharge of about 40 million cubic meters of water a year from natural resources. With return irrigation and the reuse of treated sewage effluent, the present safe yield is estimated at about 50 million cubic meters a year. Qatar's average

Table 12-2. *Water Produced for Irrigation in Qatar, 1980*

Water source	Millions of cubic meters produced	Cost in Qatar riyals per cubic meter (QR = US$1.10)
Groundwater		
Fresh	76	0.20
Brackish	3	
Distilled seawater	44	5.50
Treated sewage effluent	2	0.85
Total	125	

Source: FAO.

water balance over the fourteen-year period is shown in Table 12-1. Table 12-2 shows the source of waters produced for irrigation in 1980 and their costs.

The estimates of future water use contained in this study are based on population projections and take into account various government policies. Desirable policies include the achievement of various levels of food self-sufficiency and the expansion of light industry. The water requirements were estimated for various levels of food production; optimal cropping patterns, labor, and other factors were taken into account.

Figure 12-1 is a flow chart for a systems analysis approach to Qatar's water situation. Table 12-3 presents the results of a series of linear program formulations for various assumptions. Run 1 shows that importing all food is one of the three least costly alternatives. This option is at

Table 12-3. *Linear Programming Options for Water Resource Use in Qatar*

Parameter	Run 1	Run 2	Run 3	Run 4	Run 5	Run 6	Run 7	Run 8
Local produce (tons)								
Milk	—	—	5,600	5,600	5,600	5,600	5,600	—
Beef	—	—	240	240	240	240	240	—
Mutton	—	150	660	750	2,400	4,500	300	—
Cereals	—	—	—	18,000	23,000	23,000	—	23,000
Dates	—	—	—	—	3,300	3,700	3,700	—
Winter vegetables	—	18,000	18,000	18,000	18,000	18,000	18,000	—
Summer vegetables	—	8,100	8,100	8,100	8,100	8,100	8,100	—
Fruit	—	—	—	—	1,700	3,700	3,700	—
Cucumbers	—	—	—	—	—	650	650	—
Water used (millions of cubic meters)								
Groundwater	—	19	19	21	29	29	23	24
Treated sewage effluent	—	—	35	35	35	35	11	20
Desalinized water	—	—	—	50	142	242	35	23
Total	—	19	54	106	206	306	69	67
Costs (millions of Qatar riyals)								
Farm costs	—	40	80	160	250	350	90	95
Water costs	—	5	35	350	940	1,580	230	165
Imports	230	140	115	95	50	—	90	215
Total	230	135	230	605	1,240	1,940	410	475
Area farmed (hectares)	—	1,900	2,900	9,100	13,800	17,600	4,800	7,700

—A zero entry.

Figure 12-1. *A Systems Analysis Approach to Water Resource Use in Qatar*

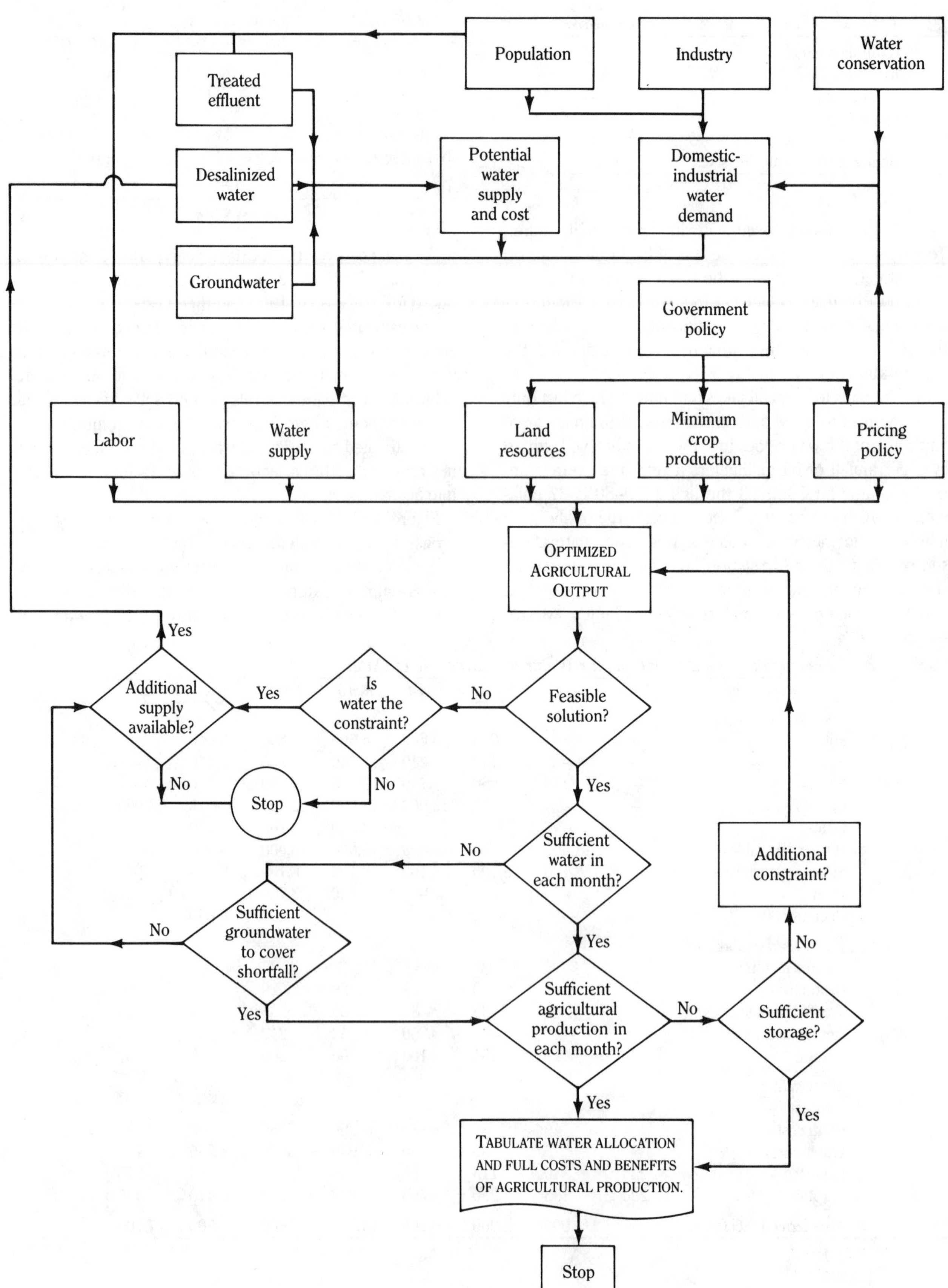

present completely unsatisfactory to Qatar's government. Run 2 limits water to the safe yield of groundwater. Run 6 is for self-sufficiency as complete as the soils will allow by the year 2000. The large increment of water cost is of course due to the large amount of desalinized water needed. These data are summarized in a different way in Figure 12-2.

Complete food self-sufficiency is clearly not possible. No studies have been made of the costs of food security, a problem different from self-sufficiency. Table 12-3 can be interpreted in different ways. For example, most of the increase in water needed between runs 5 and 6 is to produce 2,100 tons of mutton a year. The difference in water cost is roughly 600 million Qatar riyals a year, or about 700 for a ton of mutton.

Another run was made for a population projection increase of roughly half that used for most runs. At present there are about 3.8 foreign workers for each Qatari national. The capitalized cost of desalinization plants needed for nearly full self-sufficiency in food supply in 1980–2000 is roughly $1,000 a person a year. Government policies to reduce the number of foreign workers in the future could clearly save a large amount (my estimate is just under $4,000 million) with nearly full food self-sufficiency.

Importing water from Saudi Arabia is physically feasible, but current costs suggest that the water would be nearly as costly as desalinized water, although such analyses remain to be made. Increasing crop yields and reducing agricultural and domestic water losses is another possibility. (A fresh but contaminated lens of water has formed from sewage system leakage under Doha and is causing problems where the water table is at or near the surface.) Improving crop production with less water requires great effort from the farmers, but the farmers have little incentive because of the land tenure system. If the effort were to be initiated, a politically difficult decision would be required. Reducing losses in domestic use is of course partly an engineering problem but it could also be affected by policies on user charges.

Qatar has one of the simpler water problems of the world, yet the real choices are difficult. The United Nations Food and Agriculture Organization (FAO) has tried over the years to help responsible leaders understand their choices and to understand that not doing anything is also an action as far as water is concerned. In this regard, the FAO project staff has done as good a job as possible.

The final report from the FAO recommended the limited agricultural objective of improving irrigation practices and fully using treated sewage effluent. Preliminary designs of projects were also prepared, should the government wish to proceed with extensive use of desalinized water for irrigation to reach a higher degree of food self-sufficiency. This would involve 170 million cubic meters a year for pumping, transport, and use of groundwater storage during the non-growing season. Preliminary designs for

Figure 12-2. ***Qatar's Water Use Options in Order of Increasing Demand***

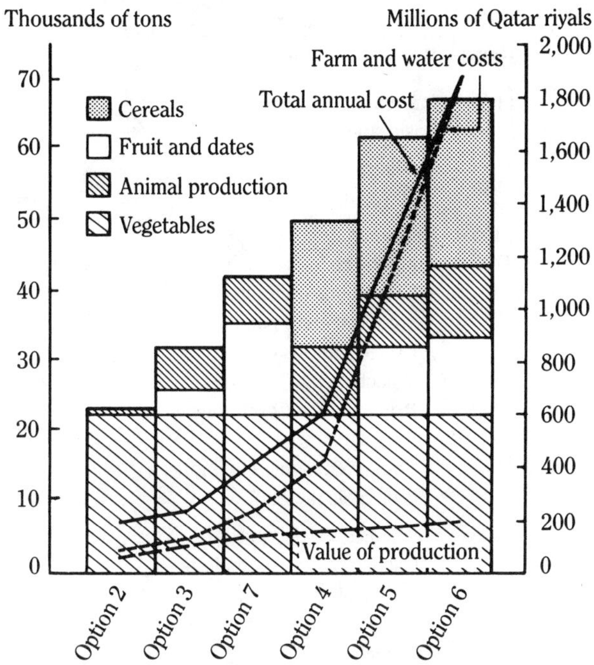

Source: FAO.

transport of reclaimed sewage effluent have also been prepared. The results should answer many of the questions and provide a better basis for future work and decisions.

Libya's Gefara Plain

The official name of Libya is the Socialist People's Libyan Arab Jamahiriya. Libya covers an area of 1,759,400 square kilometers and has a population of about 3 million people. Projected population for the year 2000 varies from 4.6 million to 6.5 million. Most of Libya is desert and receives an annual rainfall of less than 100 millimeters. However, rainfall in the Cyrenaica and Tripolitania regions reaches a maximum average of about 600 and 300 millimeters a year, respectively.

The Tripolitania area in the vicinity of the city of Tripoli has an extensive plain along the coast known as the Gefara (Jifara) Plain. The plain is backed up by an upland area called Jabal Nafūsah which reaches a height of about 600 meters. The Gefara Plain is the most important agricultural area of Libya because of its better soils and its concentration of population. Although the water problems of the plain must eventually take into account national agricultural policies, relatively little work has been done on a national scale to provide data for future decisions. Although many areas of Libya have been studied in detail, most studies are from the point of view of local use.

The Gefara Plain has also been studied in detail for

Table 12-4. *Budget for Water Management Plan, Gefara Plain*

Code	Activity	*Schedule* Present, 1980	First stage	Second stage	Third stage		Totals
1000	*Water Management*		5 years	10 years	20 years	30 years	
1100	*Planning*						
1110	Basic studies	—	—	Routine	—	—	—
1120	Projects	—	—	Continuous	—	—	—
1130	Control	—	—	Permanent	—	—	—
1200	*Execution*						
GW	Aquifers, collective wells	580	200	185	165	160	160
	Extractions (millions of cubic meters per year)	(580)[a]	200	(185)	(165)	(160)	(160)
	Projects (thousands of hectares)	—	36.0	2.5	—	1.5	40
	Irrigated area (thousands of hectares)	(96.6)	36.0	38.5	36.6	40.0	40
	Volume consumed (millions of cubic meters)	(2,900)	1,000	1,850	1,650	160	4,660
SWG	Reclaimed sewage water	—	60	85	140	220	220
	Capacity installed (millions of cubic meters per year)	(45)	60	25	55	80	220
	Projects (thousands of hectares)	a	10.9	6.8	13.4	23.9	55
	Irrigated area (thousands of hectares)	a	10.9	17.7	31.1	55	55
	Volume consumed (millions of cubic meters)	a	300	850	1,400	220	2,770
TS	Transported from Fezzan	—	175	350	350	350	350
	Capacity installed (millions of cubic meters per year)	—	175	175	—	—	350
	Projects (thousands of hectares)	—	31.8	41.1	4.9	9.7	87.5
	Irrigated area (thousands of hectares)	—	31.8	72.9	77.8	87.5	87.5
	Volume consumed (millions of cubic meters)	—	875	3,500	3,500	350	8,225
SS	Desalinized seawater	—	—	70	—	—	70
	Excess from municipal demand (millions of cubic meters per year)	—	—	70	—	—	70
	Projects (thousands of hectares)	—	—	14.5	—	—	14.5
	Irrigated area (thousands of hectares)	—	—	14.5	—	—	14.5
	Volume consumed (millions of cubic meters)	—	—	350	—	—	350
O	Other sources	—	20	30	60	80	80
	Capacity installed (millions of cubic meters per year)	—	20	10	30	20	80
	Projects (thousands of hectares)	—	3.6	2.6	7.1	6.7	20.0
	Irrigated area (thousands of hectares)	—	3.6	6.2	13.3	20.0	20.0
	Volume consumed (millions of cubic meters)	—	100	300	600	80	1,080
	Total	(580)	455	720	715	810	810

Code	Activity	*Schedule* Present, 1980	First stage	Second stage	Third stage 20 years	30 years	Totals
	Capacity installed (millions of cubic meters per year)	(580)	455	280	85	100	920
	Projects (thousands of hectares)	—	82.3	67.5	9.0	43.7	202.5
	Irrigated area (thousands of hectares)	(96.6)	82.3	149.8	158.8	202.5	202.5
	Volume consumed (millions of cubic meters)	(2,900)	2,275	6,850	7,150	810	17,085
1300	*Operation and Maintenance*						
1310	Water supply (operation; hectares)	—	82,500	149,800	158,800	202,500	202,500
1320	Maintenance	—	—	Continuous	—	—	—
1330	Technical assistance	—	—	Routine	—	—	—
3100	*Administrative Support*						
3110	Clerical	—	—	Permanent	—	—	—
3120	Supplies	—	—	As required	—	—	—

Notes: Figures under year headings refer to targets to be achieved at that time.
a. Included in GW so far. (No cost for installed capacity.)

many years, but many dimensions of its water problems have remained very fuzzy. For example, natural recharge estimates varied from 100 million to 400 million cubic meters a year in 1979. Although future water requirements for domestic supply have been predicted, they are not at all clear, since future water requirements for agriculture depend on a variety of policies, resources, and factors. Furthermore, no cost values had been assigned to possible large engineering works. Thus several years ago the FAO started to assist the Libyans in a comprehensive analysis of the Gefara Plain.

The Gefara Plain covers about 20,000 square kilometers. Rainfall occurs between November and May and varies from 100 to 300 millimeters a year. An extensive area of older dune sand is gently rolling and vegetated; it forms relatively good soil but tends to limit recharge to aquifers below by retaining much of the rainfall. Unfortunately, most of the rainfall tends to be low in intensity, further limiting groundwater recharge. Surface runoff of less than 90 million cubic meters a year occurs in the plain from the uplands to the south. Most of this water evaporates, however, or is percolated to groundwater; only a few million cubic meters reach the sea during the occasional wet year.

The latest and more reliable estimate of average natural groundwater replenishment from surface streams and rainfall is about 195 million cubic meters a year. Of the two aquifer systems, the shallow, essentially unconfined system is the one mostly under production. The deeper aquifer system is in contact inland with the shallow aquifer but tends to have poor water near the coast, where it is separated from the shallow aquifer.

Groundwater use by agriculture and by towns has increased from 210 million cubic meters a year in 1959–62 to 532 million cubic meters in 1978. Sewage water which might currently be reclaimed for agricultural use is about 41 million cubic meters a year; it is estimated to increase to about 145 million cubic meters a year by 2010.

Plants recently under construction for the desalinization of seawater will produce only 18 million cubic meters a year. Additional plants are planned which will produce 350 million cubic meters a year by 2010.

In 1983 there was an estimated deficit being drawn from groundwater storage of about 435 million cubic meters a year, with a total demand of 635 million cubic meters a year. By 2010 the demand could increase to over 1,300 million cubic meters a year if all plans are carried out for municipal and agricultural development in the Gefara Plain. The current overdraft has been forced to the people's attention by seawater intrusion which is moving inland at a maximum of about 500 to 1,000 meters a year.

The basic engineering options are clear. An outside source of freshwater must be provided either by desalinization or by pipeline from large desert aquifers to the south, while reclaimed sewage water is used to the maximum extent possible. Costs of desalinization are estimated to be about $1.80 a cubic meter; transported water will cost about $0.75 a cubic meter. Some additional water could also be provided to agriculture through the use of water-harvesting techniques, but the amount will be relatively small.

Table 12-4 presents the proposed budget for a water management plan, which is shown graphically in Figure 12-3. This plan is under discussion and clearly needs much further work as conditions change.

Table 12-5. *Summary of Results of the Linear Programming Runs*

Item	*LPO– 1990*	*LPOB– 1985*	*LPOB2– 1990*	*LPOB3– 2000*	*LPOB4– 2010*	*LPOT2U– 1990*	*LPOT2L– 1990*	*LPOT3U– 2000*	*LPOT3L– 2000*	*LPOT3U2– 2000*	*LPOT4– 2010*	*LPOS2U– 1990*
Population of Gefara Plain (millions)	2.084	1.807	2.084	2.864	3.950	2.084	2.084	2.864	2.864	2.864	3.950	2.084
Consumption pattern (kilograms per year per capita)												
Cereals	154	96	94	90	86	106	106	105	105	105	103	106
Vegetables	142	143	143	143	143	178	178	179	179	179	180	178
Fruits	46	41	41	41	41	56	56	53	58	58	60	56
Olives (green and for oil)	33	63	66	70	74	105	105	105	105	105	105	105
Meat and eggs	68	38	41	43	45	54	54	53	58	58	62	54
Milk (equivalent)	162	67	71	75	78	43	53	72	72	72	91	53
Total cultivable land (hectares)	557,000	557,000	557,000	557,000	557,000	550,000	550,000	550,000	550,000	550,000	550,000	478,500
Cultivated area (hectares)	406,546	358,191	263,535	263,637	340,655	253,362		287,572	340,323	447,082	298,134	423,732
Irrigated area	145,007	138,671	120,591	111,624	228,267	111,177		131,751	70,865	146,000	130,425	138,618
Food self-sufficiency ratio (percent)												
Cereals	82	20	20	10	20	10	10	10	10	10	10	10
Vegetables	87	100	100	100	100	100	100	100	100	100	100	100
Fruits	100	100	100	100	100	100	100	100	100	100	100	100
Meat and milk	80	80	80	50	80	40	40	40	10	20	20	20
Available water (millions of cubic meters)	1,650.0	573.5	730.2	747.1	846.7	612.1		615.4	281.4	615.4	708.4	619
Total water consumption (millions of cubic meters)	1,644.3	582.4	507.5	451.5	782.2	447.2		527.7	281.5	567.4	519.7	619.0
Pumped	213.7	326.2	199.5	199.5	199.5	235.8		172.4	172.4	172.4	172.4	172.4
Desalinized (and other)	1,155.0	20.0	100.0	60.0	80.0	0.0		0.0	30.0	0.0	0.0	30.0
Transported	147.6	175.0	123.0	49.0	280.0	169.1		276.2	0.0	315.9	215.1	334.0
Treated	128.0	61.2	86.0	143.0	222.7	42.3		79.1	79.1	79.1	132.2	79.1
Water per irrigated hectare	4.800	4.200	4.200	4.050	3.450	4.020		4.005	3.970	3.880	3.990	4.470
Manpower (millions of man-days)[a]	10.752	6.903	5.206	6.020	8.475			9.666	13.025	10.388	9.937	14.156
Total cost of production (millions of Libyan dinars)	784.764	175.858	165.560	133.356	230.609	138.786		85.649	149.719	213.630	181.349	253.246
Cost of water (millions of Libyan dinars)	655.018	97.594	108.717	74.154	144.992	57.033		86.389	22.339	96.314	78.789	118.768
Cost of water (percent)	83.5	55.5	65.7	55.7	62.9	41.0		46.5	15.0	45.0	43.5	46.9
Cost per hectare (Libyan dinars)	1,930	491	628	516	667	548		646	440	478	608	598
Average cost per irrigated hectare (Libyan dinars)	2,218	922	1,118	879	886	834		1,001	689	921	948	1,174
Average cost per nonirrigated hectare (Libyan dinars)	319	218	216	224	251	323		345	374	262	344	317

a. With more highly skilled labor, productivity increases 25 percent per year from 1985 to 1990 and 1 percent per year thereafter. With more efficient use of all input factors, the quantity of input per hectare decreases 2.5 percent per year until 1990 and 1 percent per year thereafter. A breakdown of the manpower results in terms of permanent farmers versus seasonal labor would be available with an additional printout program.

Figure 12-3. *Water Balance Development, Gefara Plain*

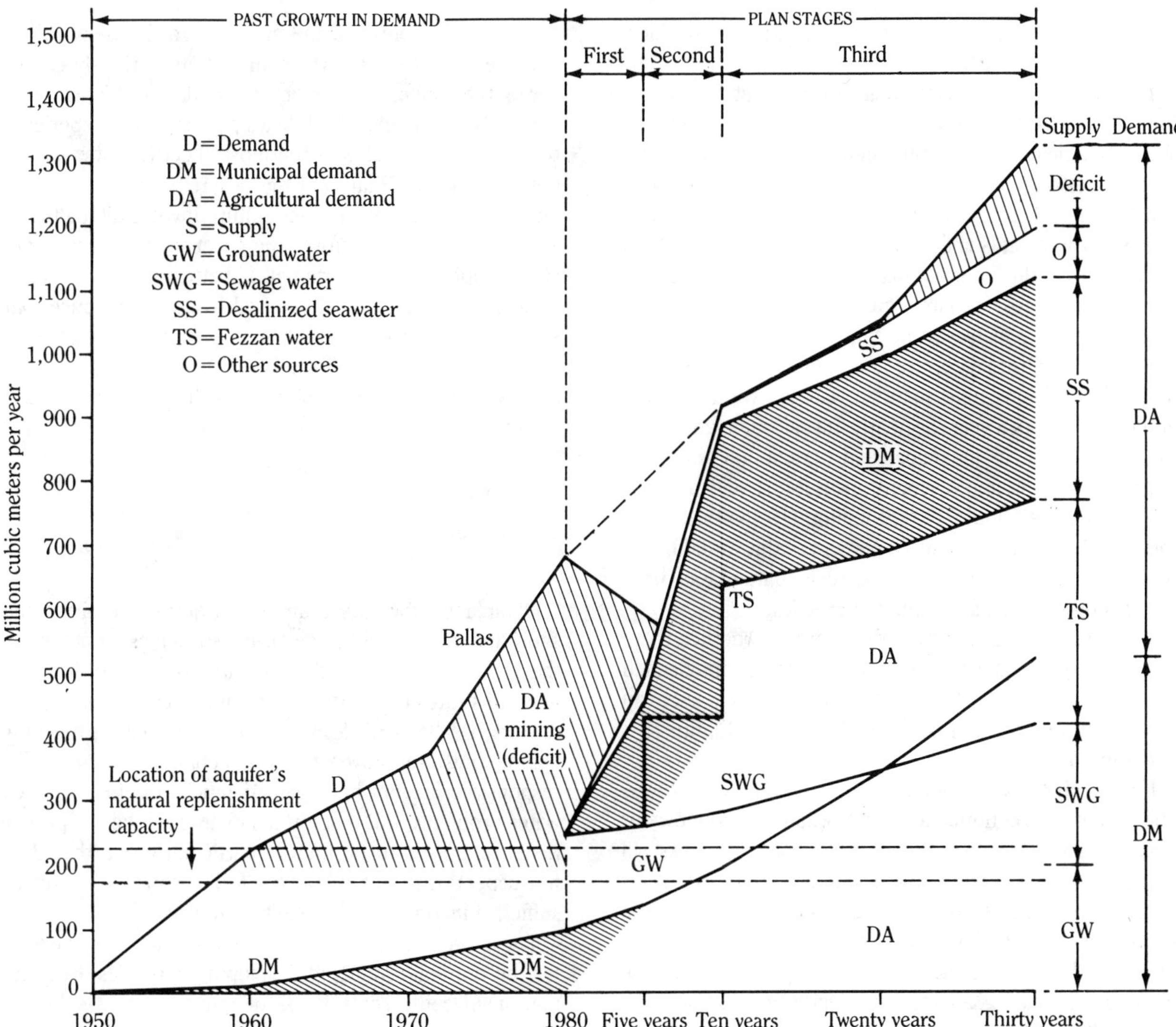

Several policy matters may be difficult to resolve. One of these is whether to transport the groundwater from the desert to the coast, where crop production is much higher per unit of water, or whether to grow crops in the more difficult desert areas where the groundwater is located and then transport the crops to the coast. A great deal is known about some areas, but for the areas to the south a greater amount of technical information on groundwater and agricultural soils is required. There may be some areas where water should be used locally and other areas where water would best be transported to the coast. So far there has been little awareness of the importance of knowing the details about the large and geologically complex freshwater aquifers south of the Gefara Plain. However, a project in the eastern part of the Gulf of Sirte that would transport water to the coast for irrigation is already under consideration, so the idea of transporting water is obviously acceptable.

In the long run, the extraction of groundwater from the desert aquifers will clearly increase costs of pumping. As long as oil revenues are available, increasing costs may not be a serious concern, but the government should have information on the long-term consequences of groundwater mining.

An important engineering point is whether desalinized water will be used for agriculture, since water distribution and seasonal storage of transported and desalinized water require different structures.

A series of linear programs were run for the Gefara Plain (see Table 12-5). The least costly alternative is to limit agriculture to the safe yield in the Gefara Plain and to import the rest of the agricultural products. It has not been possible at this stage to complete all the linear programming models which are desirable. But it is clear that the costs are sensitive to the population, the degree of food self-sufficiency, and the interest rate; they are not sensitive to small variations in crop water requirement. Labor requirements may turn out to be a serious future constraint (6 million to 14 million worker days a year), and the level of agricultural skills would need to be considerably improved.

Total annual water costs range from about 85 million Libyan dinars ($23 million) for self-sufficiency only in fruit and vegetables to nearly 800 million ($220 million) for 80 percent self-sufficiency in cereals, meat, and milk and nearly full sufficiency in fruit and vegetables. Full self-sufficiency in cereals and meat would thus cost nearly 700 million Libyan dinars a year ($190 million). It is interesting that the total cost of large quantities of desalinized or transported water is nearly the same.

Total agricultural water requirements range from just under 300 million cubic meters a year to over 1,600 million cubic meters a year for nearly full self-sufficiency. The total investment in municipal and agricultural water for engineering, farm improvement, institutions, and so on, is estimated to be about $14,000 million (1981 prices) over the next thirty years. About 70 percent of this would be for water facilities alone. The large-scale parameters have thus been estimated for planning purposes.

The most critical point for the future is whether there will be the political will to provide the institutions needed to carry out this work. The FAO has outlined such an institution, with an approximate schedule (see Figure 12-3 and Table 12-4).

If no decisions are made by the government, the towns and farmers will continue to extract whatever groundwater they can, and the options for future engineering works will be reduced. Fortunately, the groundwater in storage is considerable, and the time pressure is not critical. With a maximum of 500 meters a year of seawater intrusion, however, clearly no more than ten to twenty years should be allowed for making appropriate decisions and initiating major actions if the coastal aquifer in inhabited areas is not to be lost. Intrusion of seawater can of course be combated, but even with large amounts of freshwater and costly recharge facilities, pushing back seawater is a slow process.

The FAO has outlined the problems in the Gefara Plain and the means to resolve them without seriously endangering the aquifer. A large amount of data still needs to be collected and analyzed if Libya's people are to experience the long-term benefits of the most effective water use for both the Gefara Plain and the desert to the south. Nonengineering solutions for agriculture involve difficult decisions and concerted effort. The government has for some years emphasized that rainfall is better used to augment agriculture production. Improved irrigation practices will save relatively little water since most farmers already tend to use as little as possible. Water will be saved only by sophisticated irrigation systems, which will in turn require increasingly skilled farmers.

Conclusion

Qatar and Libya are examples of areas that are water-short and money-rich. Engineering strategies for achieving the ideal of food self-sufficiency are costly, and nonengineering strategies are difficult from a social point of view. A start has been made, however, in evaluating the nonengineering alternatives in terms of engineering costs. The responsible leaders have gained much insight into their water problems, but much remains to be done if a permanent solution is to be found—particularly one that does not depend on the availability of oil resources to provide sufficient income to subsidize agriculture.

These two examples may be inappropriate in some ways for the non–oil-producing developing countries. But they may involve the same kinds of questions and solutions faced by those regions where recharge is being exceeded and where surface water resources are not available.

Comment

Gerald T. O'Mara

Thomas explores the policy choices for two arid countries where both petroleum and freshwater are found underground. The existence of significant petroleum deposits extends the range of policy options for water to include seawater desalinization—a technical option too expensive to merit serious consideration in most countries. By examining two countries that are boundary cases, Thomas is able to reveal clearly how water use policy depends on national objectives. In exploring the policy alternatives for Qatar, Thomas develops a programming model of the country's agricultural sector and simulates scenarios that define the costs of some of the policy objectives articulated by national decisionmakers. For example, one of the lowest cost scenarios for the national food supply was to import all foodstuffs—an option that the government rejected. Another scenario achieved putative food self-sufficiency, but with a tenfold increase in cost over the least-cost alternative and only by mining the groundwater aquifer. The estimated annual cost per capita for food self-sufficiency achieved by using mostly water from seawater desalinization was about $1,000. This high level of cost was required because of the projected increase in population (almost four foreign workers for each Qatari national) as a consequence of petroleum development. Thomas notes that if the government opted for a 50 percent reduction of the projected population increase by reducing dependence on foreign workers, near food self-sufficiency was feasible at a large savings in cost (approximately $4,000 million). Analysis of water policy options thus revealed alternatives that exist only if overall development and population policies are modified.

In both countries the least-cost solution for the provision of the food supply consists of constraining agricultural water use to the sustainable yield (or less) from groundwater sources and relying on imports for the remainder of domestic supply (Thomas errs in reporting sustainable yield for Qatar at 50 million cubic meters, since his data clearly indicate that it is 22.5 million cubic meters). Thomas reports that neither country accepts the lowest cost solution. Instead, the value both place on near self-sufficiency in food supply makes it a viable alternative despite significantly greater cost. Whether these preferences would remain the same if petroleum export earnings drastically decreased is debatable, but the link between water policy and such broad national objectives as self-sufficiency is quite clear. Planning for water resource use is a process that must start from the national objectives articulated by national political leaders. Thomas has provided us with some instructive examples of this process. It will often require a dialogue that presents water policy alternatives to policymakers and identifies operant national objectives through the preferences revealed by water policy choices.

13

Irrigation Management and Scheduling: Study of an Irrigation System in India

D. N. Basu and Per Ljung

The study presented in this chapter focuses on developing an irrigation schedule for conjunctive use to improve the equity and efficiency of a traditional irrigation system in India. The study does not take into account all the complexities of the real system. But the approach is more effective in a shorter time and requires less investment than plans biased toward modernizing the physical infrastructures of older irrigation systems. Another advantage is that the approach is based on a detailed study of an existing canal delivery system and on groundwater use at the lowest level of the distribution system; it therefore takes into account the farmers' normal response to groundwater irrigation under different conditions of canal water availability.

The approach draws its rationale from the following observations that explain inefficiency in a majority of traditional surface irrigation systems in India and other developing countries:

- Water losses in the conveyance system from the main canal head to the farmgate have resulted from both physical and operational factors.
- More attention has been paid to the physical improvements in the irrigation system—for example, the lining of canals—than to management and economic alternatives that would more equitably allocate water among user groups or regions and would achieve higher overall efficiency.
- Few clearcut objective strategies or schemes have been developed for conjunctive use, even though there is significant recharge to the groundwater system through conveyance and field application losses.

It is appropriate to define the context in which the concepts of equity and efficiency are used. Surface irrigation projects whose surface water resources are highly variable and are limited by the size of the command often resort to management practices that produce inequitable water distribution and less than expected returns from the total water available. The phrase "less than expected returns" refers to either higher losses in water conveyance and the operations of the system or a lower crop yield for each unit of water released at a given level of the distribution system. In a larger context of efficiency, surface and groundwater resources are considered as two interdependent subsystems.

Equity is measured by surface water allocation for each unit of cropped area among different reaches of the command and also among farmers located at the head and tail outlets in the command of the canal system at any given reach. In a more general sense equity applies not only to canal water allocation but also to groundwater access. Many irrigation projects, particularly those in semiarid areas, typically use more groundwater if their canal water supplies are limited and variable. In such cases a measure of equity has to take into account both relative access to groundwater in different reaches of the command (or among farmers in the same reach) and the cost (to the farmers) of groundwater as a supplement to canal irrigation. Achieving equity and efficiency in such a context involves a strategy of conjunctive use in water allocation policy and irrigation scheduling. In this study the concept of equity is more relevant at the lowest level of the distribution network—that is, the canal outlet.

This study, which represents a typical surface irrigation project in a semiarid zone with high seepage losses recharging the groundwater aquifers, has the following limited objectives:

- To develop an irrigation schedule that applies to the lowest level of the distribution network and achieves higher overall efficiency in water use by supplementing canal irrigation with groundwater.

- To estimate the marginal returns from exploiting additional groundwater through an irrigation schedule for a conjunctive system.
- To maximize the favorable effects of externalities (for example, exploiting the groundwater storage of a traditionally unlined conveyance system) to achieve higher agricultural production and more equitable water distribution. This objective is only a part of the larger problem of water allocation, which takes into account both surface and groundwater resources and the inefficiencies caused by the externalities of the system.
- To identify the operational constraints and policy issues for conjunctive irrigation management and scheduling.

Our findings are intended to evaluate the feasibility of developing an alternative for irrigation management and are not conclusive. They should serve as a basis for discussion and for subsequent improvements in the development of appropriate approaches.

Project Overview

This study presents four elements of a comprehensive evaluation of a traditional surface irrigation system in a state of India. First, the study highlights the development of an appropriate irrigation management system through conjunctive use, which we believe to be the most effective approach. Three important findings of our evaluation exercise are:

- Significant year-to-year fluctuations in canal water availability and the absence of a consistent policy for canal water allocation in different parts of the command, for different types of inflow, and for variations in rainfall.
- Evidence of very high conveyance losses in the system owing to both seepage and operations.
- Positive response by farmers to using groundwater under varying conditions of canal water availability.

Second, a simple model demonstrates the potential benefits of irrigation scheduling. A soil-moisture balance model was developed for major crops in the study area, with a day-to-day accounting of water input and output for the crop-soil system. This model was based on data generated for the specific crops in the study area and was restricted to a few parameters and inputs only. It enabled us to examine the relationship between water application and yield index and thereby to study the implications of alternative irrigation schedules for canal irrigation or conjunctive water use.

Third, the study examines in detail the demands made on the groundwater system by the adoption of the appropriate irrigation schedule for a conjunctive system (as obtained from the model). The economic advantages of the suggested conjunctive use system are briefly analyzed, particularly in the context of the inefficient and inequitable water use observed in the existing system. Our study was restricted to the sample study areas, and our findings indicate only the feasibility of applying the approach on a wider scale.

Fourth, the study identifies the policy issues and constraints involved in developing the approach.

Project Profile

The irrigation project consists of a masonry structure 327 meters long with an earthen dam extending 4,550 meters. The gross and live storage capacities are 464 million and 444 million cubic meters respectively. The main and branch canals are 48 and 32 kilometers long respectively, with discharge capacities of 1,100 and 370 cusecs at offtake points. The distributaries—both minor and subminor—add up to 512 kilometers.

The project command covers 81,000 hectares, 45,000 of which are arable. The command represents a typical semiarid tropic zone with a combination of high temperatures (maximum and minimum monthly means of 41.3° C and 10.4° C respectively and an annual mean of 26.9° C) and low rainfall (an average of 629 millimeters a year), with 93 percent of rainfall confined to the monsoon season (June–September).

The dominant soil texture is sandy loam (66.1 percent); the remaining soils (33.9 percent) range from loam to clay. The soils throughout the command are very deep and are not threatened by salinity or alkalinity. The moisture-holding capacity ranges from 10.6 to 49.7 percent, and infiltration rates from 0.86 to 10.35 centimeters a hectare. The land irrigability classification (LIC) puts 29.6, 67.9, 1.7, and 0.8 percent of the command area under classes 1, 2, 3, and 4 respectively, which indicates soils suitable for irrigation. The soils are low in organic matter and nitrogen, low to medium in phosphorous, and medium to high in potash constituents.

The main water-bearing units (aquifers) in the project area occur in thick, alluvial deposits of Pleistocene or more recent age. The oldest of these deposits unconformably overlie granitic rocks of Archean age. The thickness of these deposits increases from northeast (50 meters) to southwest (600 meters). Groundwater occurs under confined as well as unconfined conditions.

The depth of the water table in open wells varies between 6 and 16 meters below ground level, and the water table represents a composite figure of piezometric and phreatic surfaces. During the past decade the water table has risen 0.14 meter a year, and the average rise during the monsoon is about 1 meter. The water table gradient is from northeast to southwest and varies from 1 to 3 meters per kilometer.

With an average permeability of 20 meters a day and a specific yield of 16 percent, the aquifers are productive and can supply vast amounts of water if properly exploited. In general, the quality of water is fit for irrigation (total dissolved solids [TDS] < 2,000 parts per million).

The major crops grown in the command during kharif (June–October) include sorghum (jowar), pennisetum (bajra), and clusterbean; mustard and wheat are the major crops during rabi (October–March). The potato is grown intensively in the head reaches of the command. Castor and lucerne are the important two-season, perennial crops of the command. Depending on the availability of irrigation, pennisetum and guar are also grown during hot weather (March–June), while cumin and fennel, which cover a considerable area during rabi, are reduced to almost negligible proportions.

Project Performance

Historical data for the past fifteen years show wide deviations in actual project performance (compared with design criteria) for reservoir yield, area irrigated, seasonal irrigation priorities, proposed cropping pattern, distribution of water in different regions, and overall project efficiency. On average, water availability—that is, the canal release from the dam through the main canal—has been significantly lower than project estimates indicated: estimates based on historical records of 50 percent and 75 percent are only 230 million and 91 million cubic meters, compared with original project estimates of reservoir yields of 598 million and 260 million cubic meters respectively.

For the entire period of study, the average area irrigated during kharif is estimated at 2,890 hectares (maximum 9,240 and minimum 110), against project estimates of 15,800 hectares. The area irrigated by canal during rabi has fluctuated between 470 and 26,470 hectares, with an average of 13,050 hectares against project estimates of an average of 4,670 hectares and a maximum of 16,470 hectares in a very good rainfall year. In view of the observed phenomenon of low water availability, actual area irrigated, on average, has been only 53 percent of project estimates, with a range of 14 to 98 percent.

The cropping pattern has changed markedly over the years because of the uncertain supply of irrigation water. The crops affected include millet, sorghum, and pennisetum, which are gradually being replaced by clusterbean in kharif and by oilseeds in rabi. The area under wheat cultivation fluctuates depending on water availability at the end of kharif season.

Historical data reveal that distributaries in the head and middle reaches receive more or less stable supplies of irrigation water during rabi, while the tail distributaries receive hardly any irrigation water in bad years.

In a year of water scarcity, the official policy for canal water allocation is to restrict the area to be irrigated and, in some cases, the crop as well. For specified crops, head reaches (those reaches closest to the main or branch canal or the head and tail outlets within the same minor) receive a greater number of irrigations than the rest. For some of the high-value crops, distribution of water is still more uneven.

The average water application for the crop delta at the main canal head has been estimated at 3.75 feet (with minor variations between rabi and kharif). Given the project estimate of about 38 percent efficiency, water application at the field level would be about 1.4 feet (0.42 meter), but the findings from the study of conveyance losses indicate that from the minor to the plot level alone, losses owing to seepage and operations would be between 65 and 70 percent.[1] Total canal system efficiency would be barely 20 percent, which implies that in a normal year water application at the plant level would be, on average, 0.7 foot (0.21 meter)—a quantity large enough for two to three irrigations. In a year of scarcity, average water application would be even less. Such low canal efficiency, coupled with interyear and intrayear variability in water supply, has resulted in coverage of only 45 percent of the total area irrigated either by canal water alone or by groundwater supplemented with canal water. Farmers have naturally responded by supplementing canal water with groundwater in tail distributaries and outlets characterized by low and variable canal supplies.

The above findings suggest two steps to developing an appropriate strategy for irrigation management and scheduling. First, examine in detail the potential for, and constraints to, groundwater development; second, understand, in greater detail, the existing irrigation and agricultural practices of farmers at the micro level (say, at the command of an outlet) and apply the approach in this situation.

Exploitation of Groundwater

The primary water balance of the study area indicates a gross recharge (from irrigation and rainfall) of 140 million cubic meters in the unconfined aquifers and net pumpage by shallow dug wells and borewells of 44 million cubic meters (31 percent). Subsurface storage receives 20 million cubic meters, which leaves 76 million cubic meters for storage accumulation (an increase to the water table) and residual loss (water lost through subsurface drainage or the evapotranspiration of deep-rooted trees).

Estimates of the recharge to deep aquifers are not available. But a lowering of piezometric surface by two meters can yield up to 20 million cubic meters—that is, 200 deep tubewells could be installed. At present there are only twenty-two deep tubewells (state-owned) of one cusec each which are in operation and which produce a quality of water fit for irrigation.

With proper management, additional groundwater—say, 50 million cubic meters—can be exploited from confined as well as unconfined aquifers. Thus there is potential for almost doubling the irrigated area with an ensured supply of water from either private or public sector tubewells.

A program of groundwater exploitation in keeping with the various constraints of the public and private sectors is envisioned as follows:

- A battery of deep tubewells would be installed in different parts of the command areas to augment canal flows and ensure that farmers receive at least three irrigations. These tubewells would tap the deeper aquifers (those below 50 meters) and would not pose any danger to shallow wells.
- Farmers would be encouraged to have their own shallow wells to tap only the upper aquifers (those above 50 meters) and would be provided with the necessary guidance and financial resources. The additional water would increase the crop yields and allow for multiple cropping in the area.
- If canal water were not available and the water table were deep, the state would provide water from deep tubewells for at least three irrigations.

Description of the Study Outlets

The approach is illustrated by detailed findings at the lowest level of the distribution system—the canal outlet and its command. The outlets chosen represent distribution at two different reaches from the main canal offtake point and at head and tail reaches within the same distributary. The general features of the outlets are presented in Table 13-1. The outlets are assumed to represent the irrigation project command in area irrigated per outlet (17.24 to 55.12 hectares), in cropping pattern, and in intensity of irrigation and available groundwater resources. Outlets HD/HOL and HD/TOL fall into LIC class 1, while MD/HOL and MD/MOL fall into LIC class 2; both these classes together constitute 97.5 percent of the command.[2] Outlets HD/HOL and HD/TOL represent soils with light texture (sandy loam), high permeability (an infiltration rate of 10 to 12 centimeters an hour), and low moisture (7 to 9 percent), while outlets MD/HOL and MD/MOL represent medium texture (loam to clay-loam), medium permeability (an infiltration rate of 3.0 to 5.6 centimeters an hour), and moderate moisture (12 to 15 percent).

Table 13-1 reveals some interesting features of current irrigation in these representative outlets. Outlet HD/HOL represents an extreme with very high canal water input (and correspondingly low groundwater use) and adoption of high-value crops such as potatoes, which achieve almost the highest yield in the state (about 38 tons per hectare). Outlet HD/TOL is also characterized by high-value crops, but with relatively low canal water input, its groundwater use is more extensive. Outlets in the medium reaches present a distinct picture—a low input of canal water and a high use of groundwater for the cultivation of low-value crops.

Steps to Conjunctive Use

The following list is a sequence of steps for developing our approach to irrigation management and scheduling in

Table 13-1. ***Characteristics of Canal Outlets and Current Irrigation Status***

Outlet and position	*Distance from minor offtake point (meters)*	*Area irrigated (hectares)*	*Design discharge (cusec)*	*Delta at outlet (meters per irrigation)*	*Number of wells*	*Crops (in order of increasing acreage)*	*Number of irrigations*		
							Canal	*Well*	*Canal + Well*
HD/HOL (head)	959	17.24	0.032	0.30–0.36	1	Mustard	3–4	5	—
						Castor	3–4	—	—
						Potato	7–8	12	17
						Wheat	4–5	—	—
HD/TOL (tail)	2,700	36.98	0.033	0.09–0.15	4	Mustard	2–4	3–7	3–4
						Castor	3–5	4–6	4–5
						Potato	—	15–17	7–20
						Wheat	4	8–9	8
MD/HOL (head)	381	21.18	0.051	0.12–0.15	4	Mustard	2	3–4	—
						Wheat	—	2–8	—
						Castor	2	3	—
MD/MOL (middle)	2,804	55.12	0.051	0.05–0.08	5	Mustard	2–3	2–5	2
						Wheat	—	3–5	5
						Castor	—	2–4	—

—Not applicable.

a conjunctive system. Each step is illustrated by the actual conditions of the cultivated area, the soil types, and the cropping patterns.

• Use the model to obtain relevant results for the major crops in the study area under varying frequency and depth of total irrigation input.

• Given the constraints of total canal water availability and the existing conveyance losses, work out a feasible canal schedule which may be considered efficient and equitable—that is, which ensures a minimum yield for major crops throughout the entire command or a large part of it. The suggested schedule has a fixed pattern of delivery at a given quantity of canal water for a unit of cropped area.

• Analyze the marginal relationship between water application and yield index and use the results to generate a groundwater schedule for the outlet, subject to total water availability through recharge from the canal system and rainfall. The schedule assumes that each outlet represents certain conditions that balance canal water deliveries with recoverable recharge. Given the conveyance losses of the canal system, recoverable recharge (from lower or upper aquifers, as the case may be) is predetermined and sets a limit to total water use. The net balance of subsurface flows into and out of the aquifers represented by a single outlet or group of outlets is not significantly positive or negative. The schedule also assumes that irrigation schedules for conjunctive use will not significantly change the cropping pattern of the study area.

• Convert the groundwater schedule for each outlet into specific physical magnitudes—for example, the total water to be delivered each week from the groundwater system and the capacity to be installed.

Table 13-2. *Parameters Considered for Irrigation Scheduling Model*

Parameter		*Crop*					
		Castor		*Wheat*		*Mustard*	
1. Date of planting		14th day (from July 1)		16th day (from November 1)		16th day (from November 1)	
2. Crop duration (days)		210		120		120	
3. Fortnight evapotranspiration values (millimeters per day) starting from July 1 or November 1		6.0, 6.0, 5.5, 5.5, 6.2, 6.2, 6.3, 6.3, 4.6, 4.6, 3.6, 3.6, 3.9, 3.9, 5.0, 5.0		4.45, 4.45, 3.85, 3.85, 4.15, 4.15, 4.5, 4.5, 6.65, 6.65, 8.1, 8.1		4.45, 4.45, 3.85, 3.85, 4.15, 4.15, 4.5, 4.5, 6.65, 6.65, 8.1, 8.1	
4. Crop factor (*Kc* value) and root depth in feet for percentage growth period		*Kc value*	*Root depth*	*Kc value*	*Root depth*	*Kc value*	*Root depth*
	Start	0.20	0.3	0.25	0.2	0.23	0.2
	10 percent	0.35	0.8	0.30	0.5	0.23	0.6
	20 percent	0.65	1.6	0.55	1.2	0.37	1.3
	30 percent	0.80	2.4	0.76	1.8	0.65	1.9
	40 percent	0.90	3.0	1.03	2.3	0.80	2.3
	50 percent	1.05	3.5	1.10	2.4	1.05	2.6
	60 percent	1.05	3.8	1.10	2.5	1.10	2.8
	70 percent	1.05	4.0	1.10	2.5	1.10	3.0
	80 percent	1.05	4.0	1.00	2.5	1.00	3.0
	90 percent	0.74	4.0	0.87	2.5	0.80	3.0
	Maturity	0.59	4.0	0.48	2.5	0.52	3.0
5. Duration of physiological growth state and corresponding yield response factor							
	I	Establishment 25 days; 0.30		Establishment 15 days; 0.2		Establishment 25 days; 0.20	
	II	Branching 40 days; 0.30		Tillering 25 days; 0.6		Branching 35 days; 0.30	
	III	Spike emergence 65 days; 0.60		Jointing and boot 30 days; 0.4		Flowering 25 days; 0.55	
	IV	Spike development 30 days; 0.55		Flowering and milk 30 days; 0.5		Grain development 25 days; 0.60	
	V	Spike maturity 50 days; 0.20		Dough and maturity 20 days; 0.2		Maturity 10 days; 0.10	
6. Soil moisture depletion factor for reduced evapotranspiration		75		50		75	

Table 13-3. *Common Parameters Used for Irrigation Scheduling Model*

1. Daily rainfall data in millimeters represent actual series for a nearby meteorological station. (Rainfall of less than 2.5 millimeters is considered ineffective.)
2. Runoff curve numbers (*CN*) convert daily values for rainfall and irrigation into daily values for rainfall excess (TSURF).

Soil moisture content	*CN value*
Wilting point	47
75 percent depletion	57
50 percent depletion	67
25 percent depletion	76
Field capacity	83

   CN values correspond to land use (row crops), treatment (straight row), hydrologic condition (good), and hydrologic soil group A (high rate of water transmission).
3. Irrigation schedule choices: rainfall, schedule irrigation, and on-demand irrigation
4. Soil moisture capacity: 44 millimeters per foot
5. Percolation rate: 1 percent

• Estimate the marginal net returns from the above schedule and from total water application through conjunctive use.

Future analysis would address the issue of water allocation more explicitly, particularly when the different regions in the study area have highly interdependent subsurface flows and modes of groundwater exploitation. Under such circumstances, externalities caused by the interdependent physical components must be explicitly taken into account in the strategy for equitable water allocation.

Structure of the Model (IRRYM)

The interrelations between crop, climate, water, and soil are extremely complex. But an analysis of a crop's response to different irrigation schedules requires only a couple of significant variables. The IRRYM is a simulation model that incorporates the key values and answers such questions as: Given a certain crop variety, cultivation practice, and agroclimatic condition, what is the impact, if other factors remain the same, of different irrigation regimes on crop yield? Tables 13-2 and 13-3 summarize the values of the parameters used to specify the factors held constant in the irrigation scheduling model. The model does not address how farmers respond to shifts in cropping patterns or changes in cultivation practices. However, the output of IRRYM can be fed into traditional economic models of farmer behavior. By running IRRYM for a series of years or for one year with an irrigation schedule that contains a random element, risk measures can be estimated and directly incorporated into more sophisticated behavioral models.

The structure of the model is straightforward and consists of five modules: rainfall infiltration and runoff, soil moisture balance, evapotranspiration and moisture uptake by the crop, irrigation schedule, and yield response to moisture stress. Figure 13-1 depicts the essential elements of the model and their relationships.

Figure 13-1. *Irrigation Scheduling Model*

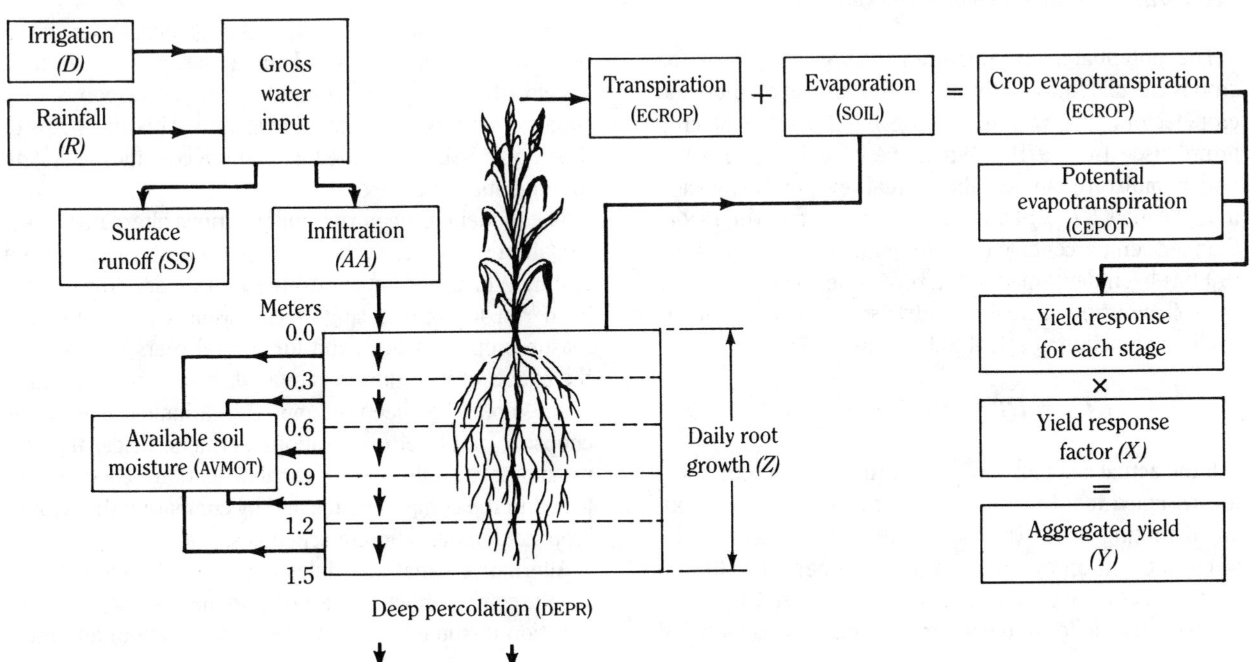

Soil Moisture Module

The soil profile consists of layers one foot thick. When rain or irrigation occurs, moisture moves from one layer to the next when the field capacity has been reached in the upper layer. The soil moisture module also has small correction factors that account for a slow percolation downward and for a slow movement of moisture upward from a shallow water table. For one meter of soil, the total available soil moisture—that is, the amount of water stored between field capacity and wilting point—is normally assumed to be 200 millimeters for heavy soil, 140 millimeters for medium soil, and 60 millimeters for coarse soil. The mathematical structure of the modules, which are described below, can vary depending on the amount and type of research data available.[3]

Rainfall Infiltration Module

The U.S. Soil Conservation Service has developed a formula for estimating runoff and infiltration from rainfall. Infiltration is a function of the hydrologic condition of the field and the previous moisture content in the top one-foot layer of soil.

$$I = P - \frac{[P - 0.2 \cdot (25{,}400/CN - 254)]^2}{[P + 0.8 \cdot (25{,}400/CN - 254)]}$$

$$\text{or } I = P \text{ if } P \leq 0.2 \cdot (25{,}400/CN - 254)$$

where I = amount of infiltration in millimeters, P = amount of precipitation in millimeters, and CN = coefficient which is a function of previous soil moisture, soil type, and cultivation practice.

Evapotranspiration and Moisture Uptake Module

The potential daily evapotranspiration (ET_{pot}) of the crop is calculated using the Modified Penman Method and crop factors that vary depending on the stage of crop growth (see FAO 1979a). When the crop is not growing under moisture stress, the actual evapotranspiration (ET_{act}) equals ET_{pot}. Stress is assumed to occur when more than a given percentage (p) of total available soil moisture (S_a) has been developed in the root zone. Below the critical soil moisture, actual evapotranspiration is a linearly declining function of actual soil moisture (M):

$$ET_{act} = ET_{pot} \cdot M / [(1 - p) \cdot S_a]$$

If the actual soil moisture is uniform over the root zone, it is assumed that the water uptake by the crop is triangular. If rooting depth is four feet, then 40 percent of the water is taken up by the roots in the upper foot, 30 percent in the second foot, and so on. If the soil moisture profile is not uniform, the water uptake profile is modified so that a larger percentage is absorbed in the wetter soil layers and a relatively smaller percentage in the drier ones.

Irrigation Scheduling Module

Depending on the type of problem that is analyzed, a great variety of decisions can be introduced in the irrigation scheduling module. In the present version of this model, canal and groundwater irrigation have been treated as separate variables with flexible operating rules regarding groundwater irrigation.

Yield Response Module

This module follows the methodology described in FAO 1979b. The actual and theoretical evapotranspiration is summed up for each stage of crop growth. If actual evapotranspiration is less than the theoretical, crop yield is reduced according to the following formula:

$$\frac{Y_{act}}{Y_{pot}} = (1 - k_1) \cdot \left(1 - \frac{\Sigma ET_{act}}{\Sigma ET_{pot}}\right) \cdot \ldots \cdot$$

$$(1 - k_n) \cdot \left(1 - \frac{\Sigma ET_{act}}{\Sigma ET_{pot}}\right)$$

where Y_{act} = actual yield, Y_{pot} = potential yield for the ensured package of cultivation practices if no water stress occurs, and k_i = yield response factor for growth stage i.

Sample Results from the Scheduling Model for Canal Irrigation

To illustrate the schedule and its significance for developing an appropriate schedule under a supply-constrained system, the first set of computer runs corresponds to a situation with exclusive canal irrigation. This also helps to develop an irrigation schedule under a conjunctive system in sequential stages.

The model has been run under various alternative canal irrigation schedules for three crops—wheat, mustard, and castor. Wheat and mustard—the rabi or dry crops—have been run for only a single year, while castor—the two-season crop—has been run for several years to represent the effects of variation in rainfall distribution. Calibration of the model has been attempted in a limited manner by choosing plot-level observations (farmers' fields) that approximate the alternative canal irrigation schedules selected in the computer runs and by comparing the relative crop yield under different schedules.

Alternative canal schedules have been developed by a simple combination of the two parameters—depth of application (assumed constant for each irrigation) and number of irrigations—to define the particular schedule. The interval between two successive canal irrigations is fixed, reflecting the growing practice of rotational systems in many irrigation projects. The range of canal irrigation schedules is intended to represent variations in total irrigation input, depth of application, and the interval between two irrigations.

The yield index computed from such an irrigation schedule will provide an idea of both the marginal effect on crop yield for increasing levels of irrigation inputs and also the relative efficiency of a particular schedule compared with an on-demand schedule. An on-demand schedule represents an ideal situation in which the stress conditions at different stages of crop growth are eliminated and a yield index close to 1.0 is achieved. Naturally, the on-demand schedule implies a varying interval of irrigation, and the depth of application is kept as small as feasible to improve the efficiency of the irrigation application.

The model also analyzes how different rainfall conditions affect the crop yield index for a given canal schedule, which indirectly reflects how efficient a particular schedule is for crops such as castor which are sensitive to the rainfall distribution pattern.

Results of alternative canal schedules for the three-crop study are summarized in Table 13-4. Detailed results of the model include relevant quantities such as actual effective rainfall, potential evapotranspiration (EPOT), actual moisture uptake by the crop (ECROP), surface runoff (TSURF), and yield index. They are presented in Tables 13-5, 13-6, and 13-7.

The most efficient schedule is represented by on-demand irrigation, which minimizes the stress condition in each stage of crop growth. The closer the fixed irrigation schedule (which predetermines the dates of irrigation) is to the on-demand irrigation, the more efficient it is. For example, although the best schedule (highest water application) for castor (schedule A in Table 13-4) closely approximates on-demand irrigation, for wheat and mustard the best fixed schedules chosen are less efficient—either the total water applied is much more or the yield index is significantly less than for on-demand irrigation. For wheat, each schedule up to schedule B (next highest water application) shows a steady increase in the yield index, but schedule A compared with schedule B results in a very small increase in the yield index. For mustard, it would appear that a schedule with water application higher than

Table 13-4. *Model Results from Alternative Canal Irrigation Schedules for Three Crops*

Schedule	*Characteristics*	*Yield index* *Average*	*Maximum*	*Minimum*
	Castor			
A	$N = 11, D = 50, TD = 550$	0.9159	0.9922	0.7812
B	$N = 6, D = 75, TD = 450$	0.7822	0.9063	0.6040
C	$N = 8, D = 50, TD = 400$	0.6730	0.7927	0.5535
D	$N = 4, D = 75, TD = 300$	0.5840	0.7044	0.4397
E	$N = 3, D = 75, TD = 225$	0.4965	0.6137	0.3652
F	$N = 4, D = 50, TD = 200$	0.4474	0.5560	0.3315
On demand[a]	$N = 12–14, D = 50,$ $TD = 600$ (year 2) – 700 (years 3 and 4)	0.9887	0.9917	0.9837
	Rainfall (millimeters)	483.1	826.1	203.9
			Yield index	
	Wheat			
A	$N = 11, D = 75, TD = 825$		0.9452	
B	$N = 11, D = 60, TD = 660$		0.9452	
C	$N = 11, D = 75, TD = 525$		0.8764	
D	$N = 7, D = 60, TD = 420$		0.7861	
E	$N = 4, D = 75, TD = 300$		0.5163	
On demand[a]	$N = 11, D = 50, TD = 550$		0.9962	
	Mustard			
A	$N = 6, D = 75, TD = 450$		0.8321	
B	$N = 6, D = 60, TD = 360$		0.6752	
C	$N = 4, D = 75, TD = 300$		0.5017	
D	$N = 4, D = 60, TD = 240$		0.3723	
E	$N = 3, D = 75, TD = 225$		0.2798	
On demand[a]	$N = 8, D = 50, TD = 400$		0.9982	

Note: N = number of irrigations, D = depth per irrigation in millimeters, and TD = total depth in millimeters.

a. On-demand irrigation assumes a varying interval of irrigation; other schedules assume a predetermined interval irrespective of the crop's stage of growth.

Table 13-5. *Results from Irrigation Scheduling Model for Castor, 1977–81*

Schedule details	*1977*	*1978*	*1979*	*1980*	*1981*
Set A					
$N = 11, D = 50, TD = 550,$ $IS = 60, 74, 88, 102, 116, 130, 144, 158, 172, 186, 200$					
Total rainfall (millimeters)					
Presowing[a]	217.0	53.3	16.7	0.0	91.9
Postsowing	609.1	537.4	306.5	203.9	379.5
Effective rainfall	509.2	483.3	301.7	178.7	392.2
TSURF	341.1	129.0	24.2	26.2	100.8
EPOT (millimeters)	805.8	805.8	805.8	805.8	805.8
ECROP (millimeters)	766.7	799.9	777.2	727.0	776.1
Yield index	0.9169	0.9922	0.9464	0.7812	0.9427
Set B					
$N = 6, D = 75, TD = 450,$ $IS = 90, 111, 132, 153, 174, 195$					
Total rainfall (millimeters)					
Presowing[a]	217.0	53.3	16.7	0.0	91.9
Postsowing	609.1	537.4	306.5	203.9	379.5
Effective rainfall	509.2	483.3	301.7	178.7	392.2
TSURF	335.0	127.9	24.1	27.4	79.8
EPOT (millimeters)	805.8	805.8	805.8	805.8	805.8
ECROP (millimeters)	742.9	750.8	684.1	622.7	735.2
Yield index	0.8333	0.9063	0.7479	0.6040	0.8195
Set C					
$N = 8, D = 50, TD = 400,$ $IS = 60, 81, 102, 123, 144, 165, 186, 207$					
Total rainfall (millimeters)					
Presowing[a]	217.0	53.3	16.7	0.0	91.9
Postsowing	609.1	537.4	306.5	203.9	379.5
Effective rainfall	509.2	483.3	301.7	178.7	392.2
TSURF	340.4	115.1	22.1	25.2	98.5
EPOT (millimeters)	805.8	805.8	805.8	805.8	805.8
ECROP (millimeters)	640.6	679.3	632.9	581.4	638.9
Yield index	0.6780	0.7927	0.6659	0.5535	0.6749
Set D					
$N = 4, D = 75, TD = 300,$ $IS = 104, 132, 160, 188$					
Total rainfall (millimeters)					
Presowing[a]	217.0	53.3	16.7	0.0	91.9
Postsowing	609.1	537.4	306.5	203.9	379.5
Effective rainfall	509.2	483.3	301.7	178.7	392.2
TSURF	335.0	127.9	22.1	25.2	79.8
EPOT (millimeters)	805.8	805.8	805.8	805.8	805.8
ECROP (millimeters)	604.7	619.1	546.7	490.6	599.5
Yield index	0.6186	0.7044	0.5466	0.4397	0.6107
Set E					
$N = 3, D = 75, TD = 225,$ $IS = 104, 149, 194$					
Total rainfall (millimeters)					
Presowing[a]	217.0	53.3	16.7	0.0	91.9
Postsowing	609.1	537.4	306.5	203.9	379.5
Effective rainfall	509.2	483.3	301.7	178.7	392.2
TSURF	335.0	107.4	22.1	25.2	77.4
EPOT (millimeters)	805.8	805.8	805.8	805.8	805.8
ECROP (millimeters)	527.2	562.5	469.2	413.3	521.9
Yield index	0.5258	0.6137	0.4589	0.3652	0.5188

Schedule details	1977	1978	1979	1980	1981
Set F					
$N = 4, D = 50, TD = 200,$ $IS = 104, 132, 160, 188$					
Total rainfall (millimeters)					
Presowing[a]	217.0	53.3	16.7	0.0	91.9
Postsowing	609.1	537.4	306.5	203.9	379.5
Effective rainfall	509.2	483.3	301.7	178.7	392.2
TSURF	335.0	125.1	22.1	25.2	77.4
EPOT (millimeters)	805.8	805.8	805.8	805.8	805.8
ECROP (millimeters)	509.3	527.0	451.5	395.0	503.9
Yield index	0.4715	0.5560	0.4130	0.3315	0.4652
Set G (on demand)[b]					
Total rainfall (millimeters)					
Presowing[a]	217.0	53.3	16.7	0.0	91.9
Postsowing	609.1	537.4	306.5	203.9	379.5
Effective rainfall	509.2	483.3	301.7	178.7	392.2
TSURF	342.4	137.0	24.0	32.8	98.7
EPOT (millimeters)	805.8	805.8	805.8	805.8	805.8
ECROP (millimeters)	802.1	796.6	802.1	796.1	801.8
Yield index	0.9913	0.9861	0.9917	0.9837	0.9908

Note: N = number of irrigations, D = depth per irrigation in millimeters, TD = total depth in millimeters, and IS = irrigation schedule.

a. Rainfall within five days before sowing.

b. 1977: $N = 13, D = 50, TD = 650, IS = 1, 14, 93, 106, 177, 127, 138, 149, 161, 174, 186, 199, 217$
1978: $N = 12, D = 50, TD = 600, IS = 1, 14, 81, 93, 110, 120, 131, 152, 165, 178, 190, 204$
1979: $N = 14, D = 50, TD = 700, IS = 1, 14, 65, 81, 94, 109, 126, 136, 147, 159, 171, 184, 196, 213$
1980: $N = 14, D = 50, TD = 700, IS = 1, 16, 51, 81, 92, 105, 115, 125, 136, 147, 159, 171, 190, 200$
1981: $N = 13, D = 50, TD = 650, IS = 1, 14, 81, 105, 116, 126, 137, 148, 160, 172, 185, 197, 214$

that of schedule A would approximate the yield achieved under on-demand irrigation, but the total quantity of water applied will be much more than with the on-demand schedule.

Marginal yield is positive for almost all the levels of water application considered, but it decreases as water application increases (see Figures 13-2, 13-3, and 13-4). In interpreting marginal relationships, depth of irrigation should also be considered. Admittedly, however, the estimate of marginal yield is only illustrative. Better-defined

Table 13-6. ***Results from Irrigation Scheduling Model for Castor, 1973–76 and 1982***

Schedule details	1973	1974	1975	1976	1982
Set A					
$N = 4, D = 75, TD = 300,$ $IS = 104, 132, 160, 188$					
Total rainfall (millimeters)					
Presowing[a]	58.8	20.7	184.1	12.0	5.0
Postsowing	601.5	43.3	842.6	786.3	469.1
Effective rainfall	447.9	60.5	772.6	671.5	325.6
EPOT (millimeters)	805.8	805.8	805.8	805.8	805.8
ECROP (millimeters)	638.1	364.1	716.0	650.8	532.9
Yield index	0.6646	0.3273	0.8246	0.7274	0.5239
Set B					
$N = 6, D = 75, TD = 450,$ $IS = 90, 111, 132, 153, 174, 195$					
Total rainfall (millimeters)					
Presowing[a]	58.8	20.7	184.1	12.0	5.0
Postsowing	601.5	43.3	842.6	786.3	469.1
Effective rainfall	447.9	60.5	772.6	671.5	325.6
EPOT (millimeters)	805.8	805.8	805.8	805.8	805.8
ECROP (millimeters)	770.6	506.0	804.8	774.6	674.1
Yield index	0.9107	0.4532	0.9988	0.9351	0.7155

Note: N = number of irrigations, D = depth per irrigation in millimeters, TD = total depth in millimeters, and IS = irrigation schedule.

a. Rainfall within five days before sowing.

Table 13-7. ***Results from Irrigation Scheduling Model for Mustard and Wheat, 1981***

Schedule	*N*	*D*	*TD*	*IS*	*EPOT*	*ECROP*	*TSURF*	*Yield index*
				Wheat				
A	11	75	825	10, 20, 30, 40, 60, 70, 80, 90, 100, 110	440	414	11.4	0.9452
B	11	60	660	10, 20, 30, 40, 60, 70, 80, 90, 100, 110	440	414	2.4	0.9452
C	7	75	525	10, 28, 46, 64, 82, 100, 118	440	402	0.1	0.8764
D	7	60	420	10, 28, 46, 64, 82, 100, 118	440	366	0.0	0.7861
E	4	75	300	10, 49, 70, 100	440	262	0.0	0.5163
On demand	11	50	550	1, 16, 31, 48, 63, 75, 86, 96, 106, 116, 126	440	439	0.0	0.9962
				Mustard				
A	6	75	450	10, 30, 50, 70, 90, 11	426	369	0.3	0.8321
B	6	60	360	10, 30, 50, 70, 90, 11	426	328	0.0	0.6752
C	4	75	300	10, 40, 70, 100	426	271	0.0	0.5017
D	4	60	240	10, 40, 70, 100	426	219	0.0	0.3723
E	3	75	225	10, 40, 70	426	197	0.0	0.2793
On demand	8	50	400	16, 35, 54, 70, 83, 99, 111, 123	426	424	0.0	0.9982

Note: *N* = number of irrigations
D = depth per irrigation in millimeters
TD = total depth in millimeters
IS = irrigation schedule.

Figure 13-2. ***Effect of Total Water Application, Depth of Irrigation, and Different Rainfall Years on the Yield Index for Castor***

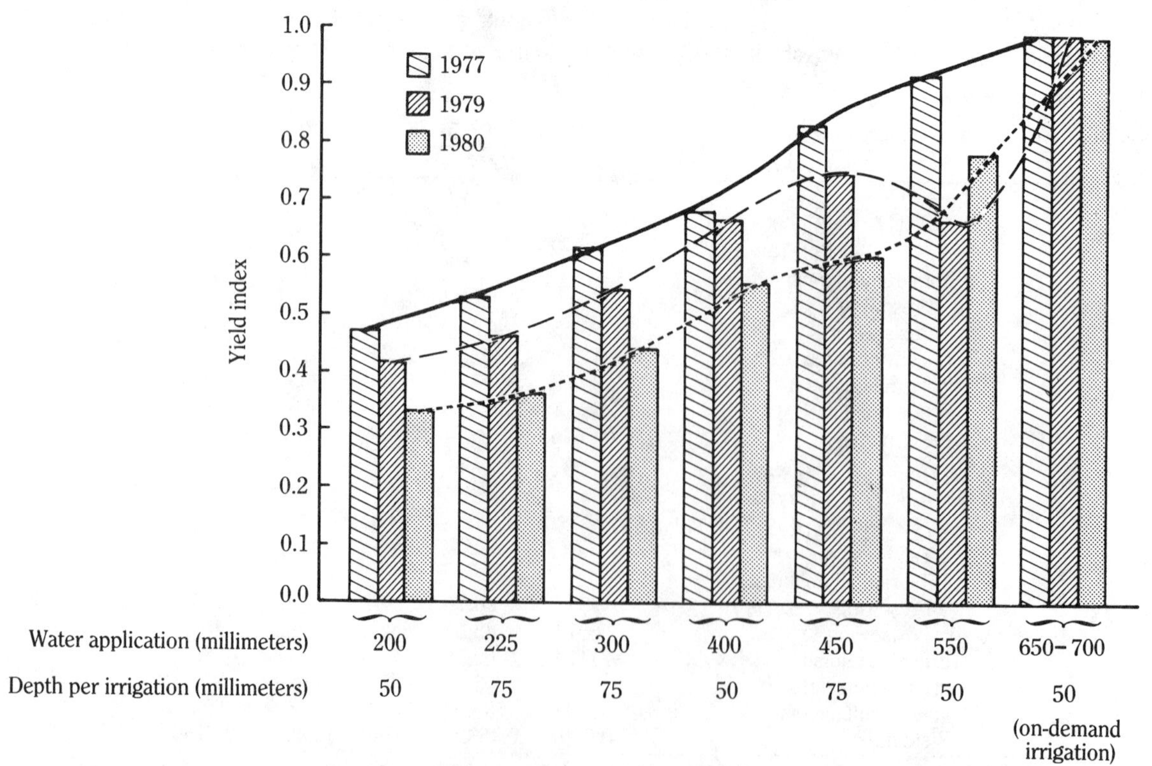

Source: Irrigation Scheduling Model.

irrigation schedules and a more refined calibration of the model are needed, as well as actual field observations under a varying set of conditions.

Development of the Irrigation Schedule

The objective of developing an appropriate irrigation schedule for conjunctive use is to deliver irrigation water more efficiently and equitably to the lowest level of the distribution network—the canal outlet. A system that supplements canal irrigation with groundwater is more equitable in situations with uncertain availability of canal water (but with dependable groundwater storage replenished by rainfall and seepages from the surface irrigation system). In actual practice, farmers are already using groundwater in various situations—such as insufficient canal irrigation in the tail outlet—to supplement canal irrigation for water-intensive crops. Groundwater is often used by only a few farmers, however, even within a particular outlet command, and the schedules followed by farmers may not be efficient or properly integrated with the canal supply system.

The irrigation schedules for conjunctive use developed in this context represent a supply-constrained system that predetermines the canal schedule for the total quantity of water to be delivered and the number and depth of irrigations. The groundwater schedule is developed in sequential stages to maximize the crop yield by minimizing the stress condition at different stages of crop growth. Such a fine integration of groundwater with a predetermined canal schedule may not be feasible or realistic. In this study, however, the total quantity of water that can be supplied at the outlet level from the groundwater system is kept within a feasible and rational limit. For example, given the command size and irrigation intensity for crops, groundwater pumping will not exceed recoverable recharge (after adjustments are made for water loss and recycling). Tentative estimates of recoverable recharge, on an average annual basis, have been made, and the imposed limit to groundwater use is on the conservative side for an average outlet. The other practical criterion introduced in developing the alternative irrigation schedule is the adjustment of dates by taking into account the stress index of the crop and the predetermined date of canal irrigation.

It is assumed that the depth of water application can be varied in the groundwater system. Usually, as can be seen from the sample results of the canal irrigation model (Table 13-4), the schedules with a lower average and varying depth of application are more efficient for overall water use.

Results of additional groundwater irrigation to supplement predetermined canal irrigation schedules have been summarized in Table 13-8 for wheat, mustard, and castor. Incremental yield is significant in all cases, but more significant in the case of mustard, where yield about doubles with a groundwater input of 240 millimeters over and above the canal irrigation input of 300 millimeters at the plant level.

What are the economic implications of an appropriate schedule for additional groundwater irrigation at the operational level (in this case command of an outlet)? To answer this question, the following data must be computed:

- Total quantity of water to be delivered after adjusting for losses from wellhead to plant, which reflects the size of the outlet command and the irrigated area under different crops (see Table 13-9 for the sample outlets).
- Pumping capacity required to meet the irrigation demand (from the groundwater system) for the peak week, which considers the entire outlet command. In computing the capacity, an assumption must be made about the type of aquifer to be tapped (shallow or deep) and consequently the type of well to be installed. In the sample study area, mostly shallow wells are being used by the farmers.
- Additional crop yield for each successive stage of additional groundwater irrigation and aggregation of yields at the outlet level. The scheduling model results are presented as a crop yield index rather than as an absolute level of output; a base level of output (preferably observed under actual field conditions) or a given level of irrigation input is then assumed to convert the relative yield index into an absolute level of crop output. This is to be done separately for each crop.

Table 13-8. *Results from Conjunctive Irrigation Schedule for Wheat, Mustard, and Castor*

Crop	*Mode of irrigation*	*Number of irrigations*	*Total irrigation at delta (millimeters)*	*Yield index*
Wheat	Canal	4	300	0.5163
	Canal and groundwater	4 + 5 = 9	300 + 305 = 605	0.8945
Mustard	Canal	4	300	0.5017
	Canal and groundwater	4 + 3 = 7	300 + 240 = 540	0.9935
Castor	Canal	3	225	0.4965
	Groundwater	3 to 5[a]	220[b]	0.7611

a. Depending on year.
b. 150 to 250, depending on year.

Figure 13-3. *Effect of Total Water Application and Depth of Irrigation on the Yield Index for Wheat*

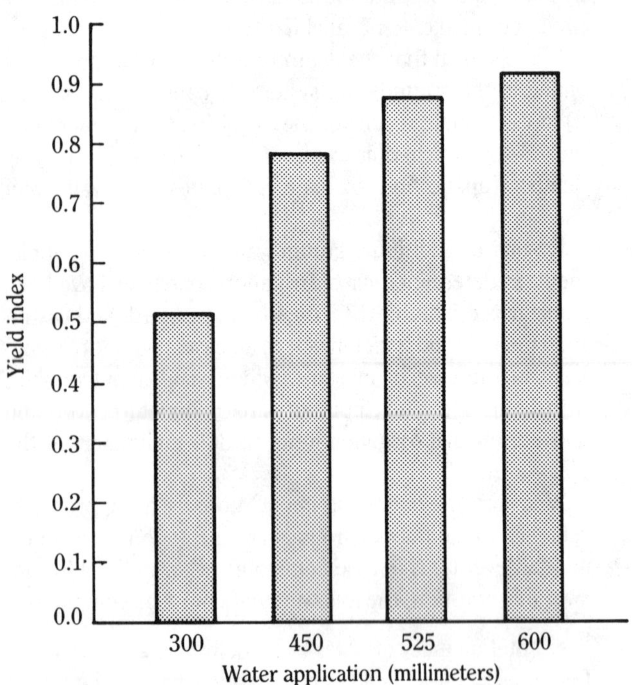

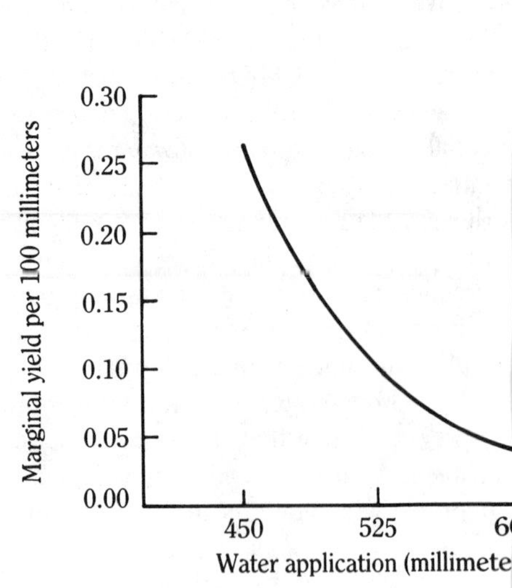

Note: Depth per irrigation = 75 millimeters.
Source: Irrigation Scheduling Model.

Figure 13-4. *Effect of Total Water Application and Depth of Irrigation on the Yield Index for Mustard*

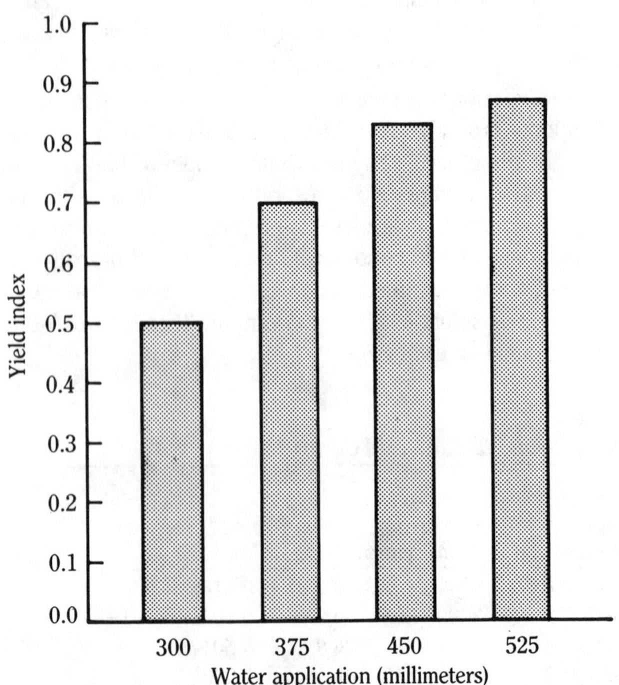

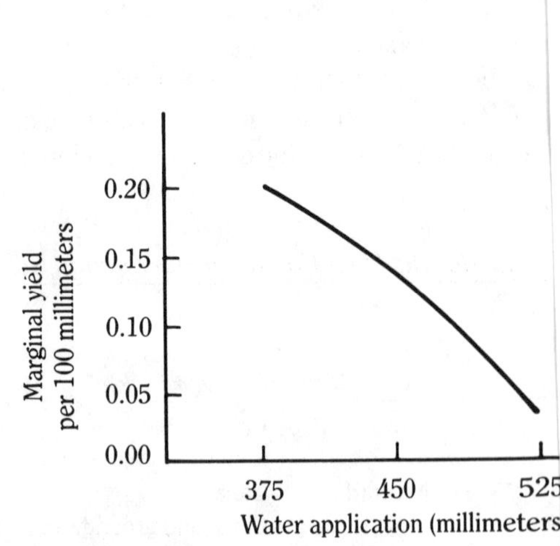

Note: Depth per irrigation = 75 millimeters.
Source: Irrigation Scheduling Model.

Table 13-9. *Irrigation Schedule and Water Requirements from Canal and Groundwater for Outlets Studied*

Outlet	Crops considered	Irrigated area for rabi and two-season crops (hectares)	Depth of irrigation at plant (millimeters)	Canal water schedule (week number)[a]	Groundwater schedule (week number)[a]	Water requirements (hectare-meters)		
						Canal	Groundwater	Total
HD/HOL	Mustard	8.49	75	7, 13, 17	9, 15, 20	3.822	2.724	6.546
	Castor	3.93	75	7, 13	1, 4, 10, 16	1.180	1.684	2.864
	Potato	2.99	50	13, 17	10, 11, 12, 14, 15, 16, 18, 19, 20, 21, 22, 23	0.598	2.580	3.178
	Wheat	1.72	60	7, 13, 17	9, 11, 15, 19, 21	0.618	0.740	1.358
	(Fallow)	0.05	—	—	—	—	—	—
	Total	17.18	—	—	—	6.218	7.728	13.946
HD/TOL	Mustard	13.32	75	7, 13, 17	9, 15, 20	5.994	4.275	10.269
	Castor	7.44	75	7, 13	1, 4, 10, 16	2.232	3.184	5.416
	Potato	3.00	50	13, 17	10, 11, 12, 14, 15, 16, 18, 19, 20, 21, 22, 23	0.600	2.592	3.192
	Wheat	1.41	60	7, 13, 17	9, 11, 15, 19, 21	0.507	0.605	1.112
	(Fallow)	8.74	—	—	—	—	—	—
	Total	33.91	—	—	—	9.333	10.656	19.989
HD/MOL	Mustard	20.73	75	7, 13, 17	9, 15, 20	9.330	6.654	15.980
	Castor	11.66	75	7, 13	1, 4, 10, 16	3.498	4.992	8.490
	Potato	5.19	50	13, 17	10, 11, 12, 14, 15, 16, 18, 19, 20, 21, 22, 23	1.038	4.488	5.520
	Wheat	1.19	60	7, 13, 17	9, 11, 15, 19, 21	0.429	0.510	0.939
	(Fallow)	19.53	—	—	—	—	—	—
	Total	58.30	—	—	—	14.295	16.644	30.939
MD/HOL	Mustard	15.76	75	7, 13, 17	9, 15, 20	7.092	5.058	12.150
	Castor	0.70	75	7, 13	1, 4, 10, 16	0.210	0.300	0.510
	Wheat	8.10	60	7, 13, 17	9, 11, 15, 19, 21	2.916	3.485	6.400
	(Fallow)	0.13	—	—	—	—	—	—
	Total	24.69	—	—	—	10.218	8,843[b]	19.060
MD/TOL	Mustard	28.44	75	7, 13, 17	9, 15, 20	12.798	9.129	21.920
	Castor	3.53	75	7, 13	1, 4, 10, 16	1.060	1.512	2.572
	Wheat	22.18	60	7, 13, 17	9, 11, 15, 19, 21	7.938	9.535	17.518
	(Fallow)	5.73	—	—	—	—	—	—
	Total	59.88	—	—	—	21.841	20.176[c]	42.017

— Not applicable.
a. Starting the week of October 1.
b. An additional 2.157 used for summer crops to arrive at balance between canal water and groundwater use.
c. An additional 1.766 used for summer crops.

- Net income estimated from additional yield (by deducting the cost of cultivation from the gross value of produce) for each successive level of groundwater irrigation.
- Cost of groundwater capacity (see Table 13-10), based on pumping capacity requirement (with appropriate conversion into annualized cost) and also the average annual operation cost for each increment of groundwater application aggregated at the outlet level.
- Net additional return or benefit (net income estimated from additional yield minus cost of groundwater capacity) per hectare for each meter of water, as well as the

Table 13-10. *Groundwater Capacity, Use, and Capital Cost per Unit of Water in Outlets Studied*
(unit of water = hectare-meters)

Outlet	Groundwater irrigation (peak week starting October 1)	Total water demand for peak week	Delivery capacity required		Actual water delivered for entire crop season	Capacity utilization factor (percent)[b]		Total capital cost for required capacity (1981–82 prices in thousands of rupees)[c]		Annualized capital cost per unit of water delivered (thousands of rupees)[c]	
			Alternative 1[a]	Alternative 2[a]		Alternative 1	Alternative 2	Alternative 1	Alternative 2	Alternative 1	Alternative 2
HD/HOL	15	1.271	2.118 (1.237)	1.495 (0.873)	7.728	15.9	22.5	74.2	65.5	1.6	1.1
HD/TOL	15	1.591	2.652 (1.549)	1.872 (1.093)	10.656	17.5	24.7	92.9	82.0	1.4	1.0
HD/MOL	15	2.694	4.490 (2.622)	3.169 (1.850)	16.644	16.1	22.8	157.3	138.8	1.5	1.1
MD/HOL	15	2.383	3.972 (2.319)	2.804 (1.637)	11.000	12.0	17.1	139.1	122.8	2.1	1.5
MD/TOL	15	4.950	8.250 (4.817)	5.824 (3.401)	21.942	11.6	16.4	289.0	255.1	2.1	1.5

Note: Numbers in parentheses indicate the capacity required in cusecs.
a. Alternative 1: dug-cum-borewell (maximum running time = 100 hours per week). Alternative 2: deep tubewell (maximum running time = 143 hours per week).
b. For alternative 1, the life of a well is assumed to be ten years; for alternative 2, fifteen years.
c. US$1.00 = Rs10.00.

Table 13-11. *Canal and Groundwater Delivered per Hectare for Outlets Studied*

Outlet	Canal water (meters)	Groundwater (meters)
HD/HOL	0.362	0.450
HD/TOL	0.275	0.314
HD/MOL	0.245	0.285
MD/HOL	0.414	0.358
MD/TOL	0.365	0.337

aggregated net benefit at the outlet level for each scheduled increment of groundwater irrigation.

Table 13-9 presents the implications of the irrigation schedule under a conjunctive system for each sample outlet studied. For the sake of simplicity, canal irrigation schedules are considered the same for all the crops studied, which is a realistic assumption under a rotational water delivery system. But the total canal irrigation requirements as well as the total irrigation application (both canal water and groundwater) for each hectare vary among the outlets, particularly because the cropping patterns vary. On average, the share of canal water and the share of groundwater delivered at the outlet level are nearly equal—a feature explained by the peculiarity of the command area under study, which has limited canal water availability, high seepage losses, and relatively large groundwater resources continually replenished by the surface irrigation system. As expected, the use of supplementary groundwater is the lowest for mustard and the highest for potatoes, which require frequent irrigation. The total amount of water delivered for each hectare (see Table 13-11) increased over the actual situation, which showed a wide variation in canal water allocation for each unit area between the head and tail outlets. The suggested conjunctive use schedule thus implies a reasonable degree of equity.

Economic justification for the suggested irrigation schedule is given in the sample results presented in Table 13-12 under a set of simplifying assumptions. The main thrust of the analysis is to answer the question: To what level of incremental groundwater irrigation is the net additional benefit positive? Two crops have been analyzed—wheat and mustard. The analysis will be quite complex for castor because of rainfall variation.

For the limited computer runs of the irrigation scheduling model, a predetermined canal irrigation input (with an appropriate schedule) of 300 millimeters at plant level has been assumed as normal for the two crops under study. The analysis is presented for only one representative outlet command. For a more refined marginal analysis, smaller incremental values of groundwater irrigation are desirable.

The model results in Table 13-12 do indicate a positive net benefit from a substantial amount of additional groundwater irrigation for wheat and mustard (from 300 millimeters of canal water to a total of 525 millimeters for wheat and 450 millimeters for mustard), if the well irrigation can be made as flexible as it is in the suggested schedule and as accessible to all the farmers in the outlet command.

Table 13-12. *Economic Evaluation of Different Levels of Water Application for Wheat and Mustard*

Total water application (millimeters) (1)	Marginal water input (millimeters) (2)	Water delivered per hectare, including losses[a] (hectare-meters) (3)	Yield index (4)	Marginal yield index (MYI) (5)	Estimated net income corresponding to MYI[b] (rupees) (6)	Marginal income per hectare-meter[c] (rupees) (7)	Marginal net return per hectare-meter, net of capital and operating cost of groundwater use[d] (rupees)
				Wheat			
300[e]	—	—	0.5163	—	—	—	—
450	150	0.21	0.7800	0.2637	1,256	5,981	3,981
525	75	0.11	0.8764	0.0964	459	4,173	2,173
600	75	0.11	0.9138	0.0374	178	1,618	−382
				Mustard			
300[e]	75	0.11	0.5017	0.2219	1,957	17,791	16,451
375	75	0.11	0.6976	0.1959	1,149	10,445	8,445
450	75	0.11	0.8321	0.1345	789	7,173	5,173
525	75	0.11	0.8668	0.0347	204	1,855	−145

— Not applicable.

a. Efficiency of 70 percent is assured at plant level from well-head.

b. Unit price is considered to be 210 rupees per quintal for wheat and 350 for mustard. Cost of cultivation (average) is 2,000 rupees per hectare for wheat and 1,250 for mustard.

c. Column 6 ÷ column 3.

d. Includes annualized capital cost (estimated at 1,200 rupees per hectare-meter on average) and 800 rupees for operating cost—a total cost of 2,000 rupees per hectare-meter for groundwater irrigation. Observed yield level at 450 millimeters = 32.2 quintals per hectare-meter.

e. Additional water increments over 300 millimeters are obtained through well irrigation.

Issues

The results presented above indicate the physical and economic feasibility of a strategy for irrigation management and scheduling which takes into account the varying availability of canal water and high losses in the system. The approach is also consistent with farmer behavior—that is, the practice of supplementing canal water with groundwater, particularly when canal water is limited (see Table 13-1). Before the implementation plan can be drawn up, however, the questions raised in the concluding paragraphs must be addressed.

How can equitable distribution of groundwater be ensured among the farmers who do not own wells in the command of the outlet? This question must be considered in the context of various constraints on the size distribution of landholdings (small holdings are predominant), the mode of financing and maintaining wells, and the price and availability of energy. How do alternative forms of well ownership—for example, cooperative or private—affect adherence to the conjunctive schedule and the equitable allocation of water?

What kinds of intervention and incentives would generate more efficient and equitable use of groundwater systems? Although this study indicates a number of factors—for example, farmers' spontaneous action to supplement canal water with groundwater—that favor private ownership of wells, there are other important considerations. Is it possible to devise a system of subsidies and price controls for groundwater use that, on the one hand, motivates well owners to exploit groundwater and, on the other hand, does not force farmers who are not owners to buy groundwater at very high rates during critical stages of crop growth?[4]

How effective are physical and legal controls (for example, statutory provisions for groundwater use) and economic and fiscal instruments? The use of aquifers with highly interdependent subsurface flows may lead to externalities in the system. In that case, various kinds of economic and fiscal instruments can be explored, such as subsidies or financial support in areas that need a higher rate of groundwater development and levying a cess on existing land revenue or a water tax in areas that need groundwater conservation. As far as possible, such controls and incentives should be implemented within the broad legal framework that controls land and water resources in the state.

How might compatibility be ensured between the apparent economic motivation for conjunctive use and the desired level of resource efficiency? For example, the existing practice of tapping the shallow aquifers in the project area is likely to be more advantageous for the private owners but may not be the right alternative for exploiting the external advantages created by significant recharge to the system through canal conveyance losses. It may be feasible to develop two types of groundwater use—one through the public system (deeper aquifers would be tapped and pumped directly into the canal system) and the other through the private system (owners would tap the upper aquifers for supplementary groundwater irrigation).

Could this plan for conjunctive use, which simultaneously takes into account the economic motivations of farmers and overall resource efficiency through the exploitation of appropriate aquifers, serve as a basis for a more objective and equitable canal water allocation policy among different regions and groups of farmers? Should such a policy be dynamic in nature? In the context of high year-to-year variability in canal water, to what extent can the groundwater system stabilize supply and lead to long-term equitable water allocation?

Finally, to develop practical solutions to the problem of inefficiency, detailed assessment of existing formal or informal institutions and their operations must be made. The underlying hypothesis for developing the approach outlined in this study has been to rely on existing institutions that have been found to be effective under local conditions. The modifications that are necessary in the structure or operation of these institutions would be introduced only over a longer period.

Notes

1. It is estimated that about half of these losses is accounted for by operational losses that are explained primarily by the lack of any definite irrigation management system.

2. One more outlet (HD/MOL) has also been studied as an example of an average outlet in the medium reaches of the system. All the details for this outlet have not, however, been worked out.

3. The present formulation of IRRYM is based primarily on three widely used and accepted publications: FAO 1979a, FAO 1979b, and U.S. Bureau of Reclamation 1978.

4. In the state under study, farmers owning wells within the area served by a canal outlet are charged fees over and above their canal water rates. This additional charge can be waived or the rates progressively increased, depending on the extent of use or control of groundwater desired.

References

FAO. United Nations Food and Agriculture Organization. 1979a. *Crop Water Requirements.* FAO Irrigation and Drainage Paper 24 (revised). Rome.

———. 1979b. *Yield Response to Water.* FAO Irrigation and Drainage Paper 33. Rome.

U.S. Bureau of Reclamation. 1978. *Drainage Manual.* Washington, D.C.

Comment

Nathan Buras

Basu and Ljung take the position that externalities—that is, the effects of a project outside its area of service or beyond its boundaries—are not necessarily measured by costs to be charged against the project. In irrigation projects, drainage is such an externality. Depending upon definitions, externalities can have nonnegative effects; hence they should be incorporated in the decisionmaking process of managing and operating development projects. Seepage losses from irrigation distribution systems could be beneficial in the sense that they could be pumped to supplement and augment surface water supplied by a canal network. The authors emphasize this point so strongly that one could easily gain the impression that irrigation inefficiency is justifiable.

Another term specific to this chapter is equity. The authors use it to define the uneven distribution of water in an irrigation project. In many parts of India, farmers at the lower end of irrigation distribution networks almost never receive any water, while those nearer the head seem to enjoy ample supplies.

Basu and Ljung present a study of an irrigation scheme in an unnamed state in India, which seems to be typical of many similar irrigation projects. The study involved four steps: (1) justification of an approach to the irrigation problem that uses groundwater through a conjunctive use system, (2) development of a model that assesses the potential benefits of conjunctive use, (3) examination of demands on the groundwater system which would result from conjunctive use, and (4) identification of policy issues generated by conjunctive use.

The water use of the project is highly inefficient. The authors estimate that the overall canal system efficiency is at most 20 percent, which means that at least 80 percent of the water delivered to the main canal is lost by evaporation, seepage, and return irrigation flow. These substantial water losses, a large part of which may eventually reach the regional aquifer, could be turned to advantage if pumped and made available to farmers—that is, if water quality is not significantly impaired. The authors assume, without offering any hydrological proof or justification, that wells deeper than 50 meters will have no influence on shallower wells.

To develop a conjunctive use approach, the authors specify five steps. First, it is necessary to develop production functions for water used for major crops which would take into account the frequency and depth of irrigation. Second, a schedule of surface (canal) water supply is specified that will produce a minimum yield of the major crops in most of the area commanded by the surface outlet. This baseline is considered by the authors to be both efficient and equitable. Third, a pumping schedule for groundwater is set up to take account of the aquifer recharge by both rainfall and canal seepage. This schedule of pumping is based on the assumptions that the aquifer properties and groundwater gradients are such that under undisturbed conditions the difference in aquifer storage during the time intervals under consideration is essentially zero, and that cropping patterns will not change as a result of groundwater being supplied for irrigation. Fourth, the capacity of the system to implement the groundwater pumping schedule is estimated. Finally, the result of exploiting groundwater in this manner is evaluated in terms of marginal net returns.

The model is a detailed simulation which essentially accounts for the soil moisture in the root zone and its effect on crop yields. There is no indication that the model parameters were verified or validated for the hydrological conditions prevailing in the irrigation project under study. It seems, however, that calibration of the model was attempted, but the results were not presented.

The results of the study involving only canal (surface) irrigation were based on the assumptions that depth of application was constant for each irrigation and that intervals between irrigations were fixed. The combination of constant applications at fixed intervals is known to result in irrigation inefficiency. As one can expect, the study has shown that on-demand irrigation was considerably more efficient. The study has also shown that the marginal yield of crops decreased with increased water applications under a rigid operation schedule.

The use of groundwater through a conjunctive use system is intended to ensure equity—that is, all users along a canal should be supplied with water, including the last ones downstream. The authors state that the pumping of groundwater should be limited to the recoverable recharge, but do no specify over what period of time. They also do not indicate how the size of the recoverable recharge is determined. This point is rather important, because the monitoring of groundwater for the acquisition of hydrological information is essential to effective management decisions, and its cost should be considered in the economic analysis. The authors do mention that an advantage of groundwater use is that farmers would come closer to the efficiency of on-demand irrigation. Here too, however, conjunctive use seems to exhibit diminishing economic returns.

The authors conclude that for a fixed amount of 300 millimeters of canal irrigation water supplied under a rigid schedule, the supplementary groundwater is beneficial if it is delivered under a flexible schedule.

A number of important issues are identified, many of them institutional. The land tenure pattern indicates a majority of small holdings, so that many farmers are in no position to develop and operate wells. Another crucial issue is that under those conditions, the aquifer is a "common pool," and its management requires physical, legal, economic, and fiscal control mechanisms. Finally, the authors allude to the possibility of surface water–groundwater interaction in a dynamic setting.

Basu and Ljung treat as externalities the inefficiencies in irrigated agriculture prevalent in many parts of India. They focus on the possible beneficial use of these inefficiencies when seepage losses are pumped and used for irrigation. Such use of groundwater under a flexible time and quantity schedule to supplement the surface (canal) water supplied under a rigid schedule is termed erroneously by the authors as conjunctive use. True conjunctive use considers the scheduling of the surface and groundwater components of a regional hydrological system in a way that the interaction of these components, including the management of storage in aquifers, is part of the overall development and operation policy.

The most recent World Bank publications are described in the catalog *New Publications,* which is issued in the spring and fall of each year. The complete backlist of publications is shown in the annual *Index of Publications,* which is of value principally to libraries and institutional purchasers. The latest edition of each is available free of charge from Publications Sales Unit, The World Bank, 1818 H Street, N.W., Washington, D.C. 20433, U.S.A., or from Publications, The World Bank, 66 avenue d'Iéna, 75116 Paris, France.